Cross-Cultural
Survey Methods

Cross-Cultural Survey Methods

JANET A. HARKNESS
FONS J. R. VAN DE VIJVER
PETER PH. MOHLER

WILEY-INTERSCIENCE

A JOHN WILEY & SONS, INC., PUBLICATION

Library of Congress Cataloging-in-Publication Data:

Cross-cultural survey methods / edited by Janet Harkness, Fons Van de Vijver, Peter Ph. Mohler.
    p. cm. — (Wiley series in survey methodology)
    "A Wiley-Interscience publication."
    Includes bibliographical references and index.
    ISBN 0-471-38526-3 (cloth : alk. paper)
    1. Social sciences—Research.   2. Social surveys—Methodology.   3. Cross-cultural
studies.   I. Harkness, Janet, 1948–   II. Vijver, Fons J. R. Van de.   III. Mohler, Peter Ph.,
1945–   IV. Series.

H62 .C73 2002
300'.7'2—dc21   2002027030

Printed in the United States of America.

10 9 8 7 6 5 4 3 2

CONTENTS

Part IV ANALYSIS OF COMPARATIVE DATA

Part V   DOCUMENTATION AND SECONDARY ANALYSIS

CONTRIBUTORS

JAAK BILLIET
Department of Sociology, University of Leuven

MICHAEL BRAUN
Centre for Survey Research and Methodology, Mannheim

MICK P. COUPER
Survey Research Center at the Institute for
Social Research, University of Michigan

EDITH D. DE LEEUW
Department of Methodology and Statistics, Utrecht University

JOHNNY FONTAINE
Department of Psychology, University of Leuven

SIEGFRIED GABLER
Centre for Survey Research and Methodology, Mannheim

SABINE HÄDER
Centre for Survey Research and Methodology, Mannheim

JANET HARKNESS
Centre for Survey Research and Methodology, Mannheim

JOOP J. HOX
Department of Methodology and Statistics,
Utrecht University

TIMOTHY P. JOHNSON
Survey Research Laboratory, University of Illinois at Chicago

KNUT KALGRAFF SKJÅK
Norwegian Social Science Data Services, Bergen

PETER PH. MOHLER
Centre for Survey Research and Methodology, Mannheim

WILLEM E. SARIS
Department of Statistics and Methodology, University of Amsterdam

NORBERT SCHWARZ
Department of Psychology/ Institute for Social Research, University of Michigan

TOM W. SMITH
National Opinion Research Centre, University of Chicago

ROLF UHER
Central Archive for Empirical Social Research, Cologne

JAN W. VAN DETH
Department of Political Science and International Comparative Social Research,
University Mannheim

FONS J. R. VAN DE VIJVER
Department of Psychology, Tilburg University

PREFACE

December 6th is the day on which children in some European countries are traditionally rewarded by St. Nikolaus for good behavior (or reprimanded by his helper, Knecht Ruprecht, for bad). In 1998, this was the day an international group of scholars decided that this book was not only needed but that it was, also, do-able. Now that it is finished, the editors have much to be thankful for and we are indebted to many people across countries and continents for their unusual commitment to our common goal.

Our thanks go to the scholars who attended the *Third ZUMA Symposium on Cross-Cultural Survey Methods*, filling an old country manse for four days with a wealth of knowledge on manifold aspects of comparative survey research and thereby setting the foundations for the book. We thank:
Antonio Alaminos (Spain), Duane Alwin (USA), Jaak Billiet (Belgium), Ann Bridgewood (Britain), Michael Braun (Germany), Siegfried Gabler (Germany), Sabine Häder (Germany), Patrick Heady (Britian), Hanneke Houtkoop-Steenstra (Netherlands), Timothy Johnson (USA), Knut Kalgraff Skjåk (Norway), Mary McIntosh (USA), Tom Smith (USA), Catherine Sim (China), Norbert Tanzer (Austria), Rolf Uher (Germany), Linda Wray (USA).

Our goal was to produce an integrated treatment of methods for comparative survey research. Given the wealth of literature on comparative research in general and the rich stock of survey methods research, it may not be immediately obvious how demanding an undertaking this was. A problem repeatedly faced was that the evidence and literature available to explain one or the other procedure, effect, or standard was not cross-cultural in perspective or treatment. Thus, enthusiastic as we were when our contributors accepted the challenge, we must now thank them for their unflagging team spirit and for sticking with us when it became clear how demanding the task really was. We wish to express our gratitude to all the contributors to the volume.

We are particularly grateful to Duane Alwin for his advice, wisdom, and invaluable support for a long stretch of the book. Two contributors, Michael Braun and Tim Johnson, helped us enormously on multiple tasks related to the book. We would also like to thank Phil Gendall (New Zealand), John Kochevar (USA), and Marta Lagos (Chile) for their valuable input, as well as two anonymous reviewers for their comments and suggestions.

We thank ZUMA (Zentrum für Umfragen, Methoden und Analysen, Mannheim, Germany) for funding the symposium, for providing financial support for editorial meetings, and for taking over the greater part of the technical preparation and production of the book. In more general terms, its ongoing support of cross-national survey research since the mid-nineteen eighties and the comparative methods research program underway at the institute since the early nineties provided a framework for much that follows here.

Many women and two men were the people immediately behind the scenes. The final versions of the manuscript were produced by a wonderfully able and patient organiser, Maria Kreppe-Aygün, aided by Julia Khorshed, Elvira Scholz, and Sabine Klein and abetted by research assistants Carmen Daramus, Cornelia Hausen, Charlotte Reinisch, and Marie-Renée Afanou. Rinus Verkooijen (Tilburg University, the Netherlands) and Tanja Langer (ZUMA) compiled the references. Carol Cassidy and Heiner Ritter (Computers Division, ZUMA) made cross-national production and editing possible. The book's preparation was greatly aided by the attention to detail paid by Tanya Hart, Lisa Kelly Wilson, Elizabeth Leimbach, and Margaret Sinnott.

Finally, we wish to thank Stephen Quigley at John Wiley & Sons and Heather Haselkorn, his assistant, for their considerable support throughout the project.

JANET HARKNESS
PETER PH. MOHLER
FONS J.R. VAN DE VIJVER

PART ONE

INTRODUCTION

Chapter 1

COMPARATIVE RESEARCH

JANET A. HARKNESS
PETER PH. MOHLER
FONS J. R. VAN DE VIJVER

Social scientists seek to understand complex realities and the usefulness of comparative research for generating and testing social theories is well established (cf. Nowak 1989; Kohn 1989a). Comparing groups, cultures, nations, or continents is an essential means of distinguishing between local conditions and universal regularities (e.g., Roth 1971; Apter 1971; Kohn 1989a). Not surprisingly, therefore, survey research has a clearly delineated international and cross-cultural tradition (e.g., Almond and Verba 1963; Prezeworski and Teune 1966, 1970; Barnes and Kaase 1979; Krebs and Schüssler 1987).

In contrast to the vast majority of publications on comparative survey research, more space is devoted in this volume to methodological concerns than to substantive issues. The book does not explore, for example, the various typologies of analysis hitherto advanced for comparative research (cf. for classic examples, Scheuch 1968; Frey 1970; Smelser 1972; Kohn 1989b). Similarly, readers are referred elsewhere for accounts of the development of survey research (e.g., Bulmer, Bales, and Sklar 1991) and of comparative survey research (e.g., Rokkan 1969; Scheuch 1973). Treatments of problems faced in one or the other single 'foreign country' in the course of a specific survey are also much better covered elsewhere (e.g., Bulmer and Warwick 1983).

The present chapter is intended as an introduction to what follows in the volume. We begin with an explanation for the focus of the book—comparative social survey research—then indicate why a book-length discussion of 'comparative' survey methods is an appropriate undertaking. This is also where we introduce, however briefly, certain key concepts and terms used throughout the volume. The chapter closes with an overview of topics covered in the four other Parts of the book.

As Kohn points out: 'As with any research strategy, cross-national research comes at a price: It is costly in time and money; it is difficult to do; it often seems to raise more interpretive problems than it solves' (1989b, 77). Kohn continues, however, by pointing out that cross-national research is potentially invaluable and grossly underutilized. Similar acknowledgements of the value of and need for cross-national research can be found throughout the comparative (survey) research literature, as can comments on the growth in research volume and the challenges comparative research presents (e.g., Holt and Turner 1973; Grimshaw 1973; Øyen 1990; McDonald and Vangelder 1998).

Survey-based data collection procedures have become a standard means of collecting comparative data systematically at the level of individuals and households Nonetheless, comparative survey research as we know it today is a fairly recent development, in particular, the variable-oriented surveys we focus on in this volume.

Up to the late 1970s, the majority of comparative surveys consisted of behavioral and demographic studies (e.g., the World Fertility Study and the Multinational Time Use Studies; cf. Gauthier 2000). Many surveys before that time were not comparative in *design, implementation* and *intention*. They were, instead, what Gauthier (2000) calls '*ex post* harmonized survey agglomerates,' that is, national survey data which were recoded according to a 'comparative' scheme. As Rokkan (1969, 20-21) pointed out:

> Only a small minority of the total number of questions asked in similar fashion across two or more countries have been part of *deliberately designed cross-national surveys*. Such studies are still a rarity: they are costly, they require a great deal of organization, they are not surefire investments. But they are methodologically and strategically much more important than the other studies: they offer a much better basis for serious and systematic consideration of comparability and equivalence issues, of questions about the logic of cross-national and cross-culture research designs, and of the organizational options in such undertakings. (italics original)

Following intensive efforts in the 1960s (cf. Rokkan et al. 1969; Armer and Grimshaw 1973), by the early 1970s, a new kind of multinational survey project had appeared. The Eurobarometer, an ongoing attitudes and value orientation survey series, is a prime example. The series has been conducted among European Community (Union) members on behalf of the European Commission at least twice a year since 1970. The World Value Survey (WVS), a decennial attitudes and values survey, began in 1981. It was followed in 1985 by the International Social Survey Programme (ISSP), an annual series of attitude and behavior surveys that began with four members countries and by the turn of the twenty-first century had grown to almost forty. More recent additions are extensions to the Barometer 'family' – the Latinobarometer in South America, the Afrobarometer in parts of Africa, the Central and Eastern European Barometer, and the newest member, the Asiabarometer, in which China also participates. The most recent multinational addition to date is the European Social Survey, which focuses on social and political values measured as attitudes, opinions, and behavior.

As often noted, cross-national survey research has burgeoned. Many political barriers to collecting data or publishing data collected have also been lowered in different parts of the world. Thus although surveying remains difficult in some regions (e.g., certain parts of Africa), other territories, such as China, have relaxed some access restrictions. Access to data and the availability of analytical software have also grown apace. Researchers once excluded from comparative analysis by the high cost of conducting surveys or the high price of data now have an extensive array of inexpensive 'public use' data from which to choose.

The increase in numbers of publications using comparative survey data reflects this growth. For example, the yearly ISSP Trend Report (on publications using ISSP data from at least two countries; www.issp.org/trends.htm) marks an increase in publications from a total of 10 in 1986 (a year after the program began) to 55 in 1990, to almost 200 publications in the year 1999. (The extensive literature using ISSP data on one country is not included.) A similar growth can be noted for publications using survey data in cross-cultural psychology (Van de Vijver 2001).

Reflecting the increased significance accorded cross-national and cross-cultural research, an increasing number of disciplines have coined special disciplinary nomenclature for multicultural studies; 'comparative sociology' (cf. Kohn 1987), 'comparative political science' (cf. Dogan and Pélassy 1984), 'comparative education' (cf. Altbach, Arnove, and Kelly 1991), 'intercultural communication' (cf. Ting-Toomey 1999), 'transcultural psychiatry' (cf. Al-Issa 1995), and 'cross-cultural psychology' (cf. Berry 1997). Irrespective of their labels, each of these fields of research aims to increase our understanding of the universals and particulars of a research area or topic and to sharpen our awareness of our own perspectives as essentially limited and culture bound.

1.1   COMPARATIVE SURVEY RESEARCH METHODS

Comparative methods cover a wide field of quantitative and qualitative research methods and perspectives, ranging from methods as different as qualitative case studies, historical narratives of political systems, and hermeneutic analysis of political events, to systematic quantitative measurement at the individual (person) and case (nation) levels.

We make no attempt to cover all these approaches and their associated strategies here; our focus falls squarely on methods used in cross-cultural and cross-national survey research. In addition, the majority of our examples are taken from the fields we and our contributors know best – comparative attitude and behavior research and cross-cultural psychology and psychological testing.

The contributions in the book cover major concerns pertinent for these fields of research. We see this focus as a particular strength in a volume that explicitly endeavors to go beyond a diverse collection of papers on a variety of issues relevant for comparative research. At the same time, much of what the volume has to say about problems and procedures in these fields is, we feel, both relevant for and applicable to other fields.

The standard literature on comparative survey methods for the social sciences is of vintage date. Editors repeatedly have found it worthwhile to reprint material already published elsewhere (e.g., Bulmer and Warwick 1983; Øyen 1990; Inkeles and Masamichi 1995). Researchers seeking guidance and practical information will continue to find valuable insights in this literature (see, too, Rokkan et al. 1969; Vallier 1971; Verba 1971; Scheuch 1973; Kohn 1989a). At the same time, this rather suggests that many of the issues to be addressed have remained the same and unresolved over several decades. Calls continue, for example, for sound and robust methods for comparative survey research (e.g., Jowell 1998) and for adequate documentation (e.g., Harkness 1999).

The present volume indicates that progress has in fact been made but that a great deal remains to be done. One important change we note as editors is a move towards an evidence-based paradigm. Form's comment from the seventies illustrates the predicament "Probably no field has generated more methodological advice on a smaller data base with fewer results than has (cross-cultural) comparative sociology" (Form 1979, cited by Kohn 1989b). Contributors to this volume, however, repeatedly insist on the need for evidence and the study documentation which provides proper access to this evidence. Asking for 'more', as contributor Tom W. Smith suggests, will bring us greater returns.

1.2   CROSS-CULTURAL AND CROSS-DISCIPLINARY CONCERNS

A covert theme of the book is that findings and strategies from other disciplines are essential for comparative survey research but that interdisciplinary exchange is badly lagging behind within-discipline progress. A second, less covert theme is that monocultural survey research and cross-cultural survey research stand to learn from one another, and a third is that comparative survey research is practiced differently in diverse disciplines. Each of these disciplines can stand to learn from the others and has, in turn, something of value to offer.

Research fields of relevance for comparative survey research that merit more attention are, for example, the translation sciences, cross-cultural communication studies, linguistic pragmatics, and social cognition research. In addition, quality measures pursued with vigor in best practice monocultural survey research point up deficiencies in cross-national survey research. At the same time, these raise questions about how best to transfer the *application* of standards into different cultures and also about the extent to which best practice *in practice* will prove to be universal or culture bound. Meanwhile, the statistical testing and pretesting that are used in developing psychological instruments monoculturally and for cross-cultural implementation serve to draw attention to the neglect of such procedures in cross-cultural social science surveys.

We welcome the fact that a number of contributions in the volume reflect what other fields and disciplines have to offer survey methods for comparative research. Contributions also highlight the need for more exchange across 'survey' disciplines – the transfer of research knowledge and standards from monocultural to comparative areas of survey research, and vice versa. Thus while certain key developments in social science survey research—the impact of cognitive research on survey methods and the increased systematic attention paid to survey process features with the goal of enhancing survey data quality—have altered the Western survey research landscape, these have gone largely unnoticed in cross-national research. By the same token, relevant developments in research procedures in cross-cultural psychology and comparative testing of older and more recent date (Berry 1980; Smith and Bond 1998) have had little or no uptake in cross-cultural social survey methods. In the monocultural survey context, too, the discussion of the culture-bound nature of perceptions and practices is only just beginning to gain momentum.

This is perhaps also the point at which to acknowledge one important way in which *this* book is culture bound. As editors, we plagued our contributors with requests for evidence-based examples and a doubtlessly irritating insistence on a consistent comparative perspective. Both acts (and our contributors' reactions) reflect our shared cultural dilemma as writers and as survey researchers. Diverse as the origins of contributors are, and extensive as our cumulative experience with different cultural groups may be (an early reviewer referred to us as 'a lot of foreigners'), we all come from Western-based cultures and work within a Western frame of reference; our fields of research are dominated by Western theory, by Western conceptions of design and implementation requirements, and by Western research findings. We can access some of the factors guiding our perceptions and preferences consciously, but much of what guides our reasoning commonly remains unconscious – as much a part of our culture as our language. As Bond, focusing on America, recently commented:

> How ironic that our field must struggle so unwittingly to shed the chains of American dependence! This dependence takes the disciplinary forms of relying upon measures, concepts, and theories developed (mostly) in America, by Americans, about Americans, and for Americans (Featherman 1993). Given that 'all Men [sic!] are created equal', there is surely much of value in what has been produced from our American heritage. This hunch is more likely to be validated if America turns out to be less culturally extreme on our measures . . . than America appeared on Hofstede's (1980) measure of cultural individualism . . . The question, however, is how we are ever going to know with epistemological confidence where the American cultural system fits? Or any other cultural system? I propose that an answer will emerge only if we do science in a way that allows each cultural voice to be heard in our colloquy. (Bond, Fu, and Pasa 2001, 24)

1.3 A COMPARATIVE METHOD

Discussions of methods in comparative research often begin by noting that all social science research is comparative (e.g., Lipset 1986) and that there is nothing essentially different about doing cross-national or cross-cultural research from doing any other social science research (cf. discussion by Grimshaw 1973).

Acknowledging that all social science research is based on comparison does not resolve the question of whether we need different methods for different kinds of this research. In truth, the issue is more one of whether we agree or not that comparative survey research is essentially just like monocultural survey research, only perhaps more difficult and more expensive. There is something to be said for this standpoint and indeed a lot has been said (cf. Armer 1973; Grimshaw 1973). Those who adopt this view can point out that the rules of inferential statistics apply everywhere and that analyses of variance and reliability are just as relevant for national surveys as for cross-national.

However, as could be expected of editors of a book about comparative survey methods, we clearly feel there *is* a need to discuss certain methodological issues pressing for comparative research, that is, research *designed* and *destined* for implementation across cultural groups. At the same time, our readers do not need to come down clearly on one (or our) side of the fence to benefit from the volume. Grimshaw (1973, 4) elegantly bridges the divide as follows:

> My argument is that while the problems involved are no different in kind from those involved in domestic research, they are of such great magnitude as to constitute an almost qualitative difference for comparative, as compared to noncomparative, research.

Verba highlights a major difference between monocultural research and the demands placed on multinational research when he notes that "one more easily ignores dialect differences within a single language area and assumes, *perhaps without validity*, that one is dealing with the same interview across the whole sample" (1971, 311; emphasis added).

Issues that might be able to be ignored in monocultural contexts cannot, however, be ignored in cross-cultural research. Comparative researchers have no grounds to *assume* identity of meaning across social, linguistic, or cultural groups. Experienced comparative researchers are also less likely to assume that the use of similar instruments administered under similar conditions is truly sufficient to ensure that respondents from different cultural groups will arrive at the same interpretations of questions or that responses collected actually also 'mean' similar things.

As a result, cross-cultural survey research is required to pursue strategies that try to come to terms with the fact that concepts may not be identical or comparable and that an instrument appropriate and adequate in one context (temporal or spatial) may not be adequate in another.

The most commonly adopted approach in conducting comparative research is to decide on a design and to replicate/implement this as best possible in each of the populations involved in the project. Where this applies to questions asked in questionnaires, this is referred to here and elsewhere in the volume as an 'Ask-the-Same-Question' model. Where it applies to replicating other design features, such as mode, fielding procedures, or data management, it is termed a 'one-size-fits-all' or a 'keep-things-the-same' approach. One of the consequences of seeing all survey research as essentially the same is, we suggest, a commitment to the one-size-fits-all perspective. Implicit in this perspective is that designs, instruments, and strategies developed according to best practice principles in one location will also be valid across cultures. Best practice is then seen as a universal, not just in terms of goals and underlying principles, but in rules of application and implementations. If we accept this perspective, there is every reason to believe that comparability and quality will be enhanced by keeping things the same. However, a basic assumption in promoting standardization of implementation is that the standard identified as 'best' in one context a) *can* be implemented and b) *should* be implemented c) that it will also be *best* across contexts.

In contrast to this view on cross-national work, monocultural research is turning to tailoring aspects of the survey process so as to reduce respondent burden and enhance respondent reward, with a view to enhancing response. The design issues under discussion in monocultural research are therefore *deliberate* differences (in mode, in question format, in questions asked, and in interview compensation, for example) in order to reduce nonresponse and to enhance data quality in general.

Several contributions to the volume question the ultimate suitability of a keep-things-the-same approach as a quality and comparability enhancer in comparative research. At the same time, however, too little methods research has been undertaken to be able to recommend any other course *on the basis of evidence*. One of the first tasks for future research, therefore, is to establish the evidence needed to be able to decide on best practices for comparative contexts.

A further complication is that the variation of contexts in which multicultural surveys are embedded means that salient alternative explanations and hypotheses multiply, as do the sources of error and bias that complicate and hinder understanding. In addition, working across cultures involves working beyond one's own cultural frame of reference. We lose much of the common ground (Clark and Schober 1992) taken for granted as a basis for research and analysis on home territory, without fully realizing that this is so. This is not a new insight.

> The primary methodological implication of foreign settings is that theoretical problems and concepts, strategies for gaining access and cooperation, sampling methods, measuring techniques and instruments, data-collection and analysis procedures, and other aspects of the research process that are appropriate for researching one's own culture will often *not* be appropriate and valid for research in foreign cultures. Indeed, it should be assumed that research methods will have to be adapted or newly devised for each culture. Thus a major methodological problem facing comparative sociological research is the *appropriateness* of conceptualizations and research methods for each specific culture. Appropriateness requires feasibility, significance, and acceptability in each foreign culture as a necessary (but not sufficient) condition for insuring validity and successful completion of comparative studies. (Armer 1973, 50)

Our perceptions remain framed, to a greater or lesser degree, by our cultural background and experience, supplemented by whoever acts as informant or mentor for the unfamiliar territory. Agar's (1994) depiction of his encounter with a social outcast as informant is an intentionally dramatic illustration of the problems which can result from accepting information of uncertain reliability.

There is no magic recipe for acquiring a less culturally framed perspective. Deciding how best to design, implement, and interpret a multicultural survey thus involves dealing with complex theoretical considerations about the validity of hypotheses, on the one hand, and determining how best to measure and interpret social phenomena across cultures on the other. Monocultural survey research can look for differences and similarities assuming *ceteris paribus*. Comparative research cannot. It must thus make explicit statements about the comparability, validity, and reliability of measurement.

In order to do so better than to date, comparative research needs also to become more thoroughly interdisciplinary. In this way it can take advantage of knowledge spread over a number of only loosely related disciplines including cognitive psychology, cross-cultural test psychology, general survey methods, translation science, anthropology, communications theory, linguistics, political science, economics, and, not least, sociology.

Our collective experience in cross-cultural research suggests a core set of issues require to be considered for each comparative study. The most important points, as we see them, are as follows:

The quality of cross-cultural measurement depends on factors as diverse as appropriate theory, instrument design, sampling frame, mode of data collection, data analysis, and documentation across all the cultures involved. The total quality is the net result of the combination of outcomes of these factors. Exclusive reliance on appropriate instrument design, or statistical analysis, or some other factor in the process, may challenge the overall quality of a study, as can concentrating on best or worst case outcomes. Neither an exclusive focus on theory nor an exclusive reliance on the repair capacities of advanced statistics can guarantee survey quality.

Quality is not an intrinsic property of an instrument, sample design, or interview technique, but a property of a specific application. The reliability of an instrument may have been high in previous studies; nonetheless, the reliability relevant for a given study is the demonstrated reliability for that study.

General guidelines on 'best research practice' are useful for researchers. However, current best practices are culturally anchored practices. Cross-cultural studies differ in many respects from one another and also from monocultural research. As a result, the cross-cultural aspects of a study need to be considered carefully before specific rules for 'best practice' are implemented.

Researchers have to demonstrate the quality of their comparative research. They have to document how instruments were designed or selected, which sample designs were chosen and which fielding procedures implemented (such as number, mode, and timing of contact attempts or response enhancement measures). Documentation habits and views on what count as 'facts' are, incidentally, likely to differ cross-culturally. It is counterproductive to simply assume that techniques standard in one culture can automatically be exported with success to a new cultural context. It is equally counterproductive to simply assume that no instrument can be used in new contexts or that standards that are unfamiliar cannot be met.

1.4   CORE TERMINOLOGY

The most common (and essential) words in a language often have long dictionary entries which try to cover all the 'meanings' and uses acquired over time. In similar fashion, key terms in disciplines tend to gain multiple meanings. In addition, across disciplines, it is not uncommon to find the same terms used to mean different things. The Glossary provides short definitions of key terms used throughout the book. At this point, it is useful to introduce just three central sets of terms:

(1)   'indicators-constructs-concepts' as a general methodological scheme that reflects the association between theory and measurement;
(2)   'error, bias, and equivalence' as terms that serve to clarify the relation between measurement disturbances or nuisance factors and comparability of data;
(3)   'validity' as the litmus test for a survey.

1.4.1 Indicators, Constructs, and Concepts

While the reasons for asking questions such as "How old are you?" may be infinite, the nature of the semantic information requested in the question is fairly clear. Survey research in the Western context usually attempts to make the question even clearer by not asking for age, which changes, but for birth date. How the information received on age is then used depends, of course, on the goals of a survey. Age can be taken as an indicator of multiple constructs, for cohort membership, for example, or for position in the life cycle.

In contrast, asking a respondent to indicate how happy he/she is on the basis of an "All in all, how happy would you say you are these days . . ." kind of question cannot be expected to provide a detailed picture of how 'happy' the person is nor what 'happiness' means to the respondent. Questions that do not ask directly about 'happiness' but ask instead for information on features theoretically associated with 'happiness' may be the culturally necessary way to acquire the information needed. Even in contexts where asking about happiness makes sense and is socially acceptable, multiple indicators are needed to gain insight into a respondent's relative position on whatever construct the questions are intended to measure.

Following Bohrnstedt and Knoke (1988; cf. Bollen 1989; Bollen and Long 1993; Marcoulides and Schumacker 1996; Tabachnick and Fidell 2000) we adopt the following scheme:

(manifest) Indicators → (latent) Constructs → (theoretical) Concepts

In this terminology, indicators (which can be measured) provide access to (latent) constructs (which cannot be directly measured). The constructs, in turn, represent underlying (theoretical) concepts. The scheme introduces the notion of mapping; first how the indicators of an instrument map on the underlying construct, and then how the construct is related to the underlying concept. Indicators are directly related to the questions in a questionnaire. Constructs relate to the domain covered by the indicators. The term often has strong statistical connotations and refers to the factors, clusters, or dimensions as found in multivariate statistical analyses. The third notion, concepts, refers to the theoretical definition of constructs and does not have a measurement perspective.

Under ideal circumstances, concepts are represented adequately in constructs and these, in turn, are well captured in indicators. However, in cross-cultural survey research, this close relationship between levels may be difficult to achieve for two principal reasons. On the one hand, it is not uncommon in social survey research to measure complex constructs with only a few questions, which may then not provide adequate representation of the constructs involved. Poor construct representations result in poor estimates of means and weak correlations of the items with external measures. This problem is compounded in cross-cultural research by the fact that small differences in question characteristics (such as small differences in means or correlations with latent constructs) may be amplified to major cross-cultural differences.

Religiosity questions can help illustrate the distinctions just made between the three terms and notions. Suppose that we want to measure religiosity across a variety of cultures and ask five questions about belief in a Supreme Being and the impact of this Being on a respondent's life. The five questions asked are then the indicators. A multivariate statistical technique that addresses the dimensionality of the questions (such as factor analysis, cluster analysis, or multidimensional scaling) gives us access to the latent construct(s) of the scale formed by the questions. The dimensionality may be identical or different across cultures.

Comparing constructs is of immense importance in cross-cultural survey research. If the data show that the indicators measure the same latent construct in all cultures, how do we know whether that means that the concept of religiosity is identical across cultures? The answer depends on the adequacy and comprehensiveness of the indicators for the cultural groups. For instance, in some religious denominations, attendance at a place of worship is an essential element of religiosity. This could then be tapped by a question about frequency of [church] attendance ('How often do you go to [church]?'). For other religions, attendance at services is a less relevant characteristic of religiosity, whereas observation of rites at home or in the workplace is important. These cultures would need indicators to measure these aspects.

A common problem in measuring broad concepts such as religiosity is whether the indicators used cover all the relevant aspects of the concept in each culture. If indicators of religiosity require to be tailored to a particular religion or denomination, rather than shared across cultures, comparisons of responses from individuals belonging to different confessions/religions are bound to be incomplete or incomparable. In such cases, the concept of religiosity differs, despite the statistical similarity of the underlying dimensions of the indicators. As a result, identity of latent constructs is a necessary but insufficient condition for the identity of concepts.

1.4.2 Error, Bias, and Equivalence

Different cross-cultural disciplines and fields of research have developed individual terminology for describing issues that are particularly pertinent to comparative research. The result is that different fields talk about related or the same phenomena using different terminology. Both the terms used and the distinctions made across these disciplines may in turn overlap with or be different from distinctions and terms familiar in related but monocultural fields. Comparative survey research of a given discipline may thus use terms not familiar in monocultural research of the same discipline or used differently in monocultural research. The terms 'error,' 'bias,' and 'equivalence' and the typological distinctions to which these terms point in different disciplines are good examples of both difference and overlap.

For example, 'total survey error' is a term used in monocultural survey research to refer to a multitude of (possible) sources of error in the survey design and process. The term includes a combination of systematic and unsystematic (supposed random) errors that may lead researchers to draw partially or wholly incorrect conclusions about the characteristics of a population with regard to the construct under study. Groves (1989) distinguishes four sources of error: sampling error, coverage error, nonresponse error, and measurement error. In recent social survey methods research, the first three sources of error in particular have received considerable attention.

Cross-cultural psychology tackles quality and error from a different perspective. It focuses more on problems related to instruments and implementations and gives sampling, coverage, and nonresponse issues far scanter consideration. The grouping of nuisance factors ('error' in survey terminology) in cross-cultural psychology concentrates on 'bias' (e.g., Van de Vijver and Leung 1997a). Bias refers to the presence of nuisance factors that challenge the comparability of measurements across cultural groups. Index scores based on item batteries or scales are an example. If scores of batteries or scales are biased, their 'meaning' or their relation to a specific concept is culture dependent, making *direct* comparison of different cultures impossible. Part Three of the volume discusses the various sources of bias outlined briefly below.

Van de Vijver and Leung (1997a, b) identify three types of bias. The first, called *concept bias*, refers to nonidentity of theoretical concepts across groups. A second type of bias, called *method bias*, involves all the sources of bias arising from methodological aspects of a study, including sample incomparability, instrument differences, interviewer effects, and the mode of administration. The third kind of bias, *item bias,* refers to anomalies at the level of items.

The overlap between the (monocultural) survey research perspective on error and bias and the cross-cultural psychology view of bias is considerable, but they nevertheless differ in emphases and in typological distinctions. In survey research, measurement error is classified in less detail than its cross-cultural psychology counterpart of measurement bias; on the other hand, the psychological typology has a much rougher classification for method bias. An integration of these research traditions is, however, not the goal here. The aim instead is to present empirical findings, explore common problems, and evaluate suggested solutions from both typologies.

The presence of bias and error impacts on the comparability of measures (and index scores derived from these measures) across cultures. The measurement implications for comparability are addressed in the concept of *equivalence*. Equivalence refers to the comparability of measures obtained in different cultural groups. Johnson (1998a) reviews uses of equivalence in survey-related literature and identifies some 40 different definitions, not counting variations within major definitions of equivalence. *Functional equivalence*, one of the most widely used terms, is also the term with the lowest degree of consensus on use. Not surprisingly, therefore, contributions in this volume also use the term differently; we alert our readers to this.

We outline here one systematic classification of hierarchically ordered types of equivalence that has proved useful in cross-cultural psychology (cf. Van de Vijver and Leung 1997a, b).

(1)   *Conceptual equivalence* is the identity of theoretical concepts across cultures. If this type of equivalence is not present, comparisons will not be possible at all.
(2)   Instruments show *measurement unit equivalence* if their measurement scales have the same units of measurement but a different origin (taken as interval and ratio scales in survey research).
(3)   *Scalar or full score equivalence* assumes the same origin across all cultures (i.e., a well-defined zero point). When scalar equivalence is present, direct comparisons can be made using *t* tests or analyses of variance.

1.4.3 Validity of comparisons

Most cross-cultural research is undertaken to compare countries, cultures, or groups on some characteristic. When these characteristics are physical attributes, the likely significance of the findings is clear. Suppose, for example, that a manufacturer of blue jeans measured the height of a representative group of Koreans and a representative group of Norwegians and that the latter group showed a higher mean. Assuming appropriate sampling, the measured difference in average body length can be attributed to actual differences in height between the two populations from which the samples were drawn. The manufacturer could then use such findings to fine-tune manufacture and sales logistics for the two populations, hopefully also taking into account any cultural differences in dress in the two countries. Suppose, too, that questions on empathy were administered to the same Korean and Norwegian respondents and that the Norwegians again showed a higher mean. The inference from the mean to the underlying construct, empathy, is less obvious than in the case of leg length, since various rival explanations may be put forward. Differences in means need not point to differences in actual degrees of empathy. The populations could, for example, differ in the social desirability of demonstrating empathy (for example, in one population showing empathy is a cultural norm, in the other, not).

Problems of competing interpretations are typical for studies in which the relationship between the measurement operations and the underlying construct is only of statistical (probabilistic) nature. Because latent constructs such as empathy cannot be measured directly in the way that height can, there is no manifest external criterion on the basis of which to test the validity of the measurement. Instead, validity assessments are carried out on the basis of extensive statistical testing for nuisance factors. Many methodological issues discussed in the volume arise from the use of probabilistic models in comparative studies.

1.5   OUTLINE OF THE BOOK

The remaining Parts of the book focus on design and implementation issues (Part II), error in the comparative context (Part III), issues relevant to analyzing data (Part IV), and issues related to data access and secondary analysis (Part V).

Part II: The first chapter discusses general approaches to designing questionnaires for comparative projects (*Harkness, Van de Vijver, and Johnson*) and is followed by a discussion of questionnaire translation (*Harkness*). The next three chapters discuss cognitive (*Braun* and *Schwarz*) and crafting (*Smith*) issues of relevance in designing questionnaires and in understanding respondent behavior. The sixth chapter discusses design and harmonization problems with background variables (*Braun and Mohler*) and the Part closes with a chapter on sampling design requirements for comparative projects (*Häder and Gabler*).

Part III: The first chapter provides an overview of types of survey error and their causes, sketching the typological divide between survey research 'error' and the 'bias' framework of psychological research (*Braun*). The second chapter takes bias and equivalence as its starting point and considers instrument design and administration as sources of error (*Van de Vijver*). The three chapters that follow discuss specific sources of survey error. *Couper and de Leeuw* provide the first comprehensive comparative treatment of unit nonresponse, while *Kalgraff Skjåk and Harkness* discuss data collection methods from a comparative perspective. In the concluding chapter, *Johnson and Van de Vijver* provide an overview of cross-cultural research on social desirability.

Part IV: The first chapter explains a number of basic techniques for analyzing bias and equivalence, followed by an overview of basic statistical techniques for substantive analyses (*Van de Vijver*). The second chapter presents an overview of the use of multidimensional scaling to analyze both bias and substantive research questions (*Fontaine*). The third chapter is a study of national identity in Belgium using structural equation modeling (*Billiet*). The fourth chapter demonstrates a cross-cultural multitrait–multimethod study (*Saris*); this is followed by a cross-national study of response functions by the same author.

Part V: The first chapter discusses the advantages and disadvantages of analyzing existing survey data as a secondary analyst, with an emphasis on the comparative context (*van Deth*). The second chapter considers data documentation needs and problems for users and archivists (*Mohler and Uher*), and the final chapter provides an overview introduction to meta-analysis in comparative survey research (*de Leeuw and Hox*).

The last authored contribution to the book consists of a Glossary of Terms used in the book that are important for the comparative context (*Johnson*).

PART TWO

DESIGN AND IMPLEMENTATION

Chapter 2

QUESTIONNAIRE DESIGN IN COMPARATIVE RESEARCH

JANET HARKNESS
FONS J. R. VAN DE VIJVER
TIMOTHY P. JOHNSON

2.1 INTRODUCTION

This chapter examines available methods of questionnaire production for use in multiple cultures. It does not discuss general crafting aspects or cognitive research findings relevant to all questionnaire design but focuses instead on how questionnaires are produced for use across languages and cultures. The chapter is divided into four sections: classification of cross-cultural questionnaire design options (e.g., working with an existing questionnaire or developing a new questionnaire); issues encountered in using existing questionnaires; development of new questionnaires; and implications for designing questionnaires for cross-cultural implementation.

Survey designers have at their disposal a rich literature on the general principles and specific details of survey questionnaire design and crafting (e.g., Sudman and Bradburn 1982; Converse and Presser 1986; Foddy 1993; Dillman 2000; Frazer and Meredith 2000; Peterson 2000) as well as cognitive research on questionnaire design (e.g., Jobe and Mingay 1989; Tanur 1992; Sudman, Bradburn, and Schwarz 1996; Schwarz and Sudman 1996). Other contributions to this volume address questions from these fields that are relevant to questionnaire design. The present chapter treats the general framework of design options for research. An arbitrarily chosen description of the procedure used to develop a cross-cultural quality of life measure (for migraine patients; Patrick and Hurst 1998) lists general steps which are standard in cross-cultural questionnaire design:

(1)   Review of literature and existing instruments
(2)   Establishment of a conceptual framework
(3)   Elicitation of items
(4)   Evaluation of cross-cultural equivalence
(5)   Development and refinement of draft questionnaire
(6)   Evaluation of psychometric properties

(7)   Evaluation of responsiveness
(8)   Preparation of users' manual and, if relevant, a scoring scheme
(9)   Submission to a supervisory council (or other signing-off procedure) and
       distribution.

At each of these stages, questionnaire designers need to consider specific cross-cultural issues. For example, a review of existing instruments should include the countries these have been fielded in and information on the psychometric properties for each individual study.

An obvious difference between mono- and cross-cultural surveys is the multilingual nature of the latter. Several strategies are used to develop instruments for multilingual studies. The scheme presented in Table 2.1 refers to the development of question scales and questionnaires; individual item development can be approached in a similar manner.

TABLE 2.1. Schematic Overview of Questionnaire Design in Cross-Cultural Surveys

Use Existing Material?	Type of Development	Commentary
Yes		(Usually monocultural) questionnaire exported.
	Adopting	(Close) translation (sequential, i.e., source then versions).
	Adapting	Increasing cultural appropriateness by changing part of the item or questionnaire (sequential, i.e., source then versions). Adaptations can be country-specific, become part of source questionnaire, or be shared by several countries.
No		No existing instrument or existing instrument not suitable; new questionnaire developed.
	Sequential	Source questionnaire developed and tested by a small group and then translated in target languages. Also involved when source questionnaire not originally intended for cross-national use.
	Parallel	Source questionnaire developed by multicultural group to be appropriate for all target cultures.
	Simultaneous	(1) Decentering: basically a two-culture procedure. Development moves between two languages and cultures; cultural specifics eliminated to maximize adequacy for both. (2) Common core of concepts implemented in country-specific fashion. (3) Emic–etic: common set of questions supplemented by country-specific sets. Goal is to cover a shared set of concepts more fully.

2.1.1 Development Decisions: Old or New, Adopted or Adapted, and Sequential, Parallel or Simultaneous

The scheme in Table 2.1 is based on three dimensions. First, are old or new questions used? (Strictly speaking, new questions may be either entirely new or modifications of existing questions.) Second, are questions (new or old) modified for one or the other culture or identical everywhere in the project? Usually, if identical questions are used in all locations, one questionnaire functions as a *source questionnaire* and the other versions are translations of this questionnaire. Third, how and by whom are instruments developed? These dimensions determine whether development focuses on finalizing one questionnaire before turning to cross-cultural considerations (sequential approach), whether cross-cultural input significantly shapes development from the beginning (parallel approach), or whether instruments for several cultures are developed more or less together (simultaneous approach). Many of the procedural options are not mutually exclusive. New and modified questions in particular may be developed via various approaches, as indicated next.

Old or new? Because much cross-cultural survey research is based on replication of sections or whole instruments from previous studies, the first decision is whether to adopt existing questions. Existing questions are often Western in origin and are most frequently implemented in 'Ask-the-Same-Question' (ASQ) studies, in which questions are exported verbatim or via translation. Sometimes research specifically sets out to replicate. In other cases, however, researchers have the choice between using an existing instrument (possibly modified) or developing a new one. Various considerations influence the approach chosen, including the type of instrument needed, the degree of cultural anchoring present in available instruments, and locations and languages in which the new questionnaire will be used. Further considerations include the theoretical model of equivalence or comparability that researchers hope to achieve in a comparative project (e.g., structural equivalence or metric equivalence) and, practically speaking, the tension between available funding and what different strategies cost. Proper development and testing of new instruments is an expensive undertaking and comparative testing exponentially increases the costs.

Although the choice between existing and new instruments is presented here as dichotomous, in reality mixtures are not uncommon. For example, 'tried and tested' questions from an existing instrument may be used alongside new questions.

Adopt or adapt? Once the decision is made regarding a new or existing instrument, researchers decide which form of question development to employ. If an existing instrument is used, three options are available. The first option, *adopting* an instrument, consists, essentially, of translating items. This widely employed approach presupposes that closely translated questions will produce an adequate instrument in the target languages (a prototypical ASQ model). The second option takes existing questions as the basis for development, but modifies or *adapts* these somewhat. Adaptation may affect content and/or formulation in order to render an instrument appropriate for fielding in other cultures.

Start from scratch? When no suitable instruments exist, researchers develop a new instrument. This may happen when the research topic is essentially new or existing instruments are fundamentally unsuitable. Instruments already available might contain so many country and/or culture-specific topics, references, or expressions, for example, that designers decide to construct new items designed to avoid systematic sources of cultural bias.

2.1.2   Sequential, Parallel, and Simultaneous Development

Three basic approaches to developing new study materials are briefly presented here and discussed in detail in Section 2.3.

The most common approach is *sequential*. In sequential development the source questionnaire is developed by a group, without deliberate inclusion of multiple cultural perceptions. The questionnaire is usually tested in the context and language in which it was developed, then exported abroad via either translation or translation and adaptation. A second approach, the *parallel* approach, incorporates cross-cultural input from all target cultures and languages while the source questionnaire is still under development. Parallel development is time-consuming and requires considerable effort to coordinate feedback and version changes. At the same time, it is likely to result in instruments that function more effectively across cultures. A third approach, *simultaneous* development, aims to produce different language versions at the same time. Questions in each language may be developed simultaneously through a *decentering* procedure of iterative translation (Werner and Campbell 1970), or a core of common concepts are generated, then questions are individually formulated to suit each culture. In an *emic/etic* implementation (Przeworski and Teune 1970), a set of common questions (etic) are produced through sequential or parallel methods for use in combination with a set of country-specific questions (emic).

Decisions about which approach to use are related to the larger research framework and depend, for example, on the survey research goals and analytic purposes. While some instruments are imported simply for national use, others will be part of a cross-national project. In the first case, analysts may be interested in national results and unconcerned with cross-country comparisons. The unit of analysis (structure- or level-oriented), the statistical analysis required, and the availability of existing instruments and data on their quality also determine which design options are open, as do cost-effectiveness, available funding, and the countries involved.

The approach taken has important implications for the remainder of the project, including the scope for statistical analysis. ASQ models, for example, allow for potential comparison at the highest levels of equivalence (see Chapter 10). At the same time, the ASQ approach requires designers to establish questionnaire adequacy in all pertinent cultural contexts before fielding. The replicated material frequently used in sequential ASQ approaches poses a further problem: questions may be culturally anchored and difficult to export via translation alone (see Section 2.2.1).

Table 2.1 provides an outline of the points introduced here and discussed in the two following sections. For practical purposes, the remaining discussion is organized as "working with existing questionnaires" and "developing new questionnaires." In practice, however, cross-cultural studies may use existing questions alongside new questions.

2.2 WORKING WITH EXISTING QUESTIONNAIRES

2.2.1 Adopting an Existing Instrument

Cross-cultural questionnaire design, as in the monocultural context, routinely begins not by writing survey questions, but by identifying literature, concepts, and available instruments relevant to the research topics. At the same time, many disciplines demonstrate a clear preference for existing instruments, that is, questions that have been tried and tested through previous use. These are then exported verbatim (if translation is not necessary) or on the basis of close translation. Single items, item batteries and scales, and entire questionnaires are adopted for multilingual and/or multicultural contexts following this approach.

Standard assessments and tests in the fields of health, education, and psychology are adopted and implemented across countries with greater frequency than is customary for social science surveys. The number of instruments adopted and the languages in which instruments are implemented have also grown considerably in recent years (Hambleton 1994; Van de Vijver and Hambleton 1996; Jeanrie and Bertrand 1999; Tanzer and Sim 1999). Thus, examples of extensively tested and documented instruments abound in health, education, and psychology research, the considerable financial investment being recouped in part through commercial marketing. Examples include the State–Trait Anxiety Inventory (Spielberger, Gorsuch, and Lushene 1970), the Beck (1961) Depression Inventory—both of which have been translated into more than 30 languages—and the Medical Outcomes Study Short-Form 36 (SF-36) health assessment instrument which has been translated into 40 languages (see http://www.sf-36.com). Examples from mental testing include the Wechsler Intelligence Scales for preschoolers (Wechsler 1989), children (Wechsler 1991), and adults (Wechsler 1981) and the Coloured, Standard, and Advanced Raven's Matrices (Raven 1938, 1956). The Raven tests use nonverbal stimulus materials and the short test instruction is always closely translated. The Wechsler tests, interesting from the questionnaire design perspective, are composed of various subtests. While some subtests are closely translated (e.g., digit repetition tests to measure short-term memory), some require adaptations (e.g., mental arithmetic items may need adaptation to local currencies), and one subtest measuring vocabulary knowledge is developed with a different set of words for each language.

If existing items are incorporated into a new study—especially if use in many different surveys or in longitudinal replications has gained them a documented 'survey pedigree'—they are commonly adopted unchanged and translated as necessary. It is assumed that close translation of a source questionnaire (discussed in Chapter 3) will result in a *target language questionnaire* that adequately measures the constructs of the source questionnaire. However, research indicates that even questionnaires which do not require translation may be inappropriate when used with a new cultural group (e.g., Weech-Waldonado et al. 2001).

In both mono- and cross-cultural research, the advantages of using available instruments are considerable (see Table 2.2 following). Not only does it make little sense to re-invent the survey wheel (Scheuch 1989), but previous use is also seen as the best pretest. In addition, data are available on how questions performed. Replication is also how instruments acquire a survey pedigree in the first place. Since financial investment in instrument development is considerable in some fields of research, the business of question pedigree is taken seriously. Replication of items also affords researchers the opportunity to compare their new data with data collected at different times and in different places. Moreover, by translating existing items, researchers interested in numerical score comparisons across cultural groups can keep all their statistical options open. Many researchers feel confident that— provided translation was accompanied by appropriate assessment—the translated questionnaire is suitable for cross-cultural use.

The adopting approach has a number of disadvantages, in part related to the fact that pedigreed items raise somewhat different translation and translation assessment issues than do new questions. For example, questions developed for one context are culturally anchored. This perhaps contributes to their original success, but it may detract from their suitability for other contexts. However, pedigreed questions have usually gained their tried and tested status in a given form and formulation and thus if used in a new study, the questions are usually adopted, translated, and fielded in an ASQ model. Since pretesting of translated questionnaires is often much less thorough than monocultural pretesting, the suitability of these questions for new contexts may remain untested. The same reservations hold for questions developed for specific comparative contexts (and specific times) then used elsewhere. These should also be tested for their suitability for new contexts.

Further possible limitations are that country-specific aspects of a topic may be less well addressed in this approach, since questions which are used for ASQ studies across numerous countries are often necessarily more general. A research group designing questions for an environmental study (reported in Witherspoon, Mohler, and Harkness 1995) ran into multiple problems of this kind. The multicultural team wanted to measure attitudes and behavior related to domestic use of resources such as water, electricity, and heating fuel. In order to ask respondents in different countries in detail about how they regulate energy and resource use, questions need to be framed in terms of common systems and procedures. If water is not metered or water and/or heating bills not calculated on the basis of amounts used, available and tested questions linking natural resource conservation to payment for consumption become nonsensical. Alternatively, if respondents use well water rather than piped water, if water is in chronic short supply, or if the real priority is how to find cooking fuel and not how to conserve heating fuel, the pool of common pertinent questions dwindles.

TABLE 2.2. Advantages and Disadvantages of Different Questionnaire Designs

Design approach	Advantages	Disadvantages
Existing instrument		
Adopting	• Easy and inexpensive to produce • Many statistical techniques available; affords examination of all types of equivalence • Questionnaire pretested in at least one culture	• Cultural anchoring raises problems of cultural suitability and translation • Replication of monocultural questions makes close translation likely, even if unsuitable
Adapting	• Easier and less expensive than a new instrument • Cultural context suitability enhanced • Adaptations possible at source questionnaire level or at individual country level	• Few statistical techniques available to deal with dissimilar stimuli across language versions • Adaptation documentation required • Coordination of different adaptations needed
New questionnaire		
Sequential	• Potentially high cultural context suitability • If tested, concept or item inadequacy or bias for source questionnaire unlikely • Many statistical techniques available; affords examination of all types of equivalence	• High development cost (but lower than parallel and simultaneous) • Cross-cultural suitability often not pretested and potentially problematic
Parallel	• High suitability for multiple contexts • Concept or item inadequacy or bias unlikely • Many statistical techniques available; affords examination of all types of equivalence	• High development cost and complex organization
Simultaneous	• 'Common denominator' of construct in all cultures involved • Cultural idiosyncrasy bias unlikely • Many statistical techniques available; affords examination of all types of equivalence	• High development cost and complex organization • After removal of culture specifics, remaining items may not represent full concept; items may be unidiomatic or vague • May not be practicable for multinational studies

In this way, questions central to measuring relevant constructs in one context may be irrelevant or impossible to ask in other contexts.

Recent research indicates a growing awareness that adopted and translated questions require extensive pretesting prior to full fielding (e.g., Hambleton, Merenda, and Spielberger forthcoming). Ideally, researchers should have access to pretest evidence and not just 'after the event' evidence. However, many instruments currently used in comparative research lack evidence of their suitability. Some research fields—for example, health, education, and psychology—are more actively concerned about remedying this deficit than other fields. Documentation published in the Medical Outcomes Trust Bulletin (e.g., MOTB 1997) indicates that comparability is often uncertain prior to implementation but that efforts are underway to remedy this. Recent mental health and aging research (Skinner 2001) emphasizes the need for more intensive pre-implementation appraisal and testing. In many demographic and social and political science surveys, testing of translated versions is much less thorough than that of source versions.

Questions are also used across countries without translation. For instance, English language instruments developed in the United States or Great Britain might be used in Australia, New Zealand, Ireland, or Canada. Here, too, care is required since the "same language" is seldom the same language in practice. Cross-cultural literature has acknowledged the problem for different varieties of Spanish spoken in the United States (e.g., Artiola i Fortuny and Mullaney 1997; McGorry 2000), but the point applies equally to other languages—including English, German, and French—and for other contexts. In essence, questions not developed for a given context must be tested prior to implementation and examined not just for regional color or orthography but for differences in referential or cultural meaning. Researchers working in different cultures but in the same language should be aware they might be blinded by term identicality. An ISSP question referring to *social security* was developed for the United Kingdom, but was also fielded in the United States, Australia, Ireland, and Canada using the same term, *social security*, despite the fact that the term refers to different programs and benefits in these different countries.

Good formulations can also become outdated, that is, questions or terms become unsuitable either because the (social) conditions they imply have changed (c.f., Skinner 2001) or because language usage has changed. For example, terms used to refer to minorities (e.g., ethnicity or sexual orientation) have shifted in English quite rapidly. Rates of change also differ across cultures and languages, thus a change of terminology (or question) needed in one language may not be needed in another.

These considerations apply to visual materials, too. Intelligence tests such as Wechsler's scales contain subtests with drawings of common household objects. Some of these must be periodically updated to reflect technological advances in both equipment and designs. Visual components of an instrument have, in general, to be assessed for their cross-cultural suitability (see 2.2.2).

2.2.2 Adapting Instruments

Adaptation is the second most common design approach after adoption. In the context of this chapter, adapted questions are those in which some component has been deliberately altered independent of unavoidable translation change. These adaptations may be substantive, relate to question design, or consist of slight wording modifications. Regardless of the form of change, the aim of adaptation is to render questions culturally or linguistically appropriate in a cross-national context.

If question adoption is not an option, adapting offers several advantages. For example, it allows researchers to ask a question at least similar to one that is known to be useful. Additionally, the adapted question may belong to a scale in which the rest of the items were adopted unchanged; thus an adapted item may fit better with other items in the scale. Adapting also circumvents the need to identify question topics and, possibly, question design formats, while heightening the appropriateness of the questions for new contexts.

One potential problem with adapting is that 'tinkering' with an item implies making small changes to it. Small changes, however, are not the same as insignificant changes. Therefore, adapted questions should be treated as new questions and not automatically compared with original versions and their performance. The equivalence of modified items must be tested and demonstrated, as described in Chapter 10. Note that while question changes are carefully documented in some surveys, they are rarely documented in others. A further drawback for some purposes is that adapted items provide limited scope for statistical analysis, since only statistical tools accommodating partly dissimilar instruments can be used.

Question adaptation takes four primary forms, the first of which is *terminological and factual adaptation.* Examples of terminological and factual adaptations include country- or language-specific issues such as school-leaving age, date of last general election, or name of official government legislature (e.g., *Congress*, *Parliament*, and *Bundestag*). Such alterations are necessary even when a questionnaire does not require translation but is fielded in different countries that use the same majority language (e.g., various English-speaking countries).

Language-driven adaptations, the second type of question adaptation, are adaptations made necessary by virtue of structural differences between languages. One example is how languages indicate physical and grammatical genders. Source questionnaires in English can be easily formulated to avoid gender-specific references by, for example, using plural forms and *he/she* references. Avoiding gender-specific reference in some languages is more complex, because grammatical structures automatically indicate gender. Chapter 3 discusses effects that this may have on survey question meaning.

A third type of question adaptation is *convention-driven adaptation*. For example, questionnaires and tests developed in English-speaking cultures nearly always reflect the left-to-right processing common to cultures in which hard copy language processing proceeds from left to right and top to bottom. People learn to process visual stimuli from left to right (e.g., English comic strips always follow this convention). Research on English self-completion formats has indicated the relevance of processing conventions for optimizing design (Jenkins and Dillman 1997). Cultures that do not read and write from left to right must have questionnaire components adapted accordingly (Tanzer forthcoming).

Culture-driven adaptations, the fourth type, involve changes needed because source and target cultures have different norms, customs or practices. One culture-driven adaptation in a three-country study involved an item from the Marlowe-Crowne Scale which measures social desirability (Johnson et al. 2001). The item "Have you ever taken anything (even a pin or button) that belonged to someone else?" was unsuitable for adoption in Turkey and Germany. The Turkish translator first understood *button* as a fastening device, as in *shirt button*. In the course of clarifying the intended reading the researchers also decided that "pin or button" was not suitable for adoption in Turkey (or for that matter Germany) because wearing decorative buttons or pins is not a common practice in either country. The item was therefore adapted; *pin or button* was replaced with an object that is easily lost, potentially attractive, and of low material value.

2.3   DEVELOPING NEW QUESTIONNAIRES

More development options are available when questionnaire materials are developed from scratch. Regardless of the approach used to produce the new questions, developers have the opportunity to tailor the questionnaire to fit the cross-cultural context in which it will be used. Even if the design chosen involves producing a source questionnaire, this can allow more freedom of choice than a sequential ASQ model with replicated items. In addition, an adaptation component can be added. In this way, the source questionnaire can be individually extended with functionally equivalent sections in given countries, albeit with the advantages and disadvantages discussed in the preceding section.

Alternatively, designers may agree upon constructs and dimensions to be measured, with or without developing a questionnaire in one language. Culture-specific questions that adequately measure the shared concepts or dimensions are then developed for each country. In traditional decentering models, different cultures share a questionnaire but the questions are developed simultaneously in each language (Werner and Campbell 1970).

In both these approaches, researchers are free to compare nomological networks. However, neither approach is suitable for large multinational survey research. Not only is considerable skill and comparative design expertise needed to develop and test such questions across a range of countries (cf. Johnson 1998a) but these skills must be available in each language/culture involved in the survey. In addition, decentering may result in rather vague questions (see 2.3.3).

A combined approach agrees on a set of common, universally applicable (*etic*) questions, then supplements these with country-specific (*emic*) questions. Little detailed information is available regarding implementation of emic/etic models in large-scale surveys (but see Triandis and Marín 1983; de Vera 1985; Berry 1989, 1999). Nonetheless, the procedure can be a useful method of data collection in cultures that are very different from one another. It can also be used in more countries than can the decentering approach. By using a combined approach, data collected with ASQ questions can be enriched with country-specific information. However, asking questions salient for one cultural group but without proper salience for another is problematic (cf. Johnson 1998a). In addition, questions that make sense in only one country but are asked in all, or questions that can only properly be analyzed for one country, are very costly questions.

Literature on questionnaire crafting and cognitive questionnaire design indicates that producing appropriate question wording and response categories for a given study and population is no simple task. In the comparative context—depending on the development route chosen—this task may be further complicated if the source questionnaire is destined for translation as part of an ASQ model. In this case, the source questionnaire must "get it right" for all succeeding versions. At the same time, new questions provide developers with greater opportunity to identify suitable wordings, wordings that can be tested *and altered* to enhance suitability. Thus, difficult adaptation or translation problems may be reduced and formal design components (e.g., response scales) can be better tailored. Research findings suggest that identical response scale designs may not be suitable for all cultures and languages (e.g., Hayashi 1992; Johnson et al. 1997; Skinner 2001; see, too, Chapters 3 and 5). For example, 5-point bipolar scales with strongly accented end points have been claimed to be less appropriate for Japanese populations than somewhat longer or less extremely defined response scales (Hayashi 1992). Moreover response scales interact with response styles and these also vary across cultures (Hui and Triandis 1989; Marín and Marín 1989; Javeline 1999).

A further advantage of new questions is that they allow researchers to investigate the dimension or construct of interest; researchers restricted to using available items may not be able to ask precisely the questions they would like (cf. Chapter 19). An ASQ model with new questions, moreover, allows researchers to target the highest form of equivalence (i.e., full score equivalence; see Chapter 10). If translated questions measure identical dimensions and score equivalence has been established, findings can then be compared item-for-item and scale-for-scale across a series of countries. This is the preferred option if the analysis compares both country average scores and structures of targeted concepts (see Chapter 14). In sum, well-developed and well-tested new instruments should reduce problems of equivalence. At the same time, they do not reduce the need to demonstrate equivalence.

The disadvantages of using new question go almost without saying. Development costs are high if cross-cultural testing is carried out. Testing and documentation of pedigreed questions—often extensive, even if not undertaken on a comparative basis—are not available. If new items for cross-cultural contexts are carefully tested, development effort and costs multiply greatly. Indeed, the prospect of comparative testing and documentation of a questionnaire scale—let alone an entire questionnaire—across twelve or more countries may seem insurmountable for both research budgets and available staff. However, if questionnaires are not tested, the risk of method bias is high. If, as is frequently the case, pretests are carried out in only a limited number of countries, findings from these can only be taken as a first and rough indication of an instrument's performance and suitability.

2.3.1 Sequential Development

Although various elaborations of a sequential approach to questionnaire development are possible, this chapter outlines a conservative model. Sequential development commonly consists of two distinct phases executed by (potentially unrelated) work groups. While the questionnaire may be designed for use in multiple cultures, it is usually fielded in only one or two contexts. The composition of the design team does not normally match the heterogeneity of the target populations. Upon completion, the questionnaire is exported to other contexts via translation or, less commonly, adaptation plus translation. As mentioned, even for questionnaires not requiring translation, researchers should consider language and adaptation needs when implementing in new contexts.

The sequential model is the most commonly chosen approach for both new question designs and when existing instruments are exported to new cultural contexts. The Eurobarometer is only one well-known example of a sequentially developed survey. Questions included in Eurobarometer surveys are both borrowed from other studies and designed explicitly for the Eurobarometer. Once the French and English source questionnaires are completed, other Eurobarometer versions follow via translation.

Although the sequential approach is economical and relatively simple to organize, it can also be problematic because, as already indicated, all the instruments require pretesting. In a sequential model, however, the source questionnaire is commonly finalized before other versions are produced. As a result, the design team has little opportunity to receive feedback from advance translation or from multiple pretests that could identify problems in the source questionnaire. The translating teams, on the other hand, are faced with the often rigid confines of fixed source item formulations.

One problem is that questions that seem optimal on paper in the source questionnaire language may prove unsuitable in another language and/or context. Two ISSP 1999 questions investigating respondents' views on income illustrate this well. The two English questions "Is your pay just?" (followed by response categories on the pattern of "much less than is just") and "Would you say that you earn" . . . (followed by response options on the pattern of "much less than I deserve") proved difficult to translate closely in German because a single word, *gerecht*, would have been a good translation for both *deserve* and *just*. Using *gerecht*, however, would have rendered the questions almost identical.

ISSP Japan had different problems with these questions. Japan could not field these (and other income questions) because it was not socially appropriate for Japanese respondents to assess others' earnings or to evaluate their own earnings as 'just' or 'unjust'.

In sum, neither newly developed nor existing instruments can be exported blindly. They require careful selection and testing and translated versions should be assessed in terms of linguistic and pragmatic suitability (see Chapter 3) and performance (cf. Chapter 10). Psychological instruments commonly address a topic with numerous items; analysts can thus afford to shed questions that do not 'work' in the field. Social science questionnaires with only one or two questions per topic obviously do not afford researchers this luxury.

2.3.2 Parallel Development

A parallel model seeks to "establish a basis for cross-national comparability in the original development and validation stages of an instrument" (MOTB 1997) by combining the linguistic and assessment expertise of developers or consultants from all the involved target cultures involved. The questionnaire is usually developed in one language, however, and may well contain some degree of cultural anchoring. Before the source questionnaire is finalized, it is subjected to multicultural testing. Following finalization, design teams produce final translated or translated and adapted versions.

Occasionally, the questionnaire is developed in a *lingua franca* shared by developers. Whatever the language used to develop a questionnaire, if a multicultural group is involved, some participants will be at a linguistic disadvantage. When development is undertaken in a *lingua franca*, the source questions themselves may not be properly appropriate wordings for use in the language of the source questionnaire. (These could then be 're-formulated' for fielding much like other versions, only without translation.) The crucial issue in either case is that misunderstandings about the purpose and content of questions do not go undetected.

Advance translation (Harkness and Schoua-Glusberg 1998) is a valuable tool in parallel development. It involves translating the questionnaire into other languages before the source questionnaire is finalized. Experience has shown that many translation problems linked to source questionnaire formulations do not become apparent—even to experienced cross-cultural researchers—until a concrete attempt is made to translate the questionnaire. Advance translations need not be as careful and laborious as translations for final versions, since their purpose is to highlight problems in the source language formulations which, unless modified, would cause problems later. In addition, if the source questionnaire is to be annotated (Harkness et al. forthcoming), advance translation feedback is a rich source of information on what requires annotation. This is particularly relevant for annotations for languages and cultures furthest removed from the source questionnaire language; these are otherwise unlikely to receive much consideration in annotation notes.

The ISSP uses a parallel approach for module development, combining questions borrowed from other studies and questions designed specifically for the ISSP. A multicultural drafting group develops preliminary versions of the source questionnaire, other ISSP members are called on to provide feedback, and each module is finalized at a general assembly of all members. At the same time, although advance translation is recommended in the ISSP, it is seldom employed.

A further example of the parallel approach, according to the Medical Outcomes Trust Bulletin (MOTB), is the European Quality of Life Instrument:

> The EuroQol instrument is a multidimensional health related quality of life (HRQL) profile capable of creating a general cardinal index of health. The EuroQol instrument has been translated and evaluated through the parallel approach—it was developed by the EuroQol Group, a multi-country, multi-center, and multi-disciplinary organization which convened in 1987 with the objective of developing a standardized instrument to describe quality of life. This instrument is currently available in English, Spanish, Catalan, Dutch, Finnish, Norwegian, Swedish, French, and German, and is intended for use in concert with other HRQL measures. The instrument has been available since 1990 and currently, there is preliminary support for the comparability of health valuations across the countries considered. However, there is little published regarding methods of adaptation, and similarly, psychometric properties await further evaluation. (MOTB 1997)

In sum: a parallel approach holds several important advantages over the other development schemes, but these come with a price tag. It is more time-consuming than a sequential approach, for example, and requires considerable coordination effort to document and integrate feedback into successive draft versions of questionnaires.

2.3.3 Simultaneous Development

In simultaneous development, translation and question development go hand in hand. Decentering is one of the oldest forms of simultaneous development (Werner and Campbell 1970; see also Schoua 1985; Tanzer et al. 1997).

It is a technique which begins from a *draft* questionnaire in *one* language and produces final questionnaires in, usually, *two* languages through a process of paraphrase and translation between language one and language two. In essence, the decentering process removes culture-specific elements. In the terminology of this chapter, decentering combines adaptation at draft questionnaire level with iterative translation. The technique is usually applied in two-culture projects and can be useful when language structures and cultures differ greatly. The advantage is that the 'same' questions are asked in each culture and language, allowing researchers to target full score equivalence in contexts where a sequential ASQ is unsuitable. A decentering approach has also been applied to existing instruments; Cortese and Smyth (1979) used a decentering strategy to produce a Spanish version of an English acculturation questionnaire.

Intuitively appealing as the process may be, it has limitations. The first is practical: decentering is a laborious, expensive procedure for two languages, and would be even more so for a group of languages. In addition, decentering leads to question formulations that—insofar as possible—avoid cultural specifics. However, minimization of cultural references may leave questions vague and/or strangely neutral and unidiomatic, resulting in unwanted ambiguity. Dimensions presupposing a specific context of occurrence, such as voting behavior or concrete expressions of religiosity, do not lend themselves to decentering, since decontextualized measures may not sample the range of behaviors associated with these adequately.

As an alternative to decentering, designers may combine a core set of common etic questions with an emic set of questions that allows for more cross-cultural variation (Przeworski and Teune 1970). In this way important specifics for individual countries—missed in the common core questions—can be covered by country-specific questions. This approach avoids the problem of "emptiness" that can plague decentered items and permits questions to be more closely attuned to individual cultural contexts. This fine-tuning also has its price; questions that differ across countries afford only limited statistical options for comparison.

The Medical Outcomes Trust Bulletin presents the WHO quality of life assessment instruments WHOQOL-100 and WHOQOL-BREF as examples of a simultaneous approach to instrument development. We draw attention to the modest note on which the description ends.

> The WHOQOL-100 and WHOQOL-BREF incorporate many dimensions of life and can determine impact of disease and health intervention on quality of life. Because this instrument is based on cultural norms of various groups, as opposed to the traditional method of basing item selection on one set of culturally homogeneous data, it truly is a cross-cultural measure. Furthermore, it is capable of capturing quality of life data in countries which lack validated measures. While testing of the psychometric properties of the WHOQOL instruments is ongoing, this collaborative effort is a fine example of what is possible when we start to think beyond domestic concepts of quality of life and well-being. But until resources and vision allow for the modeling of such a collaborative spirit and global philosophy, individuals interested in cross-cultural measures of quality of life and health will have to rely on translation and adaptation efforts. (MOTB 1997)

2.4   CONCLUSION

Cross-cultural research as an enterprise harbors some persistent myths (Hambleton and Patsula 1998), one of which is an almost unwavering belief that questionnaire design in cross-cultural studies is a matter of choosing and translating a questionnaire with a good (Western) pedigree. This chapter attempts to illustrate both the advantages and limitations of the most important approaches in use.

Blind reliance on instruments not tested for cross-cultural suitability is obviously a mistake in that pertinent cross-cultural differences may remain untapped, while differences noted may in fact be measurement artifacts. Several chapters in this volume indicate the wide variety of false conclusions that can result from such artifacts.

Regardless of the design model selected, a deliberated choice will serve the scientific community better than uncritical acceptance (or rejection) of particularly 'good' or well-known measures. Careful documentation of design decisions and implementations will enable cross-cultural researchers to begin accumulating a knowledge base of experience that will promote greater awareness of the relative advantages of each approach and move the community closer to establishing "best practice" recommendations with confidence.

Comparative research requires designs tailored to accommodate multiple cultural—not monocultural—requirements. A primary requirement for cross-cultural questionnaires must be their proven suitability for all the cultural groups to be studied. However, a clearer understanding of the requirements for cross-cultural instrument design and crafting must still be sought. Cross-cultural developments need, for example, to turn attention to question crafting and cognitive research findings in relation to translated versions, not just source questionnaires. The current assumption seems to be that crafting and cognitive characteristics also travel through translation. At the same time, study report documentation, in the social sciences at least, also commonly omits details of how instruments were translated or pretested.

Across disciplines, development procedures, like instruments, differ greatly. While cross-cultural psychology may focus more on assessing performance or traits with multiple measures than on crafting individual questions, social science research may over-rely on past performance in other contexts and neglect testing for new contexts. Cross-disciplinary cooperation and an intensified exchange of relevant strategies and research findings may prove to be an important catalyst in improving comparative survey design in general.

Chapter 3

QUESTIONNAIRE TRANSLATION

JANET HARKNESS

3.1 INTRODUCTION

The foregoing chapter indicates that the majority of survey translations are carried out as part of an 'Ask-the-Same-Question' (ASQ) model. The present chapter focuses on questionnaire translation and assessment within this framework and therefore assumes that the questions in the source questionnaire are known to be suitable for fielding in the cultures and languages required. Issues related to questions that have inherent weaknesses as source questions or that 'work' well in one language and culture but do not travel well are dealt with in other chapters (see Chapters 2, 4, 5, and 6).

The chapter is divided into two main parts. The first discusses the practical implementation and assessment of questionnaire translation and has three emphases: (1) important approaches and procedures available for translating questionnaires and assessing translation output, (2) the kind of people needed and tools and job specifications for these, and (3) options for the translation/review/adjudication process central to arriving at a final translated version. The second part of the chapter discusses language issues directly relevant for deciding on translation versions. To show that translation is not just a matter of finding the right words, the discussion focuses on structural problems. It considers meaning and close translation in survey translation and then presents two examples of how structural differences across languages conflict with close questionnaire translation. Before turning to these, a few preliminary remarks are called for.

Whenever possible, translation should be integrated into the study design. In practice, however, translation rarely is seen as part of questionnaire design and usually is treated as an addendum. In most instances, translation begins once the source questionnaire has been finalized. Sometimes this is unavoidable. For example, testing and diagnostic instruments that prove successful for one context often are 'exported' in translation long after they have been developed. In this case, meticulous screening of both translations and question performance is needed before the instruments can be assumed to be culturally comparable (see Chapter 2). Sometimes, especially in within-country research, the need for translation may become clear only when the sample is drawn and includes, say, minority language speakers. Even in cross-national projects, however, it may not be possible to determine exactly how many (or which) languages will be involved in fielding. In numerous regions of Africa, multiple languages are spoken within one geographical area. Thus, both regional and national samples could involve many languages. The oral 'on the fly translations' sometimes used in these contexts raise numerous problems, since researchers have no control over what is asked and what responded.

Translation issues are one of the most frequently mentioned problems in literature dealing with empirical comparative research, both qualitative and quantitative. An extensive literature dealing with diverse aspects of instrument translation is spread across different disciplines—most prominently, health research; quality of life research and related fields, such as aging research; education; and cognitive and psychological testing. This literature is too large and too heterogeneous to review here and can be best accessed through discipline-tailored searches. In addition, many of the publications referred to in the chapter include extensive references.

Useful general guidelines for survey translation are already available; psychology and educational testing have produced a series (e.g., Van de Vijver and Hambleton 1996; Hambleton and Patsula 1998; Van de Vijver and Tanzer 1997). In social sciences, Harkness and Schoua-Glusberg (1998) present a critical overview of procedures used to translate and assess questionnaires. Guidelines on how to write good questions in English are useful aids in formulating source questionnaires intended for translation. At the same time, it is important to remember that these apply to formulating in English, not in other languages. Much of the literature, however, neglects modern translation theory and research from linguistics and pragmatics; we suggest these have much to offer questionnaire translation and assessment.

3.2   PEOPLE AND PROCEDURES FOR TRANSLATION AND ASSESSMENT

We turn now to the practical implementation of questionnaire translation and assessment. The section focuses on the kind of people needed to translate and review translations; on important translation and assessment approaches and procedures; on selection, training, and job specifications for translators; and on providing the tools translators need to do their work.

3.2.1   Translators, Reviewers, Adjudicators

Three different sets of people are required to produce the final version of a translated questionnaire: *translators, translation reviewers,* and *translation adjudicators.*

Translators should be skilled practitioners who have received training on translating questionnaires. *Reviewers* need to have at least as good language abilities as the translators but also must be familiar with questionnaire design principles, as well as the study design and topic. The *adjudicating body* is responsible for making final decisions about which translation options to adopt, preferably in cooperation with reviewers and translators. Depending on the size and staffing of a project, the adjudicating body may be just one person, such as the principal investigator, who carries the ultimate responsibility for the version fielded. Adjudicators must understand the research subject, know about the survey design, and be proficient in the languages involved. In cases where the person with the final responsibility does not understand both source and target language well, one or more additional consultants with the appropriate language and survey proficiencies have to be included.

3.2.2 Approaches

Translation is necessarily a text-based undertaking; translation assessment can be based on evaluations of textual quality and also on statistical analysis. Text-based assessment checks whether, for example, a translated question linguistically and pragmatically asks the same question as the source question, whether an optimal version has been achieved (e.g., simple and clear) that is tailored to the intended audience (such as the oldest senior citizens, children, or specific linguistic groups), and whether the translation reads naturally or signals it is a translation. It also establishes whether parts of a question are missing, whether the response categories are appropriate, and whether measurement properties seem to be intact. Translation and these forms of assessment essentially are based on subjective judgments about the translated materials. This is one reason why translation and assessment is best undertaken in a team environment, as this leavens the subjective nature of decisions about which version is 'best.'

Team approaches to translation and assessment have been found to provide the richest output in terms of alternative options to choose from for translation and in terms of a balanced critique of versions (Guillemin, Bombardier, and Beaton 1993; Acquadro et al. 1996; McKay et al. 1996; Harkness and Schoua-Glusberg 1998). The team should bring together the mix of skills and disciplinary expertise needed to decide on optimal versions. Together, members of this team must supply knowledge of the study, of questionnaire design, of fielding processes (Van de Vijver and Hambleton 1996; Johnson et al. 1997), and of the target field relevant for administration; they also must have the cultural and linguistic knowledge needed to translate appropriately in the required varieties of the target language (e.g., Acquadro et al. 1996; McKay et al. 1996). In many cases, a team-based process will rule out using translation bureaus, since, for example, translators may work long distance or not be allowed to interact directly with clients or each other. It also rules out simply passing on the work to a fielding agency and letting them 'get on with it'.

Two main forms of team-based work are used in questionnaire translation: the *committee approach* (Schoua-Glusberg 1992) and what we call an *expert team* approach. In committee approaches, a good part of the work, apart from first translations, is done working in a group; in expert team approaches, members do more of the work individually, or in pairs, than as one unit. The alternative to team approaches is to have one translator produce one translation, which is then appraised in any of various ways. This also is discussed in the following section.

3.2.3 Five Basic Procedures: TRAPD

Five basic procedures are involved in producing a final version of a questionnaire:

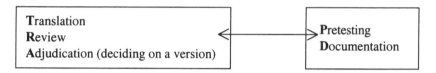

All or some of these procedures may need to be repeated at different stages. For example, pretesting and debriefing sessions with fielding staff and respondents will lead to revisions; these then call for further testing of the revised translations. Even if a committee approach is not employed, any strict compartmentalization of procedures and people involved (for example, leaving translators out of assessment stages or reviewers out of adjudication stages) is counterproductive and not recommended.

Team Application of TRAPD
To simplify the presentation somewhat, we focus here on a committee-based approach. An expert team approach would compartmentalize some parts of the process more. We also assume a simpler model, in which the adjudicator can attend at least final stages of the review process or is, alternatively, one of the senior reviewers in the team, with a dual role of reviewer and adjudicator. Other procedures and further elaborations on the basics presented here are, of course, possible (e.g., McKay et al. 1996; Hambleton forthcoming). Committee translation can be organized in two main ways: parallel translations and split translations (Schoua-Glusberg 1992).

Parallel translations: Several translators make independent parallel translations of the same questionnaire (Brislin 1980; Schoua-Glusberg 1992; Guillemin, Bombardier, and Beaton 1993; Acquadro et al. 1996, here called team translation). At a reconciliation meeting, translators and at least one translation reviewer go through the entire questionnaire discussing versions and agreeing on a final review version. In some instances, the adjudicator may attend the review process or be a reviewer; otherwise the version produced through discussion moves on to adjudication.

Split translations: Task-splitting can save time and effort, particularly if the questionnaire is long (Schoua-Glusberg 1992). At least two translators plus a third person for the review process are needed. The translation is divided up between translators in the alternating fashion used to deal cards in card games. This ensures that translators get an even spread of material from the questionnaire. Each translator translates his/her own section. At a reconciliation meeting, translators and one or more translation reviewers go through the entire questionnaire discussing versions and agreeing on a version. The adjudicator(s) may attend the review process and become involved in the review. Alternatively, the reviewed version moves on to adjudication.

Care needs to be taken to ensure that consistency is maintained across the translation. This is true for any translation, whether produced in parallel fashion, using a split approach, or by one translator. However, the modified committee approach may require particular care. For example, it is conceivable that two translators translate the same expression but come up with suitable, but different translations that might then not be discussed as 'problems.' Consistency checks can ensure that one translator's translation of, say, "What is your occupation?" as "What work do you do?" can be harmonized with another translator's rendering as "What job do you have?"

Committee translation and review followed by committee adjudication: Committee approaches often merge review and adjudication wholly or in part, depending on the expertises of the team and practical considerations such as time schedules and locations. Alternatively, two committee rounds can be used, one to review and revise a version and one to decide whether the revised version will be accepted. In this case, translations are produced following either the parallel approach or the split approach and reviewed in committee as described above. A second committee round is organized for reviewer(s), adjudicator(s), and consultant(s). This route is useful if the person officially responsible for signing off a version needs time to consult in depth before deciding on which version to adopt. This could be the case if questionnaires are complicated or different subject expertise is needed for individual parts. Two rounds also could be more efficient if the person (or persons) officially signing off on a version is not proficient in one (or both) of the languages involved. For example, questionnaires for a national population study in Spain could be needed in Castilian Spanish, in Catalan and, in some circumstances, in Basque. A similar study in Israel might need a questionnaire in Hebrew, Arabic, Russian, and English. If two rounds are used, the adjudication round needs to include reviewers, as well as documentation from the review (see Section 3.2.5), in order to benefit fully from insights gained during the review process.

One-to-One Translation

In many disciplines, it is normal to use only one translator; it also may be common to rely on the translator's reputation as a quality assurance. In survey translation, too, using one translator is not uncommon. Here it is sometimes called the *one-to-one*, *solo* or *direct* translation approach.

When multiple translations are needed for one country, costs increase and with them the pressure to use one translator for each language. However, the split committee approach described above is another option to reduce time and costs. Seen against the costs of developing and implementing a study, translation costs are usually low. On the other hand, the cost of inappropriate versions or mistakes in questionnaire translations can be high.

Using one translator may be cheaper and faster than using several translators and also can simplify organization, since work schedules for a team do not need to be coordinated. However, relying on one person to provide a questionnaire translation is problematic, in particular if no team-based assessment process follows. A translator working alone has no opportunity to discuss and develop alternatives. Regional variance, idiosyncratic interpretations, and inevitable translator blind spots are better handled if several translators are involved and an exchange of versions and views is part of the review process. Whichever translation approach is used, as indicated above, it is important to ensure that people involved in assessment provide the necessary spread of knowledge, cultural and linguistic input, and expertise.

3.2.4   Assessment Procedures Accompanying Translation-to-Adjudication

Other necessary assessment procedures that accompany the translation-to-adjudication process are presented in Table 3.1. Space restrictions prevent discussion of each of these individually. Statistical assessments of translated instruments are discussed in Chapter 10.

TABLE 3.1.   Assessment Procedures Commonly Used in Research

Qualitative on translation	Qualitative from instrument pretesting
Textual appraisal	Bilingual respondent feedback (debriefing)
Field staff feedback	Monolingual respondent feedback
(text editing)	(debriefing)
Cognitive testing	Field staff feedback (debriefing)
Think alouds	Cognitive testing
Focus groups	Focus groups
Back translation (focus then not on target	Comprehension tests (e.g., for support
translation text)	materials)
	Quantitative from pretesting
	Double ballot administrations (data analysis)
	Bilingual fielding splits (data analysis)
	Item Response Theory
	Structural Equation Modeling
	Multitrait Multimethod

Translated questionnaires should be tested as thoroughly as questionnaires designed for one context, and most of the techniques used for testing monolingual questionnaires are equally relevant for testing translated questionnaires. Assessment, incidentally, should include everything translated for a study, including hidden CAPI instructions to interviewers and any support materials, such as show cards, diagrams, etc. Attention should also be paid to any culturally anchored visual components.

In many instances, however, this is not what happens. Social science questionnaires, for example, rarely are pretested in translation with the same thoroughness afforded important monocultural questionnaires. Translated versions of psychological and health-related instruments, on the other hand, are commonly tested more thoroughly. These often are distributed on a commercial basis, which in part explains how testing can be funded.

Pretesting strategies, such as focus groups, cognitive interviews, split pretests with bilinguals and monolinguals, as well as respondent and field staff debriefing, are important sources of feedback that can help improve different aspects of the translated questionnaire versions. In addition, fielding staff can be useful in pre-editing translations before fielding, not only in providing input at pretest debriefings. Back translation assessment is a particular case dealt with in greater length below. Further text-based assessment options include rating procedures, assessment for clarity and simplicity, and task completion tests for testing instructions and visual aids, for example. (For discussion of the advantages and disadvantages of bilingual splits and double ballot administrations, see, for example, Hulin 1987; Hayashi, Suzuki, and Sasaki 1992; Acquadro et al. 1996; Harkness and Schoua-Glusberg 1998. For monolingual feedback, see Acquadro et al. 1996; McKay et al. 1996; Harkness and Schoua-Glusberg 1998. For focus groups, see Schoua-Glusberg 1988 and for different uses of think alouds, see Harkness 1996; Harkness and Schoua-Glusberg 1998. For outlines of multiple procedures, see McKay et al. 1996 for the US CPS; Teresi et al. 2001 on testing for ethnically diverse aged populations; Nápoles-Springer 2001 on procedures for testing health-related quality of life instruments; and Hambleton forthcoming on psychological instruments.)

Back Translation: The best known translation assessment procedure in survey research (sometimes mistakenly called a translation approach) is called back translation (e.g., Brislin 1970; Werner and Campbell 1970).

Back translation was devised to solve a predicament. Researchers fielding questionnaires in languages they could not read wanted some assurance that the questionnaire was indeed 'asking the same question.' Presumably they either had no one they could trust as reviewers or adjudicators or they hoped to eliminate the subjective element of textual assessment and saw back translation as a way of "gaining some insight" (Brislin 1980) into the translated text they could not read.

Back translation operates on the premise that 'what goes in ought to come out.' The procedure starts once a target language questionnaire has been produced (translated). The method is as follows: the translated questionnaire is first translated back into the source questionnaire language. Then the two questionnaire versions in the *source* language are compared for difference or comparability. The second back-translated text in the source language is taken as an indicator of the quality of the target language translation, which is not, itself, assessed. Thus, if the back-translated text is 'like' the original source questionnaire, the translated text is considered to be good (cf. Brislin 1970, 1976, 1980; Werner and Campbell 1970; Sinaiko and Brislin 1973). Early advocates of the procedure also pointed to some limitations—for instance, the fact that good back translators might automatically compensate for poorly translated texts and thus mask problems (cf., for example, Brislin 1976).

Languages are not isomorphous, however, and we cannot expect that what goes in will, indeed, come out. We also cannot assume that because something comes out 'right' this means that what we cannot read (and really want to know about) is also 'right.' A German General Social Survey item (suggested by Achim Koch, ZUMA) provides a neat example of this.

"Das Leben in vollen Zügen genießen" is an item from a battery in a Likert-statement format. Let us suppose that the German researcher team for some reason want to have the item in English but speak no English. A translator is engaged. Let us also leave open the translator's first language. (Ideally, translators should always translate into their strongest language.) He/she translates the item into English as "Enjoy life in full trains." This is a possible translation if the translator adopts a close literal approach or, alternatively, is not proficient enough in one of the languages, English or German, to notice pragmatic cues. The German team now have a translation in English but cannot tell how good this is. They organize a back translation (into German) to find out. This back translation produces exactly the same wording as in the original German item: "Das Leben in vollen Zügen genießen." The team therefore conclude (wrongly) that the English translation is good. A suitable English translation of "Das Leben in vollen Zügen genießen" would be "Live life to the full," not "Enjoy life in full trains" (in American English: "Live life to the fullest"). One reading of *Zügen* is indeed the plural of (railway) *train*; the other, idiomatic, reading is "in full draughts."

One could certainly expect, however, that some translation problems will turn up through back translation. Moreover, it is important to note that because back translation is such a well-known term, it sometimes is used to describe assessment procedures that include a back translation step but are, in fact, considerably more detailed than the basic procedure of translate, re-translate, and compare source language versions.

At the same time, a basic back translation procedure will not provide the rich detail needed to assess whether a translation is adequate, in particular where the material involved is needed for comparable measurement across countries. In addition, basic back translation makes no contribution to deciding what counts as 'like' or 'unlike' in the two source language versions. These decisions, plus any resulting changes to the translated text, lead inevitably back to subjective, text-based assessment (cf. Hulin 1987; Acquadro et al. 1996; Harkness 1996; Harkness and Schoua-Glusberg 1998; Hambleton and Patsula 1998; McGorry 2000).

Researchers would be well advised to avoid making decisions about the quality of translations they cannot have appraised by reliable people with whom they can discuss details. Proper planning can greatly reduce the need for any kind of 'blind' assessment. Even if researchers do need to assess questionnaires they cannot read and have no one suitable to take over the assessment, other procedures, such as probe interviews and think-aloud protocols, offer good alternatives and provide more detail (Harkness 1996). Nonetheless, these, too, are not sufficient for final decisions on the quality of a translation and are no substitute for the array of procedures described earlier.

3.2.5 Documentation

Translation and review decisions need to be documented for three main reasons. First, those reviewing and adjudicating need notes on options discarded or problems noted in order to decide better on the 'final' choices. Second, many important instruments have a history of versions. Longitudinal studies need to record which version of which item is involved in a particular study. Third, secondary analysts can benefit from records of unavoidable differences or adaptations and from notes on mistakes found after a questionnaire was fielded.

Changes made at each stage of questionnaire revision, unresolved difficulties and translation compromises, or any adaptations made should be recorded. This speeds up both the review and adjudication processes. Moreover, the final version documentation, which records differences between source and translations, compromises, and any adaptations (e.g., bipolar scale translated as unipolar for pressing reasons) is useful for analysts and in developing later versions of instruments.

In addition, projects using instruments with official accreditation assign official version numbers to a given final version and need to annotate changes across versions. Version documentation across multiple languages is the rule for such projects but can be a mammoth task. It is needed to standardize version updating for given languages and to ensure that any differences in instruments can be taken into account in analysis.

In social science survey projects, documentation of translation decisions and difficulties is, to date, rare. Experience gained from collecting information on translation procedures used in the International Social Survey Programme (ISSP; Harkness, Langfeldt, and Scholz 2001) indicates that records should be kept *while the work is being done*. At a later stage, problems and decisions have been forgotten. Translation protocols are helpful for this (Harkness et al. forthcoming).

3.2.6 Selection and Training

Survey translation looks rather simple. Questions are often short, vocabulary often simple unless the survey deals with a special topic with special vocabulary, and frequently repeated instructions and answer scales need to be translated only once.

Nonetheless, translation is one of the most frequently mentioned problems in comparative research. Could the translators be the problem?

Survey literature advocates that translators should be 'bilinguals,' 'professional translators,' people who understand empirical social science research, or combinations of all of these. Relying on titles is not the appropriate way to select either translators or translation reviewers. 'Professional translators,' for example, could refer to anyone who earns his or her living by translating, not the skilled practitioners we hope to employ.

In selecting translators, it is wise to look at performance and experience as well as official credentials. The people most likely to be good questionnaire translators are people who are already good translators and who learn/are trained to become questionnaire translators. Translators can be tested for their suitability for questionnaires using old questionnaire materials. They can, for example, be asked to identify problems, to provide translations or to correct translations, and to make questions more suitable for the target population (cf. Harkness et al. forthcoming).

Given the scarcity of training opportunities for survey translation, not many translators will have been trained to translate questionnaires adequately. Thus, in many cases, proven translating skills will be more important than survey translation experience. Translators who have had experience in translating questionnaires but were never actually trained to handle this kind of text may, indeed, prove difficult to (re-)train.

If questionnaires tend to favor short questions and simple vocabulary and if translators used are already working as translators, why is extra training necessary? One answer is that design problems are sometimes not addressed at the design stage and then manifest themselves in translations (see Chapters 2, 4, and 5). Poor translations can mean that respondents are not asked what they should be asked. Poorly designed questions mean they may not be able to be asked what they should be asked. Another explanation is that questionnaires are only seemingly simple texts. Questions and answer formulations, in fact, lead 'a double life' which translators without training are unlikely to perceive fully.

On the one hand, questionnaires enact the question and answer dialogue between researcher and respondent; they ask simple questions and offer (hopefully) suitable answer options. At the same time, items and response categories constitute a tool of measurement. The careful balancing of word choices and structures (e.g., the neutral formulation 'to what extent do you agree or disagree with...' instead of the somewhat more everyday '(how much) do you agree with...'), the sometimes strangely explicit repetition of noun phrases instead of using pronouns (when your *father*... did your *father*), and the symmetrical design of answer scales with what otherwise would be rather odd combinations (e.g., "a little bit unimportant/a little bit important") are all verbal traces of measurement objectives. Translators can be expected to notice them but not to intuit their purpose.

Translators who have been informed about the measurement components of questions and trained to be sensitive to design requirements as well as target audience requirements are in an optimal position to produce good versions. They are also more likely to be able to point out when one or the other requirement cannot be met and to recognize problems (Hulin 1987; Hambleton 1993; Borg 1998).

3.2.7 Tools, Training, and Job Specifications

Tools: Equipping translators properly for the task helps them perform better (Holz-Mänttäri 1984; Gile 1995; Kussmaul 1995; Gutknecht and Rölle 1996; Wilss 1996). Translators need to understand the function of target and source text to see the best possible translation options. What they produce as a translation depends not only on their ability and training but on the task specifications they receive and the quality of the material they are asked to translate. Translation sciences literature emphasizes the importance of specifying the objectives and requirements for translators in the form of a translator's brief. (If not given job specifications, translators mentally decide their own, since they cannot translate in a vacuum.) This brief includes information about what kind of translation is required, for what purposes, and for what kind of audience. A questionnaire for the oldest senior citizens, for example, might need to include older, more traditional vocabulary, one for children would call for simple words that are part of children's vocabulary.

Task specifications must thus indicate the intended audience, level of literacy and tone of text (e.g., official or more casual tone), the function of the text (e.g., a questionnaire for fielding or a gloss to describe the contents of a questionnaire), and the degree of freedom permitted in translation. This last depends, for example, on whether pragmatically equivalent components can be substituted in translated questions or not (see Section 3.3.2). It is strongly recommended that translators are given support materials, example texts, and the information relevant for their part in producing instruments. This would include briefing them on questionnaires as instruments of measurement.

Training: In the case of surveys, training and informational materials easily can be developed from available questionnaire translations. Old questionnaires can be used for training and practice (cf. Harkness et al. forthcoming). Translators could also benefit from seeing a questionnaire 'in action' at a pretest, since their work, on file and paper, is some distance removed from how questions are administered and how respondents and texts/interviewers interact. Source questionnaires can and should be annotated to help guide translators. Translators can decide better what needs to be conveyed if the source questionnaires contain notes on what reading is intended, what dimensions are to be tapped, and how, for example, response categories are intended to be used (cf. Harkness et al. forthcoming).

Performance Specifications: To date, there is no consensus in survey research on translation and assessment best practices and no task-related specifications of required performance (Harkness and Schoua-Glusberg 1998; Harkness 1999). Available standards and guidelines do not agree on procedures, particularly on the use of back translation. Survey research projects also differ in how they brief translators, if at all, and how they assess translations and document the products.

ISSP modules in recent years have included notes for translation, but there are no stipulations on how to translate or assess. The European Social Survey first round provides national teams and their translators with a starter kit on translation and assessment, an annotated questionnaire, as well as guidelines on translator selection and training (Harkness et al. forthcoming). World Health Organization projects such as the QOL studies or the Mental Health 2000 survey regularly specify general translation and assessment procedures and also may provide a question-by-question commentary on the questionnaire to aid translators and interviewers (e.g., Power, Doran, and Scott 1999).

The quality of the source questionnaire affects the quality of the translation, especially if close translation is required (discussed in Section 3.3.1). Kussmaul (1995) suggests it is the translator's job to improve poor source texts in translation. Questionnaire designers, in contrast, are likely to be adamant that questions should be translated, not 'improved.' At the same time, source questions can be awkwardly formulated or ambiguous. When source instruments can still be changed, translators can report back and thus help improve the source questionnaire. In some survey circles, however, translators may be expected to reproduce the ambiguity (and the same poor measurement properties). Apart from the fact that this will not always be possible, the benefits of carrying over dubious measurement are themselves dubious.

Well-formulated but old source questions also can create problems (cf. Skinner 2001). Questions can become outdated when reality changes and the questions then ask about things that no longer make sense. Translators who notice the illogicality (which may or may not apply for the source questionnaire) would have to negotiate the right to update (adapt) the questions. In such cases, it is also important that the documentation of the translation covers the problem or the change.

3.3   INSTRUMENT MEANING AND TRANSLATION

As indicated at the outset, the second half of the chapter turns to address linguistic aspects of questionnaire translation.

3.3.1   Referential Meaning and Close or Literal Translation

In an ASQ model, translated items must present the same stimulus as the source questionnaire items and do so by referring to the same entities (abstract and concrete) as do the source items. If a source question refers to (domestic) 'cats,' then the translated item is expected to convey domestic cats or domestic felines. It is not expected to refer to what in some other context might be a functional equivalent (such as 'dogs' or 'parrots') nor to anything like English 'panthers,' 'canines,' or 'hyenas.'

In addition to referring to the same entities (retaining referential meaning), survey translation is expected to stay close to other features of the source text, including lexical and structural components. Thus, if a questionnaire for children uses language appropriate for younger respondents (*tummy* instead of *stomach*), the translation is expected to do the same. If the source text repeats a noun within a few words of the first mention ('when your father...what did your father...'), rather than using a noun followed by a pronoun ('when your father...what did he...'), the target text is expected to do the same. If the source item uses a hypothetical construction or a past time reference, the target language item is expected to do so as well. Of course, in reality, neither words, concepts, nor structures match up neatly across languages.

Such 'close' or 'literal' translation is the rule in an ASQ framework. Close or literal translation in survey research has to do with trying to retain the measurement properties of source items. (In other disciplines, it has other and multiple functions and a variety of names.)

Depending on degree, close translation can stick so closely to the words and structures of the source text that the lexical and structural needs of the target language are neglected. However, if a translation signals it *is* a translation or simply comes across as strange in some way, this becomes part of the communication. The goal in ASQ models is also to keep the stimulus the same; consequently, translated items should read like items, not like translations of items. Respondents are cooperative in Gricean terms (Grice 1975) and can be expected to do their best to make sense of any strangeness. If close translation is indeed too close, this may mean respondents arrive at unintended interpretations of questions or an unintended perception of the study.

Thus, for questionnaires, close translation is a balancing act. Translated versions that are clear and make sense to respondents may not be as close as researchers might like; Weidmer, Brown, and Garcia (1999) indicate how relevant clarity is for translations. Conversely, over-close translations can appear stilted or be unclear for respondents. As indicated in Chapter 2, approaches that combine adaptation with translation have more freedom to change elements of items.

3.3.2 Pragmatic meaning

Pragmatic meaning is meaning *in context*—that is, what words mean in use in a given context for whoever is involved. This section explains why questionnaire translations must take pragmatic meaning into account (cf. Chapter 4).

We rarely say or write exactly what we mean or specify everything we intend to communicate. (We also are rarely aware of all of what we have, in fact, communicated.) Nonetheless, people do manage to communicate successfully; contextual information and pragmatically based principles of interpretation help us to decide what is intended (cf. Grice 1975; Levinson 1983; Sperber and Wilson 1986).

Cognitive research in the monocultural context has demonstrated the relevance of the pragmatics of communication for understanding how respondents read questions. Linguistic pragmatics is concerned with what it is that speakers do when they speak and how they manage to do this (Austin 1962) and with how people manage to communicate successfully, given that meaning is fluid and that intended meanings rarely are spelled out exactly (cf. Grice 1975; Levinson 1983; Sperber and Wilson 1986). Intercultural communication research has demonstrated that cultures differ not only in what is salient for interpretation but also what is salient to communicate in one or another context (e.g., Gudykunst and Kim 1992; Wiseman, and Koester 1993; Agar 1994; Katan 1999).

The four components central to clarifying what is meant with a given utterance in a given context are what is said, what the speaker intentions are (what the speaker/writer intends to be understood), the 'common ground' between participants (the shared information participants in an interaction have; Clark and Schober 1992), and the macro and micro contextual setting of an utterance. These are just as important for survey questions and answers as they are for any of the other things we do with language. A few 'ordinary' questions help illustrate this.

In many contexts, the questions 'Can you open the door?' and 'Can you tell me the time?' would be intended and understood as requests by the speaker to have the door opened or to be told the time. Compliance with the request would produce the help requested or a response motivating noncompliance such as, 'Sorry, my watch has stopped.' Addressed to a child, the questions could be intended and understood as questions about the child's ability to open the door or tell the time. (The second would also probably produce a demonstration.) More precise formulations would help identify a request ('Could you please open the door for me?') and an information-seeking question ('Are you able to open the door by yourself yet?') but this degree of explicitness is often unnecessary in English.

At the same time, what is made explicit and what is left unsaid differs across languages, as do the reasons for speech or silence and the interpretations of these acts (e.g., Hall 1976; Hofstede 1980; Summer Wolfson and Pearce 1983; Nakanashi and Johnson 1993; Katan 1999). Translations into other languages might need to make speaker-intended meaning linguistically more explicit. Alternatively, language systems without the double sense of *can* in these examples might automatically mark the difference between requests and, say, questions about ability. In this case, the translator would always need to be able to recognize the intended meaning in the language which does not specify. Alternatively, the language structure or social norms might dictate that certain meanings are always explicit or inexplicit in formulations.

Questions, including survey questions, only have meaning in context. It is misguided to believe that we can develop questions with a unique and literal reading based on the lexical meaning of words alone. Thus, survey translations that focus on conveying lexical meaning and exclude consideration of pragmatic meaning have an obvious Achilles heel. Pragmatic meaning necessarily gives contextual considerations more emphasis. Contextual considerations are cultural considerations. Therefore, paying more attention to pragmatic meaning in translation means paying more attention to culturally appropriate meaning (Katan 1999). To date there has been little consideration of pragmatics research in literature dealing with questionnaire translation.

3.4   TRANSLATION AND LANGUAGE STRUCTURE

Discussions of survey translation often focus on difficulties matching vocabulary or finding the right word for a concept across languages. Many issues discussed as problems of translation would be better addressed at the question design stage. The fact that words do not match up across languages is often less of a problem than that concepts do not match up across cultures. Thus what may seem to be linguistic differences are in fact very often culturally anchored differences (cf. Wierzbicka 1992). In addition, lexical problems are only one aspect of translation, albeit an important one. By way of contrast, we focus here on structural differences across languages that are relevant for questionnaires.

3.4.1 Grammatical Gender and Survey Questions

The examples discussed below involve differences in grammatical gender. Gender reference is only one example of an array of structural considerations that need to be addressed for questionnaire translation. These include tense references, pronoun systems, deictic reference, and structural reflections of social standing between interviewer/questionnaire text and respondent and many more exotic elements, such as how to enumerate using numeric classifiers in Thai.

The structural facts of a language shape how users of that language refer to themselves, others, and their surroundings. For example, languages differ in how they categorize entities as near or far, present or nonpresent, or as belonging to the I/We in-group or the Them/Other out-group. How we refer to people and things around us also is affected by codes of social behavior. In many English-speaking countries, for example, political correctness has changed how we refer to certain groups, including ethnic minorities, disadvantaged people, and women. Two examples here indicate how gender reference can affect how respondents read questions. A few remarks about languages and structures are necessary first.

People have sexes; languages may or may not have *grammatical gender*. Grammatical gender assigns animate and inanimate objects a gender. Some languages have elaborate systems of grammatical gender (Spanish and French), some simple or vestigial (English) and some none (Hungarian). The number of grammatical genders may differ: German has three, French two. Some languages indicate the gender of people talked about (Spanish), while Koasati, an indigenous American Indian language, indicates the gender of the speaker. We consider the first form here.

One upshot of grammatical gender is that in talking about, say, a doctor or a baker, the sex of the person may be indicated in the linguistic form used for *baker* or *doctor*. *Gender-neutral reference*, in contrast, leaves the sex of a person spoken about unidentified. This is often expected in English nowadays; job advertisements, for example, are expected to be gender neutral in reference. In many other public contexts, too, gender-neutral forms or inclusive reference, as in *he/she* doubling, are the norm.

English structure makes it fairly easy to avoid identifying someone's sex. The structural facts of other languages make this more difficult. Even inclusive forms on the model of *he/she* may be more complicated to construct and use, as an example from a German question illustrates. The short introduction to an English source question from a social networks questionnaire (ISSP 2001) begins:

Now think about your best friend, . . . (but not your partner).
Is this best friend . . .

German requires references to a concrete 'friend' to be marked for the gender of the friend. Thus in order to refer to a respondent's 'best friend,' the German questionnaire formulation needs to cover both possibilities, that is, that the best friend is male or female. Other linguistic considerations related to doing this (case endings) mean that the German formulation needs two forms for *friend*, two for *partner*, two for *best*, as well as two for the pronoun *your*. We have glossed the German here and underlined the elements which change.

Nun eine Frage zu dem besten Freund ..... / der besten Freundin ..... (aber nicht dem
Now a question on the best friend (male) ..... / the best friend (female) ..... (but not the

[Ehe-] Partner   / der [Ehe-]Partnerin).
[marriage] partner (male)   / the [marriage] partner (female)

Ist Ihr bester Freund   / Ihre beste Freundin .....
Is your best friend (male) / your best friend (female) .....

As this shows, inclusive forms in German are longer and more difficult to produce and read and thus can create design and layout problems in questionnaires. One traditional solution in English, as in German, is to use male forms for everyone, as in 'Everyone has to do his best,' for mixed gender recipients. This practice is, however, no longer socially acceptable in many contexts in English.

Gender-neutral or inclusive language is relevant for survey questions. In a module on gender roles, for example, noninclusive language may well affect female respondent cooperation negatively. Even more important, differences in gender reference can affect how respondents read and perceive what is asked. The following example from a self-completion paper-and-pencil questionnaire is about what people doing different jobs should earn (ISSP 1987). (There is a further question about what respondents think people actually earn; the list of occupations is reduced here.) Both in estimating what people earn and in indicating what they feel people ought to earn, it is likely that some respondents could take gender into account. They may know that women (in their country) earn less than men for the same job. In answering the "ought to earn" question, some could feel that men doing a certain job should receive a different wage from women doing that job.

Next, what do you think people in these jobs ought to be paid—how much do you think they should earn each year before taxes, regardless of what they actually get?

First, about how much do you think...
(a) a skilled worker in a factory earns?
(b) a doctor in general practice?
(c) the chairman of a large national corporation?
(d) a bank clerk? (how much does he/she earn)?
(e) a secretary?

In the English question, one job is marked for men (*chairman*), one is explicitly inclusive (*bank clerk*, *he/she*). Two are gender neutral in reference (*doctor* and *skilled worker*) and one (*secretary*) is linguistically uncertain. Men doing secretarial work often are called *personal assistants* in English. Thus, *secretary* is more likely to be used for women, even if the word does not explicitly exclude men. In contrast to the English, the German translations, with one exception, refer to men. In the last item, 'a secretary', however, the German uses the feminine form of the word. It thus refers to women working as secretaries.

Wieviel meinen Sie sollte ungefähr verdienen
   (a)   ein qualifizierter Fabrikarbeiter? (male factory worker)
   (b)   ein praktischer Arzt? (male doctor in general practice)
   (c)   ein Vorstandsvorsitzender eines großen nationalen Unternehmens? (male chairman of a large national corporation)
   (d)   ein Bankangestellter? (male bank clerk)
   (e)   eine Sekretärin? (female secretary)

For cross-national comparability this raises three questions. First, do respondents read all or any of the items in English as questions about how much people should earn or as questions about what a particular sex should earn? Second, do German respondents receive a different stimulus because of the gendered references? And third, is there a better or more comparable translation in German for 'a secretary' in this context? We are not aware of research on the first two points. For the last, a case can be made for and against the translation as it stands.

The masculine form of the word *secretary* in German (= *Sekretär*) would be used to refer to a piece of furniture (a desk) or to men who have high administrative positions. In addition, although the term in German for a *typist* (*Schreibkraft*) does not identify a person's sex and can be used for both men and women, the work and pay differ from those of a *secretary*. The same comparability problems apply for *personal assistant* in German, which could take both male and female forms (*persönlicher Assistent/persönliche Assistentin*). De-sexing *Sekretärin* and using the feminine form for both sexes would over-tax current German language practice. Thus, a man doing secretarial work would need a different descriptor. Adding in a second term for men alongside the German word for a female 'secretary' could prompt all sorts of unintended interpretations. Consequently, if all we required was a translation of *secretary*, *Sekretärin* would be fine for many contexts. Since here measurement issues are involved, the German translation may, however, be biased. The best solution—at the design stage—would be to change the occupation mentioned. Not given this option, the item as translated is potentially biased.

For Germany and Great Britain, social realities may mean the effects are negligible. Without research on whether gender references in questionnaire translations affect responses and cross-cultural comparability, however, decisions on how to handle gender aspects of translation in questionnaires remain a matter of 'gut feeling.' As such, they are automatically based on established practice. This will differ across cultures and languages and, in any case, may *not* be the appropriate practice for measurement instruments. Informed decisions call for evidence of effects or the lack of them.

3.4.2 Translation and Answer Scales

Every element of response scales can pose multiple difficulties for translation, only two of which we touch on here. Both wording and response scale design differ across translated response scales, sometimes presumably inadvertently, sometimes because of structural and lexical differences across languages, and sometimes because preference is given to the formulations/formats a fielding institute (or survey tradition) normally uses.

An extensive body of monocultural literature has discussed the effects response scale designs and scale wording have on respondent perceptions of what is asked and of how they should best reply (see overviews in Schwarz et al. 1991a; Schwarz 1996; Alwin 1997; Krosnick and Fabrigar 1997). There is far less literature on how to construct comparable response scales across languages or on how, for purposes of translation and analysis, to recognize whether response scales across languages are comparable with either the source questionnaire categories or with one another. Consequently, there is a paucity of literature that systematically discusses the problems involved in translating scales for comparative projects. In practice, of course, answer scales for source questionnaires are rarely specifically 'designed' for a comparative project. Source questionnaire designers (often substantive researchers, not questionnaire crafters) use the scales with which they are familiar, either from their own languages and survey traditions or from the language in which the questionnaire is being developed. These are automatically culturally and linguistically anchored in this language. Recent research recognizes that scale designs suitable for one population may not be suitable for other cultural groups. Skinner (2001), for example, reports on comparability problems across cultures in using unipolar and bipolar response scales.

Without established guidelines on how to handle response categories, translations presumably are based on researcher experience and intuition, informed by their perception of the function of response scales. Intuitive requirements usually include keeping the number of scale points the same across versions and staying as close to the semantic content of the source questionnaire response categories as possible. However, if an institute's standard practice runs counter to the source questionnaire design, the format simply may be altered.

Retaining the semantics in response categories can be complicated. There are commonly two basic 'meaning' components in response categories: the dimension that is measured (ability, agreement, satisfaction) and whatever quantification or negation is involved (*strongly*, *somewhat*, *a little*; *yes/no*; *dis*agree or *not* agree). In simple terms, therefore, translations have to attempt to match a scale of 'agreement' with a scale of 'agreement,' rather than with a scale of 'acceptability' and to match modifications (quantifications), such as *strongly* (agree) and *somewhat* (agree), with components in the target language felt to quantify similarly. Numerically labeled scales probably would simply be copied, since numbers technically need no translation, but this is also problematic. Cultures differ in their lucky and unlucky numbers, in their degree of numeracy, and in the rating scales (e.g., 1-5, 1-6, or 1-10) they use in everyday contexts. This can color cultural perceptions of the relative values of numbers and influence respondent selection of response categories.

Two basic approaches are taken to producing response scales 'in translation.' Either the scales from source questionnaires are translated along with the questions or, alternatively, response scales are used that are similar and already in use in the target language (and survey tradition).

Using Available Response Scales

Countries sometimes translate items but not response categories. This applies in particular to countries with well-established survey traditions and hence with considered views on what works best for their context. Instead of translating, designers or researchers take response scale formulations they consider to be the 'same' or equivalent to those in the source questionnaire. In doing so, they demonstrate the same tendency to prefer tested components noted in Chapter 2 for existing questions. However, the basis of equivalence or comparability with a source response scale is likely to be intuitive, not demonstrated. When home-grown scales are grafted onto source questionnaire items, the response scales are treated as technical measurement components, not as textual components of a questionnaire to be translated. The position taken here is that they are both and that both functions need to be considered in producing a 'translated' version.

The following example illustrates that home-grown scales may be preferred even when no translation is needed. A Likert-type agreement scale frequently used in English source questionnaires runs: Strongly agree/Agree/Neither agree nor disagree/Disagree/Strongly disagree/Can't choose (or in the United States, 'Don't know'). This scale was implemented in an Australian self-completion questionnaire as set out below. A box showing respondents how they should read the response categories was placed before the Likert-type item statements, which were listed on the left-hand side.

Yes!!	-- Strongly agree
Yes	-- Agree
??	-- Neither agree nor disagree
No	-- Disagree
No!!	-- Strongly disagree
–	-- (Can't choose)

Yes!! Yes ?? No No!! –

(The Likert-type statements were listed here.)

Following the arrow, we see that the agree-disagree scale is basically 'translated' into a scale using 'Yes-No' labels but one which defies dichotomous analysis. Respondents are presented with 'Yes!!' as equivalent to 'strongly agree' and '–' as equivalent to the label 'Can't choose.' The middle category in the source questionnaire, 'neither agree nor disagree' is presented as two question marks. If we take these at face value, the middle category becomes something like either 'undecided' or 'don't know.' The literature is not conclusive about how respondents read and use middle categories (e.g., Schuman and Presser 1981; Garland 1991; Krosnick 1991; Krosnick and Fabrigar 1997). Presumably, if respondents focus on labels, the interpretation they give the middle category also determines what they make of the minimalist rendering of 'Can't choose', rendered here as '–.'

Obviously, the equivalence of response scales across countries must be demonstrated rather than assumed; we cannot automatically assume that this version is comparable with the source questionnaire format (see also Chapter 18).

Translating Response Scales

The second option is to translate answer scales. However, translation and survey tradition preferences often go hand-in-hand, as the examples in Table 3.2 show. The table presents three different French versions of the same 'agree/disagree' scale discussed above. Two are European French versions and one is Canadian French. All three French scales differ from one another and from the English scale. Standard Canadian French is known to differ from French in France, but as the other French scales also indicate, differences here seem to have more to do with how source questionnaire response scales are treated than with differences within varieties of French.

TABLE 3.2.   Four Agreements Scales: One English, Three French

ISSP English source scale	Eurobarometer France	ISSP France	ISSP French Canada
Strongly agree	Tout à fait d'accord	Vous l'approuvez fortement	Fortement d'accord
Agree	D'accord	Vous l'approuvez plutôt	D'accord
Neither agree nor disagree	Ni d'accord ni pas d'accord	Vous êtes ni pour ni contre	Ni pour ni contre
Disagree	Pas d'accord	Vous la désapprouvez plutôt	Désaccord
Strongly disagree	Pas d'accord du tout	Vous la désapprouvez fortement	Fortement en désaccord
Can't choose	NSP (ne sais pas)	Vous ne pouvez choisir	Impossible de choisir

The English Eurobarometer agreement scale is the same as the English ISSP scale in Table 3.2. Eurobarometer documentation states that source questionnaires are developed in both English and French. Thus, while no descriptions are available of what Eurobarometer development actually entails, this French Eurobarometer scale may not be a translation.

The *Eurobarometer French* scale differs from the English scale in structure more than in semantics. That is, while both scales share a semantic dimension of 'agreement', the headwords (key semantic word in each label) 'd'accord'/'pas d'accord' are more suggestive of a unipolar scale ('in agreement/not in agreement'), even accepting that scale polarity is not decided by labels alone (Schwarz et al. 1991a). The English scale uses labels suggesting a bipolar construction agree-disagree and the formulations are linguistically symmetrical; only the end-points of the scale are modified, each with *strongly*. Thus the English scale structure is implicitly assumed to carry the information that 'agree' is a weaker degree of agreement than 'strongly agree.' The Eurobarometer scale is not linguistically symmetrical ('tout á fait d'accord' versus 'pas d'accord du tout'). The off-scale category in English is 'can't choose,' while in the Eurobarometer French version it is 'ne sais pas,' or 'don't know.' 'Don't know' and 'can't choose' are often assumed to be functionally equivalent in English contexts, provided only one of the two is used. However, this is not meant to suggest that a no-opinion response category does in fact capture no opinion (cf. Krosnick and Fabrigar 1997).

The *French Canadian* scale resembles the English scale in superficial respects quite closely. A verb pair is used that matches agree-disagree in bipolar structure and semantics, and the modifier *fortement* matches *strongly* in semantics. However, the middle category changes headwords: instead of 'neither agree nor disagree,' it uses, roughly translated, 'neither for nor against.' The off-scale category matches the 'choice' semantics of the English but seems more of a hurdle to take (literally translated 'impossible to choose').

The *ISSP France scale* differs structurally and semantically the most from the English source scale and each of the other two French scales. The scale uses bipolar headwords, as in the English scale, but these are semantically closer to approve-disapprove than to agree-disagree. In English, this would be suggestive of different dimensions. Moreover, each label is a full sentence, written from the interviewer/researcher perspective, as in 'You approve of it strongly' for the first point, 'You are neither for nor against' for the middle category, and 'You are not able to choose' for the off-scale category. The second and fourth scale points (corresponding to agree–disagree) are weakened to 'You tend to approve of it'/'You tend to disapprove of it.' As indicated above, the English source design does not modify these categories and relies on the scale structure to carry the message.

Languages do not match up on a one-to-one basis. As a result, we cannot expect to find neatly corresponding structural or semantic components across languages. If the differences just noted in the French scales were differences in English scales, they would prompt doubts about equivalence. However, we cannot simply assume that these differences in French do produce effects feared or assumed for similar differences in English. As research currently stands, we have neither an assurance of comparability nor any tested procedure for deciding which French response scale might be 'most' equivalent to the English (see also Chapter 18). We know little enough about some of these differences in English, since monocultural research on scales has not focused on comparability of designs and formulations, but more on effects of difference. We know even less about how to optimize translated scales.

3.5   CONCLUSIONS

Survey translation is on the way to establishing sound guidelines, but a number of important steps remain to be taken, both practically and conceptually. Translated questionnaires must be tested as seriously as source questionnaires, and more needs to be invested in the processes of translating and assessing translations. Training materials and training routines are needed for translators. Simply passing the work on to 'a good student' or a bilingual secretary is not an acceptable solution. Interdisciplinary input is long overdue; developments in linguistics, translation studies, and pragmatics have much to offer survey translation. Systematic research is needed on key aspects of questionnaire translation—on how to handle answer scales, how to resolve and document inevitable differences across translations, and how to refine available procedures for comparative assessment.

Chapter 4

COMMUNICATION AND SOCIAL COGNITION

MICHAEL BRAUN

Item bias detection techniques compare item characteristic curves across different countries in order to establish bias and equivalence (see Chapter 10). However, as Van de Vijver and Leung (1997b, 87) point out, "[g]iven the poor knowledge as to what kind of items can be expected to be biased, it is not surprising that bias studies have not generated new insights on recommendable practices in instrument construction for cross-cultural research." The array of statistical techniques available to assess the measurement properties of instruments and to detect bias (see Chapter. 14) is not matched by a similar array of tested procedures to develop equivalent instruments or even a list of tested and reliable guidelines to follow.

This chapter focuses on respondents' reactions to survey items.[1] Learning more about how respondents perceive and process items can guide our attempts to improve instrument design for comparative research. The discussion here concentrates on cognitive and communicative aspects; motivational factors, such as providing socially desirable answers, are treated in Chapter 13. We begin with an outline of basic psychological factors of communication and cognition and the peculiarities of the survey situation, independent of culture, then discuss how respondents interpret questions and the relevance of the cultural context for interpretation.

4.1 BASIC PSYCHOLOGICAL FACTORS

Human understanding is directed by different cognitive and motivational structures and processes which work through two general processing strategies: 'bottom-up,' data-driven processes and 'top-down,' theory-driven processes. People usually make use of both of these strategies. Top-down processes are regulated by expectations on the side of the recipients that result in active interpretation of messages using extra-linguistic world knowledge. "Prior knowledge has many effects in this interpretive process: it directs the perceiver's attention to particularly significant aspects of the information while allowing unimportant details to be ignored; it mediates inferences that permit the person to 'go beyond the information given' (Bruner 1957); it guides judgment and evaluation; and it fills in default or expected values for unobserved attributes" (Smith 1998, 404). This applies also to the exchange of information and communication.

Highly abstract schematic knowledge structures guide our focus of attention and how we interpret new information (Rumelhart 1980). Bodenhausen (1992) distinguishes four functions of schematic structures: first, information acquisition by directing attention to selected stimuli; second, interpretation of information in terms of a framework of expectations; third, selective information retrieval from memory as a function of activated schematic structures; and fourth, the generation of inferences on the basis of activated schematic structures. As a rule, schemas are activated automatically as a function of the match between a stimulus and a schema and the accessibility of the schema. Frequently activated schemas can be permanently accessible. Otherwise schemas have to be activated. Fiske and Taylor (1991) identify several kinds of schemas, among them scripts, prototypes, and stereotypes. Scripts refer to typical sequences of events (Abelson 1981), prototypes to typical characteristics of members of categories, and stereotypes to the ascription of properties of categories to their members (Chen and Chaiken 1999).

Words and utterances have potential meaning and pragmatic considerations contribute to determining how an utterance is understood in a given context. Pragmatic meaning mediates literal meaning and pragmatic considerations are based on the 'common ground' participants in the communication share (Clark and Schober 1992), such as their knowledge of one another, the intentions in the communication, and their shared knowledge of the world. In everyday communication, we try to tailor what we say or write to provide our audience with the information they need to understand what we intend to communicate. Our audience contributes to the success of communication by assuming that what we say is based on common ground and by using 'grounding procedures' to probe the adequacy of his/her interpretation (Schober 1999).

Four maxims observed in cooperative conversation (Grice 1975) contribute to ensuring successful and smooth communication: manner, relation, quantity, and quality. Grice's model was intended as a descriptive model of everyday interaction in which both senders and receivers are assumed to observe the maxims as a matter of course. The *maxim of manner* states that contributions should be clear, i.e. comprehensible and unambiguous, while the *maxim of relation* ensures that contributions are relevant to the goals of the interaction, which implies that communication partners are expected to use the context to clarify meaning. Listeners, for example, make inferences that transcend the semantic content of a contribution in order to find out the pragmatic meaning. The *maxim of quantity* stipulates that contributions should be informative, that is, that neither too little nor too much information is presented. As a result, information which can be taken as known will not be mentioned. The fourth, the *maxim of quality* relates to the factual quality of contributions; for example, only contributions 'known' to be fact should be presented as fact.

4.2   SOCIAL COGNITION IN THE SURVEY SITUATION

Respondents in surveys have a number of cognitive and communicative tasks to complete: interpret a question, generate an opinion, match the opinion to a response category ('formatting'), and edit the response in keeping with subjective needs to conform to social desirability norms (Strack and Martin 1987; Tourangeau and Rasinski 1988). Communication in standardized interviews is highly asymmetric; interviewers ask and respondents answer. If respondents, for example, ask about the meaning of questions or otherwise deviate from the interview script, interviewers have prescribed responses to ensure minimal effects on the respondent's perception of questions and response options, such as the doubtlessly perplexing response "Whatever it means to you" (Fowler 1992, 219). Formalized interviewer scripts along these lines can help to reduce the impact of variations in the conduct of interviewers. However, interviewer effects are only one source of error. The situational context of an interview, the wording of individual questions, question order and response category design may each present violations of the cooperation principle with potentially damaging consequences for the quality of the data (Schober 1999).

In essence, researchers formally violate the cooperation principle, while respondents assume it applies and act accordingly. Respondents use whatever they perceive as pertinent to come to conclusions about their expected role in the survey interaction and how they should answer, including information not considered by the researcher or questionnaire designer (Schwarz 1994; Sudman, Bradburn, and Schwarz 1996; Tourangeau, Rips, and Rasinski 2000). For instance, they try to infer from the interview context what information is required of them (maxim of relation) and do not repeat information already given or which they rightly or wrongly assume the interviewer to know (maxim of quantity). In doing so, they assume that everything which may seem a bit odd to them is in fact not odd, just part of a set-up they are still learning. "If a word seems vague, ambiguous, or strange, it isn't really vague, ambiguous, or strange, because the surveyer is confident respondents can figure out what it means" (Clark and Schober 1992, 27).

While respondents take into consideration the (presumed) intentions of the researcher which they infer from the context, the latter interpret the responses in a literal manner, i.e. they assume that respondents refer exclusively to the propositional content of the questions and ignore pragmatic meaning and context (Clark and Schober 1992). Thus, researchers find some reactions of the respondents odd and interpret them as artifacts that show the limitations of the human mind. However, these artifacts are, in fact, at least partially the result of an attempt on the part of respondents to make sense of the questions on the basis of the cooperation principle. If respondents are informed about which communicative rules apply and which do not, they react accordingly. If, for example, respondents are told that the aim of a study is to test response alternatives, they will not assume that the researcher will comply with all of the maxims. As a consequence, they will change their response behavior and not draw information on what the researcher is interested in from the response scales (see Schwarz 1994, 1996 for overviews).

When respondents comply with the cooperation principle (while the researcher does not) and try to clarify the meaning of a question from its context, context effects result. These can occur at any stage of answering a question. At the stage of interpretation, context effects are especially likely for unclear and ambiguous questions and, in particular, for fictitious issues (Bishop, Tuchfarber, and Oldendick 1986). They can also occur with questions which are ostensibly clear but which violate the principle of quantity. Context effects affect the task of opinion generation, if previous judgments and evaluative implications, general norms, or affective reactions and moods activate certain information and make them more accessible (Sudman, Bradburn, and Schwarz 1996). At the stage of formatting, previous items can have an impact on the use of answer scales (Parducci 1982; Daamen and de Bie 1992). Finally, at the stage of editing, i.e., the modification of a response to account for social desirability, previous questions can have sensitized the respondent for desirability issues.

4.3   QUESTION, EXPERIENCE, AND CULTURAL CONTEXTS

The social-cognition approach emphasizes that respondents try to derive the meaning of unclear or ambiguous questions from the context. The textual context of any section of a questionnaire is the rest of the questionnaire, in particular those parts already processed. It includes introductory texts, similar questions, the sequence of questions, and characteristics of response scales. Variables relating to the personal experiences of respondents, such as socio-demographic characteristics, respondent behavior, including psychological or physical states, and external conditions are less frequently used in explanations of how respondents try to solve the four tasks in a survey. However, these personal experiences variables also form an important context. It has as much an impact on respondent behavior in general and the interpretation of questions in particular as do characteristics of the questionnaire.

In addition to the contexts formed by other components of the questionnaire and the personal experiences of respondents, cultural contexts are of utmost importance in comparative research. All three kinds of context provide respondents with information that can have an effect on the interpretation of a question. While all three are similar with regard to this information function, this is not the case with regard to their direct impact on attitudes: only contexts created by personal experience and cultural contexts exert a direct influence on attitudes in the sense of the 'true' value.

The three kinds of contexts may also operate in interaction; those between question contexts and personal experience of respondents are the source of the conditional context effects discussed in the social-cognition literature (Smith 1992). Whether there is a context effect and what direction this takes can be dependent on the personal experience context. Schwarz, Strack, and Mai (1991) have shown that the impact of a marital satisfaction question on life-satisfaction depends on the perceived quality of the marriage. When the question on marital satisfaction precedes the question on life-satisfaction, respondents who are satisfied with their marriage report higher life-satisfaction and those who are dissatisfied with their marriage report lower life-satisfaction than is the case when no question on marital satisfaction is asked (see, too, Chapter 5). Given conditional context effects, comparison of means between different question contexts results in biased interpretations and researchers therefore have to compare correlations between both questions across the different contexts.

Respondents participating in intercultural surveys usually do not know that the questions are not local in origin. However, even if some of them guess that the questions originate from a different cultural context, they will nevertheless assume that all the questions, including those which square badly with their own societal reality and public discussion of issues, do make sense for their countries. They will assume that researchers intend the interpretation which they make automatically, along the lines of: "If the surveyer thinks this word has an obvious meaning, then it must be the meaning that is obvious to me at the moment" (Clark and Schober 1992, 28). If a relationship between a question and the situation in a given country can be established, then it will most likely be established. In interpreting responses, researchers have to consider that these are not 'context-free' measures.

4.4   CULTURAL CONTEXTS: MAIN AND INTERACTION EFFECTS

4.4.1   Main Effects of Cultural Contexts

Cultural norms, values, and experiences influence the processing of the four tasks of the response process (Johnson et al. 1997). There are intercultural differences in the saliency of different concepts and, thus, in the chronical, i.e. permanent, accessibility of the information associated with them. The more frequently events of a special kind happen in a society, the more likely is the formation and stabilization of schematic structures representing abstract depictions of such events.

In intercultural surveys, all norms of communication might be violated, even where questions make sense in some countries and do not pose problems of interpretation there. Question context effects arise through other components of a survey that are not meant by the researchers to contribute to the interpretation process. Cultural context effects, however, do not result from the embeddedness of a question in a question context, but in a cultural context. Interculturally different interpretations of a problem or task can be analyzed in terms of framing effects that are the result of different framing conditions (Stocké 2002), which are represented by the different cultural contexts. Gaps left with regard to the specification of important informational components can render people in some countries incapable of answering the question, whereas in other countries respondents will fill the gaps with their own culture-specific knowledge. Ambiguity-based framing effects will result as the following examples illustrate.

Differences in the age of first marriage and in cohabitation rates change the interpretation of questions on premarital sex. In traditional societies, respondents will automatically think of adolescents when confronted with such questions, whereas in modern societies the age group has to be made explicit in order to obtain materially comparable responses. Moreover, schemas do not have to be related to persons or events, there are also 'ideology' schemas (Fiske and Taylor 1991) which can be activated automatically. Premarital sex can be opposed for entirely different reasons—for instance, religious (divine order) or rational (to avoid pregnancy). Different ideology schemas are activated in different countries depending, for instance, on whether religion has something to say about premarital sex or whether teenage pregnancy is prevalent. Responses with different underlying ideology schemas cannot be meaningfully compared across countries.

An item in the ISSP 1988 module (Zentralarchiv 1991), "A pre-school child is likely to suffer if his or her mother works" (*Child suffers*), leaves at least the following informational components unspecified: the specific age of the child, the labor force participation of the mother and of the father, and the availability of alternative childcare arrangements and their quality (Braun 1998, forthcoming). Respondents fill these gaps, perhaps automatically, through the activation of schemata. Thus some think of infants, others of 5-year-olds. These schemata are formed by the social realities in the different societies, for instance, labor force participation rates of women, availability of part-time jobs, parental-leave regulations (in particular for fathers), availability of alternative childcare, unemployment rates of men, and public debate on a special role of mothers for their children. As a consequence, respondents in some societies will interpret this item as implying a situation of full-time employment of both parents. In other societies, however, respondents will automatically assume that either part-time employment of mothers is the issue, or that fathers will work less than full-time.

Figure 4.1 shows the results of a methodological experiment in which respondents were asked to evaluate the potential suffering of a 3-year-old child in different situations, for instance, when both parents work full-time, when the mother works full-time, but the father only part-time, and when the mother works full-time and the father does not work. The responses are presented as depending on responses to the item "All in all, family life suffers when the woman has a full-time job" (*Family suffers*). Low values with these items mean that respondents think there will definitely be suffering, while high values denote the absence of suffering. Between eastern and western Germany, even respondents who have the same values on the item *Family suffers* differ greatly in their evaluation of the effects on a 3-year-old child. In western Germany, the majority of respondents who strongly disagree with the statement "All in all, family life suffers when the woman has a full-time job", i.e. respondents with a score of 5 on the item *Family suffers*, do think that a 3-year-old child will suffer if both parents work full-time. (The average score for this group of respondents is clearly below the 'neutral' line.) This is not the case in eastern Germany. Similar values in both parts of Germany are obtained only for the situation where the father does not work. Obviously, with *Family suffers*, Germans in eastern states answer a more difficult item than Germans in western states, after both have filled in the missing informational components. In addition, the three lines are closer together in eastern Germany than in western Germany, indicating that the conduct of the father is of comparatively little importance in eastern Germany.

In addition to questions which are perceived as unclear or ambiguous in some countries, questions not relevant for the ongoing public debate or that do not cover relevant options for a topic are also problematic. Heuristic-based framing effects which generally arise under less elaborated processing of information will result. Respondents are, especially under these circumstances, not motivated to use the full information that the question presents. Instead, they pick out some salient aspects compatible with their own attitudes and use them as a basis for evaluation. Thus the specification in *Family suffers*, stating that the mother works full-time, might automatically be ignored by respondents who, while wanting to express their nontraditional attitudes, also cannot imagine that both parents work full-time, given the situation in their own location.

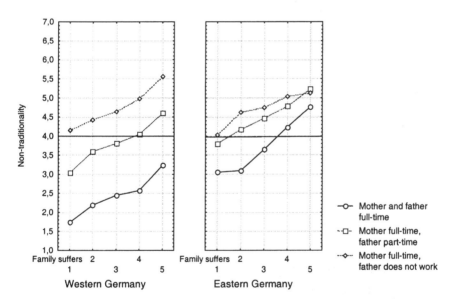

SOURCE: Thyssen Study 1998. N=2.007.
Adapted from Braun (forthcoming).

Figure 4.1. Evaluation of the potential suffering of a 3-year-old child in three situations dependent on the response to the item *Family suffers.*

Finally, schema-based framing effects are generated by an automatic activation of mental structures. This can be triggered by particular informational components in some countries. What is then important is that items holistically activate one or more images—but not on a one-to-one mapping with the item. They can also be involved with the other two framing effects—by filling in missing informational components, or in the selection of salient informational components. In the examples given earlier the item text might activate for Germans in the western states the image of a stressed mother hurrying from her part-time job to pick up her child at kindergarten. This representation automatically fills some of the gaps in the item "A pre-school child is likely to suffer if his or her mother works." The same schema could allow respondents to overlook the specification "full-time" in "All in all, family life suffers when the woman has a full-time job."

Schema-based framing effects can result in items being understood in ways not intended by the researcher. Another family-related item, "A working mother can establish just as warm and secure a relationship with her children as a mother who does not work" (*Warm relationship*), illustrates this well. Public debate on the effects on children of women working, or criticism of women who do go out to work when children are at home, determines how salient one or the other reading is (Braun, Scott, and Alwin 1994; Braun 1998, forthcoming). If debate has focussed on criticizing women, even nontraditional respondents may understand the item to mean just that, and their response would then indicate that working mothers are not to blame. Respondents in former communist countries are likely to read the item this way. In these countries, labor force participation of women was compulsory in spite of very traditional attitudes among the population; it is now necessary for economic reasons. This interpretation is also likely in some continental European countries, such as Germany, where ostracizing working mothers has a long tradition. In Scandinavian and Anglo-Saxon countries, however, the second reading is more salient. In these countries it is unnecessary to defend working mothers, because there is virtually no criticism of this group. In these cultural contexts, respondents then focus on the effects of working mothers on children. In Scandinavian countries this is due to the fact that female labor force participation is seen as a necessary means to achieve gender equality, and in Anglo-Saxon countries the family is regarded as a private matter and economic independence and prosperity are important societal values.

Thus, it comes as no surprise that in Figure 4.2 respondents in western Germany and Russia show extremely 'nontraditional' attitudes with *Warm relationship* compared to items more explicitly directed at the effects on the children (*Child suffers* and *Family suffers*). In the United States and Sweden, however, there is not much difference between the items. As seen in Figure 4.1, the items on well-being of children are, in western Germany at least, not particularly 'difficult' to endorse, because respondents do not think about extreme family set-ups. Thus, item difficulty alone cannot explain the intercultural pattern observed here.

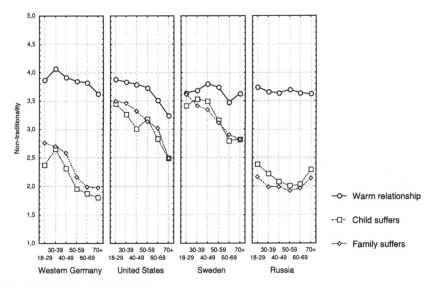

SOURCE: ISSP Study 1994. N=7.041.

Figure 4.2.   Three ISSP items designed to measure consequences of female labor force participation dependent on respondent age.

4.4.2   Interaction Effects of Cultural Contexts

Conditional context effects can also be generated through cultural contexts, by a direct impact on the effect of question and personal experience contexts and by modulating conditional context effects that are mediated by the impact of personal experience contexts on question contexts.

An experiment conducted by Strack, Schwarz, and Wänke (1991) helps illustrate the impact of cultural contexts on question context effects. German respondents evaluated the deliberately ambiguous issue of an "educational contribution" more negatively when they were asked beforehand to estimate the average tuition fee in the United States than when asked to estimate the financial support students were granted in Sweden. As a probing question revealed, respondents interpret "educational contribution" differently depending on the topic of the previous question.

The behavior of respondents in this experiment is culture-dependent and might not be replicable in other countries: it rests on the assumption that "educational contribution" is ambiguous, i.e. in Germany it can be understood both as tuition fee and financial support. This, however, is only the case in societies where public debate focuses equally on tuition fees and financial support. If only one of these topics is part of the public debate, respondents would assume that this is the topic to which the question refers. The previous question would not influence the interpretation of the concept of "educational contribution," since the question, of itself, is then unambiguous in the given cultural context. However, earlier questions can function as an 'argument' (Petty and Cacioppo 1986; Petty and Wegener 1999) and affect respondents' decisions on whether they are for or against a measure. Given the public discussion of tuition fees, the question concerning tuition fees in the United States can serve as an argument for the efficiency of such a measure and thus increase agreement. Under the same societal condition, the question about students' grants in Sweden can serve as an argument that high-quality education is available without the introduction of tuition fees and, thus, reduce agreement. The effects Strack, Schwarz, and Wänke (1991) found for Germany reflect the ambiguity of the questions in that context. Argument-based interpretation, on the other hand, would reverse their findings.

Cultural contexts can also interfere with the information function of response alternatives. A question included in ISSP 1988 (Zentralarchiv 1991) asks "which of the following ways of life" a respondent would recommend to a young person. It offers four response alternatives: "to live alone, without a steady partner," "to live with a steady partner, without marrying," "to live with a steady partner for a while, and then marry," and "to marry without living together first." The last three response alternatives are clearly ordered from less traditional to most traditional. The first is ambiguous and has potentially two most salient interpretations: either not to tie oneself down (have different partners) or not to get involved and live alone, get on with studying, etc. If respondents perceive an order in the four response categories, then the first option will be read as the least traditional living arrangement. German respondents seem to have read the response categories this way (Braun 1993). In Germany, it is predominantly the young who chose this category (9% of the under 35-year-olds versus 3% of the over 64-year-olds). The majority of Americans chose the most traditional response category ("to marry, without having lived together first"); 10% of the young, but 17% of the older cohorts prefer the first category (which would suggest a nontraditional reading). The idea that young people should study hard to gain qualifications rather than become involved in relationships is also a traditional view in America. It therefore seems likely that the response category "to live alone, without a steady partner" was read in this way, and not as the least traditional category in the set offered. In other words, what the cultural context suggests as a likely interpretation prevails in this case for (some) Americans, even despite the context of the response alternatives.

Another example is discussed in Ji, Schwarz, and Nisbett (2000; see Chapter 6): People in collectivist countries are used to monitoring their own and other people's conduct in order to make interaction smoother and they are thus less likely to be influenced by response categories in questions concerning observable behavior.

An example for the modulation of conditional question context effects by cultural contexts can be constructed from another psychological experiment. Asking American students a question on the frequency of dating before one on life-satisfaction resulted in a strong positive correlation between both variables (Strack, Martin, and Schwarz 1988). The explanation is that the question on dating activates information which the respondents are permitted to use in the following question on life-satisfaction. However, when both questions were asked in the reverse order or a joint introduction for both was used, both variables appeared to be virtually unrelated. In different cultural contexts no, or a negative, relationship between both variables might also result when the dating question precedes the satisfaction question. This could happen, for example, in cultures where a high number of dates signals failure to find a steady partner.

4.5   CONCLUSION

Poor translations of good questions mean respondents read and respond to a question they should not have been asked. Well-translated questions that have different readings in different cultures are problematic questions for cross-cultural implementation. Respondents 'read' the questions differently and answer accordingly. In order to convey the same meaning in different countries, details in the formulation of a question may require changing. This can include specifying additional informational components, eliminating elements that can activate schematic representations in only some countries, or modifying formulations that might be regarded as provocative.

Note
1. The reported research was supported by grant 921 9800 3 from the Fritz-Thyssen Foundation.

Chapter 5

DEVELOPING COMPARABLE QUESTIONS IN CROSS-NATIONAL SURVEYS

TOM W. SMITH

5.1 INTRODUCTION

Cross-national research endeavors, among other things, to construct multicountry questionnaires that are functionally equivalent across target populations. (On the different types of equivalence, see Knoop 1979 and Johnson 1998a.) Questions not only need to be valid, but must also have comparable cross-national validity. However, the very differences in language, culture, and social structure that make cross-national research so analytically valuable are the same that seriously hinder the achievement of measurement equivalency. Though the difficulty of establishing comparability is widely acknowledged, the challenge is more often ignored than met.

In a comprehensive review of major books and articles that use cross-national research, Bollen, Entwistle, and Alderson (1993) found that "[m]ajor measurement problems are expected in macrocomparative research, but if one were to judge by current practices, one might be led to a different conclusion. Issues surrounding measurement are usually overlooked ... Roughly three quarters of the books and articles do not consider alternative measures or multiple indicators of constructs, whether measures are equally valid in the different countries, or the reliability of measures."

Moreover, the less than 25 percent of comparative research addressing whether "measures are equally valid in different countries ..." sometimes made a passing reference to equal validity. This neglect arose in part because 88 percent of authors reviewed used the 'same' measures in all countries and may have felt "that using identical measures guarantees equal validity." Bollen, Entwistle, and Alderson further observed that "few authors ... discuss reliability (26 percent), which should temper our confidence in the quality of measurement and in the results ... [and] just 6 percent of books and 14 percent of articles report the reliability of their measures."

Given the value of cross-national research, the importance of comparable measurements, and the frequent failure to take measurement seriously, increased attention to issues of quality in cross-cultural survey measurement has become a necessity. This chapter hopes to contribute toward the goal of increased attention to quality by discussing 1) the development of comparable questions—focusing on both the question-asking and answer-recording sections of a question, 2) response effects that contribute to general measurement error and variable error structures across nations, and 3) steps that should be taken in cross-national survey research to enhance survey validity and comparability.

5.2   QUESTION WORDING

To begin, the way in which questions and answers are used to obtain survey information must be considered in the larger context of survey measurement. Production of high quality survey data is linked to several factors: cognitive aspects of information processing and retrieval involved in answering questions, motivational context of the questionnaire settings, and response frameworks in which information is communicated to the researcher (e.g., Oksenberg and Cannell 1977; Cannell, Miller, and Oksenberg 1981; Bradburn and Danis 1984; Tourangeau and Rasinski 1988; Schwarz 1999a; Tourangeau 1999; Tourangeau, Rips, and Rasinski 2000).

Some have argued that comparability in *question wording* is a primary consideration among survey response process issues under discussion. Kumata and Schramm (1956) noted that "the wording and translation [is] the weakest link" in achieving cross-national equivalence and, consequently, in achieving valid comparisons. The automatic link between wording and translation made here by the authors is a common assumption in survey research; Harkness (2001) discusses problems resulting from this assumption.

Cross-national studies need to examine wording of both parts of questions: the question body, which presents the substance and the stimulus, and the response scale, which records answers. This chapter examines both parts in turn.

5.2.1   Asking the Question

Consider the question's body and conceptual focus. The first challenge is to produce a questionnaire allowing the achievement of comparable questions across different language versions of the questionnaire. This requires a 'good' item that travels well and a good translation. All too often, the problem is not bad or incorrect translation (such as when the religious phrase in English, 'The spirit is strong, but the flesh is weak' is translated into Russian as something approximating 'The vodka is potent, but the meat is rotten'), but rather intrinsic differences in both the languages and the cultures of which the languages are part. For example, the English 'mental health' may be translated into Chinese as either *jingshen jiankang,* connotating spiritual health, or *xinli jiankang*, indicating psychological health. *Jingshen jiankang* in particular contains an element of meaning essentially absent from the English 'mental health.' Similarly, the English and French terms 'liberty' and *liberté* have strong historical ties closely associated with the formative revolutions that occurred in each country (e.g., such phrases as '. . . life, liberty, and the pursuit of happiness' and '*Liberté, Egalité, Fraternité!*'), though in other languages the 'equivalent' terms may carry no strong historical and revolutionary connotations. Finally, even cognates between fairly closely related languages may differ substantially. For example, for Spanish-speaking immigrants in the United States, *educación* includes social skills such as proper etiquette not included in the more academic meaning of 'education' in English (Greenfield 1997).

A related problem occurs when a concept easily represented in one language has no suitable corresponding terms in another language. For example, a study of Turkish peasants (Frey 1963) concluded that "there was no nationally understood word, familiar to all peasants, for such concepts as 'problem,' 'prestige,' and 'loyalty.'" Similarly, the Japanese concept of *giri*—having to do with duty, honor, and social obligation—has no "linguistic, operational, or conceptual corollary in Western cultures" (Sasaki 1995).

Language differences are by no means the only obstacle to measurement equivalence. Indeed, differences in contemporary conditions and existing structures also present difficulties with equivalence. Since questions, like words, have meaning not only on the basis of their constituents, but also in terms of the contexts in which they are used, the context can alter question meaning even though questions in one survey are essentially the same in format and semantics as those in another survey. As Bollen, Entwistle, and Alderson (1993) note: "Consider the young woman who has reached her family size goal. In the United States, if you ask such a woman whether it would be a problem if she were to get pregnant, she is likely to say yes. In Costa Rica, she may say no. This is because in Costa Rica, such a question may be perceived as a veiled inquiry about the likely use of abortion rather than a measure of commitment to a family size goal."

In other cases, structural differences mean that exactly equivalent objects or entities do not exist, or that terms used to describe one thing in one country describe something else in another country. For example, one can ask about approval of the national monarch in Great Britain and the Netherlands, but not in France and Germany. Though all four languages permit researchers to pose the question, only Great Britain and the Netherlands are monarchies. Thus only in these countries is the approval question relevant and sensible. Likewise, the Food Stamp Program—a major American welfare program in which participants receive script with which to purchase food—has no close equivalent in most other countries. Similarly, the phrase 'social security system,' when used in most European countries, refers to social welfare programs in general. When used in the United States, however, the term most likely calls to mind only the federal old age/retirement pension program.

Condition and structure differences mean that what and how one asks will vary across societies. This is as true for concrete behavior and demographic questions as it is for attitudinal measures and psychological construct indicators. For example, one study involving occupation in rural Mali found that in addition to the standard American occupation ratings of each job's relationship to data, people, and things, researchers needed a fourth dimension—relationship to animals (Schooler et al. 1998). Similarly, items regarding a respondent's spouse must allow for plural mates in Islamic and most African societies.

Adding such problems as linguistic and structural equivalence to the already notable monolingual challenge of creating valid measures greatly reinforces the need for multiple indicators. Even given the most careful translation, it is difficult to compare the distributions of two questions that employ abstract concepts and subjective response categories (Smith 1988; Grunert and Muller 1996). Though probably possible to ask effectively equivalent questions like "In what year were you born?" and "Did you vote in the last national election?" it is highly doubtful that responses to the query "Are you very happy, pretty happy, or not too happy?" are precisely comparable. In all likelihood the closest linguistic equivalent to 'happy' will differ from the English concept in various ways, perhaps conveying different connotations and tapping other related dimensions (e.g., satisfaction). At a minimum it probably expresses a different level of intensity (for example, where 'happiness' occurs on a scale ranging from absolute bliss to total despair). Similarly, the terms 'very,' 'pretty,' and 'not too' are unlikely to have precise equivalents in non-English languages. Even in the situation in which the English adjective 'very' is consistently (and correctly) translated into the French *très,* it is not known if its strength is sufficiently identical to cut the underlying happiness continuum at the same point.

Consider the following example of a scheme assessing whether the French or Americans have more psychological well-being:

(a)   A measure of general happiness
(b)   A measure of overall satisfaction
(c)   A scale of measures of domain-specific satisfaction

Franco-American comparisons on any one of these questions or scales would be suspect because of possible language ambiguities. Even the multi-item measure of domain-specific satisfaction would be insufficient since the survey would build all items around the shared and repeated use of the term 'satisfaction.' The repetition of 'satisfaction' would compound any nonequivalence across items since the error is correlated from one item to another. Nor would the combination of the domain-specific and overall satisfaction items contribute much to solving the comparability problem, since any French-English disparity in the meaning of 'satisfaction' would merely be perpetuated. However, asking a respondent how 'happy'/*heureux* he or she is adds a variable that is linguistically distinct from the satisfaction item and that avoids obvious problems of correlated error stemming from repeated terms. Similarly, the use of the 10-item Bradburn affect-balance scale also shares this advantage since it employs varied terminology to ask respondents how often they have experienced five positive and five negative emotions. While MacIntosh (1998) has argued that the Bradburn affect-balance scale as used in the World Value Study was not cross-nationally comparable—due to differing emotional structures and/or to translation problems—the point here is that the Bradburn scale would not replicate measurement error associated with the format or terminology of other psychological scales.

If researchers employ linguistically distinct measures, then it is possible to get unambiguous results—providing the results are consistent across items (e.g., the French leading or trailing the Americans on all measures). When using only one measure, it is impossible to determine whether any measured differences (or even a measured nondifference) are societal or merely linguistic. With the use of two measures, a consistent pattern on both items can establish a clear finding. If the measures disagree, it is possible that one is social and the other linguistic, but the researcher has no basis to identify which is which. It would be desirable to include three linguistically distinct measures of the same construct. If all three agree, the result is a clear and robust finding. If two agree but the third shows a different pattern, one must be more cautious in stating results, but there is at least a 'preponderance of evidence' toward one substantive interpretation of cross-national differences. If all three results disagree (positive, negative, and no difference), then results yield no firm evidence regarding the question and further work is needed to clarify concepts and improve items and translations. This three-item construct is similar to the 'triangulation' approach discussed by Van de Vijver and Leung (1997b). The merits of multiple indicators from cross-national research are discussed, for example, in Prezeworski and Teune (1966), Scheuch (1989) and Jowell (1998).

"Another way of thinking about the problem is that for many cross-national comparisons country and language are totally confounded. To separate out the language differences from country differences (which is what one is trying to study) additional leverage is needed. That leverage comes by using multiple measures that vary the words in specific questions. This approach would not help if the linguistic differences were persuasive across the languages. Or if by chance the different term happened to replicate the same distinctions. But they create a prima facie basis for believing that linguistic artifacts have been minimized, a presumption that can not be made with a single item or set of items using the same key terms."

Not only do translanguage comparisons add to the burden of creating valid construct measures, but they inevitably require more items to achieve the same degree of cross-language validity as compared to monolingual studies. As Jacobson, Kumata, and Gullahorn (1960) have noted, "However difficult it may be to deal with theoretical issues concerning psychological processes which intervene between observable stimuli and responses in intracultural studies, the cross-cultural research situation magnifies these problems and adds new ones."

Although comparative research compounds the general advantages of using multiple indicators, the literature review of Bollen, Entwistle, and Alderson (1993) found that "multiple indicators appear in only a small minority of the books (18%) . . . [and] in a similarly modest percentage of journal articles (26%)."

5.2.2   Response Scales

Though establishing equivalency in question concept and substance is very important, it is only half the battle. Surveys also need equivalency in response categories. Several solutions have been posited to increase the equivalency among questions (and ultimately answers) in cross-national research.

Issues addressed in the present chapter are a) whether cross-cultural measurement can rely on nonverbal scales to avoid the problem of language, b) the number of response options that should be used in response scales, and c) the extent to which determining verbal label strength will facilitate cross-national scale calibration.

Nonverbal Scales

Substitution of numerical or other nonverbal scales is sometimes suggested as a solution to problems inherent in verbal labels (Fowler 1993). These include such numerical instruments as ratio-level, magnitude measurement scales, 10-point scalometers, feeling thermometers, and frequency counts. Non-numerical, nonverbal scales include visual guides like ladders, truncated pyramids or stepped-mountains, and figures or symbols used in psychological testing. The reasoning behind these alternatives is that numerical scales reduce problems by providing a universally understood set of categories that have precise and similar meanings (e.g., 1, 2, 3 or 2:1, 3:1) and that there is no need to devise language labels denoting response category intensity. Similarly, visual questions and image-based response may reduce verbal complexity.

But nonverbal approaches are not problem-free. First, many numerical scales are more complex and harder to use than simple verbal items, and variations in cross-country comprehension could easily increase noncomparability. For example, the magnitude measurement method assigns a base value to a reference object, and other objects are evaluated by assigning values to them that reflect their ratio to the fixed item. The crime of robbing $1,000 of merchandise from a store may be selected as the reference object and assigned a seriousness score of 100. Respondents are then asked to rate the seriousness of other crimes (e.g., speeding and murder) in relation to the seriousness score of 100 pre-assigned to the store robbery. Alternatively, seriousness might be rated by showing respondents a reference line and asking them to draw shorter or longer lines expressing the relative seriousness of other crimes. One problem with this technique is that 10-15 percent of respondents (in the United States) are seriously confused by this complex, demanding task and are unable to supply meaningful responses. Moreover, it is highly likely that the confusion level would vary across countries, perhaps covarying with levels of numeracy.

Second, research shows that numerical scales are not as invariant in meaning and free of error as their simple, straightforward, mathematical nature presupposes. Schwarz and Hippler (1995) demonstrated that people rate objects quite differently on 10-point scales ranging from +1 to +10 than they do on scalometers ranging from -5 to -1 and +1 to +5. Also, on the 10-point scalometer, people routinely misunderstand the mid-point of the scale (Smith 1994). Furthermore, the 101-point feeling thermometer is not actually used as a refined measurement tool. It is rare for respondents to choose more than 10-20 values (mostly 10s and 25/75s). Also, some seem to be influenced by the temperature analogy and avoid very high ratings such as 'too hot.' (On feeling thermometers see Wilcox, Sigelman, and Cook 1989; Alwin 1997; Tourangeau, Rips, and Rasinski 2000. Doob (1968) considers whether local climate might be a factor in cross-national research—and possibly interact with the feeling thermometer.) Furthermore, no extant research establishes whether the above-mentioned patterns in numerical scale use are cross-nationally consistent.

Third, most societies have lucky and unlucky numbers (e.g., notice how many hotels have no 13th floor). These perceptions may influence numerical responses, and since lucky and unlucky numbers vary across societies, effects will vary as well.

Fourth, most numerical scales only reduce and do not eliminate word use as part of the response scale. For example, researchers must explain not only the operation of the scale itself, but also how the objects are being rated (usually with a like/dislike response). Likewise, the verbal description of the feeling thermometer scale is quite lengthy.

Finally, alternative numbering or grouping schemes influence the reporting of frequencies. Respondents are often unable (or at least unwilling) to try to provide an exact count of some behavior or possession and round their responses to an approximate number (Tourangeau, Rips, and Rasinski 2000). As explained earlier in the feeling thermometer example, people often round to tens (or hundreds, etc.). But in other cases they may round to another number such as 12—reflecting the unit of a dozen or the number of months in a year. Again, there is no assurance that cultural practices behind numerical favorings are identical across societies.

Similar problems arise with nonverbal, non-numerical questions and scales because visual stimuli are not necessarily cross-culturally equivalent (Tanzer, Gittler, and Ellis 1995). For example, the color called 'orange' in English is not clearly coded or distinctly labeled in Navajo. Thus, the Navajo have difficulty matching objects that are 'orange' (Jacobson, Kumata, and Gullahorn 1960). Likewise, Tanzer (forthcoming) demonstrates that the physical ordering of images affects survey responses. In Western-designed matrix items used in psychological testing, the missing element is placed in the bottom right corner. This is logical to those educated in languages that read from left to right and top to bottom. However, for Arab respondents reading right to left and top to bottom, the matrix should be arranged with the missing element in the lower left corner. Similarly, in several low literacy and multilingual African societies, citizens cast their votes not via written ballots, but by placing the vote in a ballot box identified by party symbols and sometimes a picture of the party leader. Africans accustomed to this system of voting could presumably follow a self-completion questionnaire item on party voting that was image-based, but would probably experience difficulty completing a questionnaire containing only written versions of party and candidate names. By the same token, respondents from developed countries could experience difficulty following the—to them—strange pictorial format.

One must ensure that visual stimuli are accurately replicated across countries. The 1987 ISSP study on social inequality included the following measure of subjective social stratification: "In our society there are groups which tend to be towards the top and groups which tend to be towards the bottom. Below is a scale that runs from top to bottom. Where would you place yourself on this scale?"

Ten response categories were listed, with 1 = Top and 10 = Bottom. This item was asked in nine countries (Australia, Austria, [Western] Germany, Great Britain, Hungary, Italy, the Netherlands, Switzerland, and the United States). In all countries, a majority of respondents placed themselves toward the middle, that is in the 4–7 range. However, the Netherlands was clearly an outlier. The range for the percentage of countries that placed themselves in the middle was 24 percentage points, from 83.8 percent in Australia to 59.8 percent in the Netherlands. More than half of the difference (12.4 percentage points) was attributable to the Netherlands. Likewise at the bottom of the scale, that is, in the 8–10 range, the spread was 31.3 percentage points. Here the Netherlands also contributed to almost half the difference (13.6 percentage points). While most other differences appeared to reflect actual differences in social structure, the Netherlands' distinctive distribution did not fit other measures of Dutch society (e.g., income distribution). Neither were Dutch responses as distinctive on other social inequality measures, for example subjective class identification (Smith 1990).

An examination of the Dutch wording indicated it was both equivalent to the English meaning and appropriate and clear in Dutch. Further analysis revealed that the visually displayed scale in the Netherlands differed from that employed in the other countries. All other scales had 10 vertically stacked squares (with the highest box labeled "Top" and the lowest labeled "Bottom"). While the Dutch scale did have 10 stacked boxes, they were in the shape of a truncated pyramid, and the bottom boxes were wider than those in the middle and on the top. Apparently, Dutch respondents were attracted to the wider, lower boxes because they viewed them as indicating where more people were. The suspected impact of these different visual images was later verified by experiments (Schwarz, Grayson, and Knäuper 1998).

Simple Response Scales

The second suggested solution to problems inherent in verbal labels is, in a sense, the opposite of the numerical approach: the 'keep-it-simple-stupid' tactic. In surveys this means using only dichotomies. It is argued that 'yes'/'no,' 'favor'/'oppose,' and other pairs of antonyms have similar meanings and cutting points across languages. The case for this tactic is that, though language differences make it difficult to determine just where along a continuum a respondent may be, it remains comparatively easy to determine on which side of the tipping point he or she is.

However, the assumption that dichotomies are simple and equivalent across societies is questionable. Consider the following example from the discipline of law: Under English law a person may be judged as guilty or not guilty in a criminal case or liable or not liable in a civil case. While both are dichotomies, they differ greatly in placement of the tipping point. In the criminal case the standard of proof is 'beyond a reasonable doubt,' while in the civil case it is 'by the preponderance of the evidence.' Theoretically, this means a person could be found liable in a civil case if the bare majority of the evidence points toward liability, but in a criminal case a guilty verdict requires overwhelming evidence against the defendant. Also, the law requires a criminal jury trial to levy a unanimous verdict (all jurors voting guilty or not guilty and a split decision resulting in a 'hung jury' and mistrial), but in many areas civil cases can be decided with less than unanimity.

Extending this legal example, the guilty/not guilty dichotomy recognized by English law is not followed even in closely related Scottish law. Scotland recognizes three verdicts: guilty, not proven, and innocent. Thus, even within Great Britain the idea that guilt and innocence is a simple dichotomy breaks down. Of course, for those unfamiliar with British legal practices, this example may not be a useful illustration. In that case, it serves as an equally important example of how institutions, such as the legal system, differ across nations and that such structural and conceptual differences are obstacles to establishing equivalence in cross-national research. Similarly, other simple, dichotomous distinctions may not hold up across languages and/or cultures.

A second drawback to this approach is, of course, its loss of precision. Dichotomies measure only direction of attitudes and not their extremity and are likely to create skewed distributions. Moreover, a survey using dichotomous items would require several questions to differentiate respondents into as many categories as one item containing five or seven scale points. (Of course, as noted earlier, on other statistical grounds a single item with multiple responses does not usually produce as valid or reliable a measure as a multiple-item, composite scale.)

Calibrating Response Scales

The third proposed solution is to attempt to calibrate the scale by determining the strength of the verbal labels used. There are several ways to measure the strength of response categories along an underlying response scale. In one approach, respondents rate the strength of terms by defining each point on the scale.

Researchers have developed three standard variants to this approach. First, one can rank the terms from weaker to stronger, from less to more, or any similar continuum (Spector 1976). This indicates only the relative position of term strength, not absolute strength or distance between terms. Second, one can rate each term on a numerical scale, usually employing 10 to 21 points (Mosier 1941; Jones and Thurstone 1955; Cliff 1959; Myers and Warner 1968; Mittelstaedt 1971; Bartram and Yelding 1973; Vidali 1975; Worcester and Burns 1975; Wildt and Mazis 1978; Tränkle 1987). This variant permits the absolute strength or distance between each term to be known, facilitating the creation of equal interval scales. It is also possible to use an alphabetical scale or unlabeled spaces, rungs, or boxes as in a semantic differential scale (Osgood, Suci, and Tannenbaum 1957). The letters or spaces are then transformed into their numerical equivalents. Finally, magnitude measurement techniques place each term on a ratio scale (Lodge et al. 1975, 1979, 1982; Hougland, Johnson, and Wolf 1992). The magnitude measurement technique assigns an arbitrary value to a reference term and has respondents rate other terms as ratios to the base term (see earlier discussion). This variant permits more precision than the numerical scale approach, since the terms are not constrained by the artificial limits of the bounded number scale.

The second of these three variants seems most useful. The ranking method fails to provide numerical precision necessary to calibrate terms across languages, and the magnitude measurement technique is difficult both to administer and to complete (about 10–15 percent of respondents seem unable to master the procedure). In addition, the extra precision provided by the magnitude measurement procedure when compared with the 21-point scale approach appears unnecessary.

The direct rating approach rates words along various dimensions. Of most interest are those that rate terms either along a general good/bad or positive/negative dimension or those that rate the intensity of modifiers (Mosier 1941; Jones and Thurstone 1955; Cliff 1959; Myers and Warner 1968; Mittelstaedt 1971; Bartram and Yelding 1973; Lodge et al. 1975, 1979, 1982; Vidali 1975; Worcester and Burns 1975; Wildt and Mazis 1978; Hougland, Johnson, and Wolf 1992). Similarly, other studies rated probability statements (Lichtenstein and Newman 1967; Wallsten et al. 1986), frequency terms (Simpson 1944; Hakel 1968; Strahan and Gerbasi 1973; Schriesheim and Schriesheim 1974; Spector 1976; Bradburn and Sudman 1979a; Schaeffer 1991; O'Muircheartaigh, Gaskell, and Wright 1993), and percentage description terms used in public opinion surveys (Crespi 1981 and RAC 1984).

The studies generally show that a) people (usually college students) can perform the required ratings tasks, b) ratings and rankings are highly similar across different studies and populations, c) there is high test/retest reliability, and d) several different treatments or variations in rating procedures yield comparable results. Thus, the general technique seems robust and reliable. While this literature is reassuring, other studies indicate that various measurement artifacts can influence responses to numerical scales (Schwarz et al. 1985; Wilcox, Sigelman, and Cook 1989; Schwarz et al. 1991; Smith 1994; Schwarz and Hippler 1995; O'Muircheartaigh, Gaskell, and Wright 1993). Additional research also indicates that vague frequency terms correspond to different absolute values that depend on the commonness or rarity of the specified event or behavior. Thus, people who 'usually' vote may vote only once a year or less (depending on the frequency of elections), but people who 'usually' dine out dine out more than once a week (Bradburn and Sudman 1979a; Schaeffer 1991).

Another approach to assessing intensity of scale terms and response categories is to measure the distributions generated when different response scales are used (Orren 1978; Smith 1979; MacKuen and Turner 1984; Sigelman 1990; Hougland, Johnson, and Wolf 1992; Laumann et al. 1994). In an experimental, cross-subjects design, one random group evaluates an object (e.g., presidential popularity or personal happiness) with one set of response categories, and a second random group evaluates the same object with a different set of response categories. Since the stimulus is constant and the subgroup assignment is random, the number of people attracted to each category depends on the absolute location of each response category on the underlying continuum and the relative position of each of the scale points utilized. With some modeling around what the observed distributions suggest as an underlying distribution, it is possible to estimate the point at which each term cuts the underlying scale (Clogg 1982, 1984).

The alternative version uses a within-subjects design in which people are asked the same question (i.e., presented with the same stimulus) two (or more) times with different response categories (Orren 1978). This differs from a test/retest reliability design in that a) the measurement instrument is not constant (since the response categories differ), and b) the two administrations are essentially consecutive without intervening time and/or buffer tasks. This provides additional information since it allows the direct comparison of responses, but initial evaluations may artificially influence responses to later scales. In a subsequent administration, respondents may feel constrained to choose the same response given in an earlier administration in order to maintain consistency. They may select responses representing the same scale position (e.g., in the middle) or using the same term.

Distributional approaches are advantageous in that they ask respondents to do only what they are normally required to do: answer substantive questions with a simple set of response categories. But these approaches also have their drawbacks. For example, it is more difficult to access a large number of response terms via distributional approaches; thus they are better suited for assessing a discrete response scale already adopted than for evaluating a large number of terms for possible use in response scales. Furthermore, the results depend on the precise underlying distribution and the modeling procedures adopted. Finally, distributional approaches create more work for analysts since the strength of terms must be indirectly estimated from the distributions rather than directly calculated from respondent ratings. Though possible to evaluate more terms using more random subgroups, this would necessitate an increase in sample size. Similarly, the same respondents could be asked many repetitions of a question and given different response scales for each. However, this would soon become tedious, and later repetitions would probably be distorted by the previous administrations.

The direct rating approach has been used in a study of terms used in response scales in Germany and the United States (Smith 1997; Mohler, Smith, and Harkness 1998), and this pilot study was very promising. Results showed many highly equivalent response terms in both Germany and the United States, as well as some notable systematic differences. In addition to the many technical challenges the approach creates, its major drawback is the necessity of separate methodological studies in each country and language to establish the calibration. Obviously, every substantive cross-national study cannot undertake this effort. Theoretically, however, once calibrations are determined they could be used by other studies without requiring any extra data collection.

In brief, considerable effort must be made to achieve cross-national comparability of questions. In addition to securing the most careful translations possible, developers must ensure that questions actually have the same meaning across societies. Likewise, developers would be wise to design response scales to achieve maximum equivalence, minimize differences in particular items and response scales, and employ multi-indicators that vary particular concept words and response scales in order to corroborate results by linguistically independent measures.

5.3   RESPONSE EFFECTS

Other challenges to cross-national comparability arise from differences in various response effects (Hui and Triandis 1985; Usunier 1999). Though response effects are a source of measurement error in all surveys, cross-national surveys are especially vulnerable to various error components being correlated with country. Thus, differences observed across countries may represent differences in response effects rather than in substance. Work by Saris (1998) across 13 cultural groups/nations indicates that measurement error is not constant. He notes that "even if the same method is used, one can get different results due to differences in the error structure in different countries." Among the more important cross-national sources of measurement variation are effects related to social desirability, acquiescence, extremity, no opinion, middle options, response order, context and order, and mode.

First, responses are frequently distorted by social desirability (DeMaio 1984; Tourangeau, Rips, and Rasinski 2000). Image management and self-presentation bias lead respondents to give responses portraying them in a positive light. The general tendency is to overreport popular opinions and underreport unpopular or deviant opinions, or—even more frequently—to overreport socially acceptable activities and underreport undesirable behaviors. In a similar vein, some respondents over- and underreport, gearing responses not to general social norms, but to the perceived values of the interviewer.

These tendencies appear to be common across social groups, but undoubtedly vary in both intensity and expression. The pressure to conform may be more or less intense, with conformist societies presumably experiencing a higher instance of social desirability effects than individualist societies. Social desirability effects may also be related to and compounded by interactions between the characteristics of respondents and interviewers. Research indicates that the race and ethnicity, gender, social class, and age of respondents and interviewers interact to alter responses. For example, the United States' largest and most consistently documented interviewer effect is the expression of more tolerant intergroup attitudes when being interviewed by a person of another race/ethnicity (Schuman et al. 1997). The increased-tolerance effect has also been documented in Kazakhstan in situations involving Russian and Kazakh interviewers (Javeline 1999). Likewise, social desirability effects are likely to be greater when there is a greater status/power differential between interviewer and respondent. These effects are likely to vary differentially across nations. In developing countries, for example, interviewers are more likely to be members of an educated elite, while in developed countries interviewers are typically of average status.

Second, the definition of 'sensitive topics' and 'undesirable behavior' varies among both individuals and cultures. As Newby (1998) noted, cultures differ from one another in definitions of private and public information. Also, legal parameters for survey topics vary across cultures. China, for example, now permits extensive survey research, but forbids many political questions, such as those inquiring directly about the Communist party. However, in most instances, the constraints stem from social conventions rather than legal regulations because societies differ greatly in what they consider appropriate survey topics. Questions pertaining to alcohol use are much more sensitive in Islamic countries than in Judeo-Christian societies. Likewise, patterns of acceptable household composition vary across cultures. For example, cohabitation of unmarried partners was a widely accepted practice in Sweden long before it became socially acceptable in the United States.

To deal with social desirability effects, survey designers can both frame questions in a less threatening manner and train interviewers to be nonjudgmental in asking items and reacting to answers. In addition, designers can introduce modes that reduce self-presentation bias (as discussed later).

A second response effect is the tendency toward acquiescence, sometimes called yea-saying bias, in which respondents lean toward over-compliance and tell the interviewer what they think the interviewer wants to hear (Tourangeau, Rips, and Rasinski 2000). Acquiescence is particularly likely to occur on agree/disagree items and others that offer clear affirm/reject responses. Acquiescence leads people to select the affirming response, though research clearly indicates that this bias can vary across cultures. For example, Church (1987) found yea-saying to be particularly strong in the Philippines; Landsberger and Saavedra (1967) reported similar effects among Spanish speakers in the United States and Chile. Javeline (1999), using experiments with reversed coded items in Kazakhstan, found not only a high level of overall acquiescence, but an even greater level among ethnic Kazakhs than ethnic Russians. Van Herk (2000) also demonstrated that Greek respondents offered more positive responses than respondents in other European countries.

How can survey designers reduce acquiescence? One solution is to balance scales so that the affirming response is half the time in the direction of the construct (e.g., six agree/disagree items on national pride with the patriotic response being represented by three agree and three disagree responses). It is hoped that such reversals force a respondent to consider the items' meaning and reply in a substantively meaningful pattern. If this does not occur, then the answers cancel themselves out and the respondent reflects a middle position—which is a more accurate response for a substantively uninvolved respondent than is one extreme or the other (where he would have been were the items consistently scored in one direction). Other solutions are to build formal reversals into an instrument to catch yea-sayers (Bradburn 1983; Javeline 1999) and, finally, to employ alternative question formats such as forced-choice items that are less susceptible to acquiescence bias (Converse and Presser 1986; Krosnick 1999). Javeline (1999) observed that "members of certain ethnic groups—in the name of deference, hospitality, or some other cultural norm [agree falsely] ... more frequently ... (T)he fact that they do must be taken into account in designing questionnaires. We cannot change the respondents, so we must change our methods."

A third response bias is found in different uses of scale extremes. Some people seem especially attracted to extreme categories (e.g., strongly agree, most important), while others avoid these and tend to favor less extreme responses (e.g., agree, somewhat important). Respondents tend to follow the extreme/nonextreme patterns regardless of their true strength of attitude towards particular items. Thus, the choice of categories may represent a response set rather than a substantive gradation of opinions. To complicate matters, research indicates that this tendency can vary across ethnic groups. For example, in the United States African-American students are more likely to select extreme responses than are White students (Bachman and O'Malley 1984). Likewise, a sample from the United States Navy shows Hispanic respondents selecting extreme categories more than non-Hispanics, although differences appeared more clearly on a 5-point scale than on a 10-point scale (Hui and Triandis 1989). Similarly, cross-national studies have documented differences in the propensity to select extreme categories. Asians in general, and Japanese in particular, are inclined to avoid extreme responses (Chun, Campbell, and Yoo 1974; Lee and Green 1991; Hayashi 1992; Chen, Lee, and Stevenson 1995; Onodera 1999). Whether these differences in category selection are tied to cultural differences (e.g., subgroup cognition norms) or explained by structural differences in other factors related to extremity preference (e.g., education, age, and income) is not yet clear (Greenleaf 1992; Greenfield 1997).

Experts propose several steps to compensate for extremity-related response styles. First among these is the implementation of a multitrait, multimeasurement design (van Herk 2000). For example, the response scales used can be varied (as in the 5- and 10-point scales cited in Hui and Triandis 1989) to see if effects occur across measurement instruments. Some suggest ranking, rather than rating, items. While this formally eliminates the possibility of an extreme response effect, it forces respondents to complete a more difficult task, loses measurement differentiation, and assumes no cross-object ties (van Herk 2000).

A third remedy involves targeting functional equivalency on the scale rather than matching translated terms. Given the assumption that the Japanese are predisposed to avoid response categories with strong labels, this approach softens the labels for Japanese categories, so that 'strongly agree' and 'agree' would be rendered in Japanese more along the lines of 'agree' and 'tend to agree' instead. Others suggest that the problem is found in a disconnect between translation equivalence and response-scale equivalence. For example, a study of English, Chinese, and Japanese students (Voss et al. 1996) indicates that a quantitative comparison of terms used in typical survey responses and deemed equivalently translated were not rated as similar in intensity. These results suggest that part of the problem is that equivalence (at least in terms of survey responses) is not achieved via standard translation.

In either case, the issue at hand is whether noncomparability in one aspect is needed to establish comparability in a more essential area. Traditional rules mandate that methods should be identical across surveys, but the challenge is to identify cases in which methods identical on one level are not identical on other levels affecting measurement. In these instances, identical structure does not establish equivalency. For example, graduate-level language examinations at the University of Chicago require a typical passage translation from French of about 700 words, while its German equivalent is only 450 words. The difference in word length is intentional, providing an equivalently difficult translation task achievable within equal lengths of time.

Several improvements can also be implemented at the analysis stage. Researchers could first determine whether response extremity is similar across countries and then analyze items collapsed into dichotomies to determine whether conclusions are appreciably altered. An analysis of ISSP items regarding scientific and environmental knowledge used one summary scale with 12 items that merely counted the number of correct responses, while another employed 5 response categories (definitely true, probably true, can't choose, probably false, and definitely false). The two scales produced similar findings (Smith 1996).

A fourth measurement variation occurs in the selection of "no opinion" and similar nonresponse categories related to extremity response sets (i.e., "Don't Knows" (DKs)). Across countries DKs are higher among the less educated (Young 1999), but DK levels still vary by country, even when education is controlled for. While part of this cross-national variation is undoubtedly real and reflects true differences in the level of opinionation, some appears to stem from different response styles. As Delbanco et al. (1997) suggested, "Attitudes about responding to surveys (e.g., a tendency for individuals to say they do not know an answer or to refuse to answer) may differ across countries."

For example, Americans have a lower tendency to refuse to supply personal income information than do many European respondents (Smith 1991a). Also, people in some countries apparently show a greater willingness to guess about questions pertaining to scientific knowledge rather than admit their ignorance by giving a DK response (Smith 1996). Another common assumption is that DK responses will be more frequent in developing countries due to lower education levels and different social norms (e.g., a reluctance to share private thoughts with strangers).

There is considerable debate in survey research as to whether surveys should generally encourage or discourage DKs. From Converse's nonattitude perspective, people tend to express opinions even when they do not have any, and surveys should use full filters in an attempt to discourage the false expression of opinions by nonattitude holders (Smith 1984). But Krosnick and others argue that explicitly offering DK options does not improve data quality and probably assists satisficing, and therefore Krosnick favors not offering an explicit DK response option (Krosnick 1999). However, little evidence is available on how such tendencies vary across countries.

Closely related to the issue of 'no opinion' responses is the matter of whether 'neutral' middle options should be offered. A no-middle-option item might ask one to "strongly agree, agree, disagree, or strongly disagree," while a middle-option version would ask one to "strongly agree, agree, neither agree nor disagree, disagree, or strongly disagree." Research from several countries indicates that by providing ambivalent respondents with a clear response option, the middle-option scale produces more reliable results (Smith 1997; O'Muircheartaigh, Krosnick, and Helic 2000). However, this middle-option advantage may depend upon the number of response options involved. (See, for example, Alwin 1992 regarding the finding that two- and four-category response scales are more reliable than three-category scales in attitude measurement due to the ambiguity of the middle category.)

Sixth, response order can also influence the distribution of answers. Under various conditions, respondents will tend to favor the first offered response (i.e., primacy effects) or the last response (i.e., recency effects). Many American studies find that when questions "are presented visually, respondents are likely to begin by processing the first response option presented; when the items are presented aurally, respondents are likely to begin processing the final option they heard" (Tourangeau, Rips, and Rasinski 2000). Thus, the former leads to primacy and the latter to recency effects. Response effects have also been found to interact with both cognitive ability and respondent motivation (Krosnick and Alwin 1987; Sudman, Bradburn, and Schwarz 1996; Krosnick 1999), though it is unknown how robust these patterns are across cultures and languages.

Seventh, variation in surveys also arises from question context and order (Smith 1991b). Broadly stated, context-and-order effects occur when previously asked questions influence responses to later questions. The questions fail to remain independent of each other and the prior stimuli or one's response to these join with stimuli from subsequent questions to affect responses to subsequent items. The many variations of context-and-order effects depend on how previous questions influence later items. Tourangeau and Rasinski (1988; Tourangeau 1999), for example, describe four steps in answering a question: 1) interpretation of question meaning and intent, 2) retrieval of relevant and necessary information from memory, 3) judgment as to relationship between memories and question, and 4) suitable response selection from the offered response categories. Each stage may contain carryover or backfire effects from content encountered earlier in the survey. A carryover effect at the interpretation stage might involve using a definition supplied in an earlier question to understand a later question. A backfire effect at the same stage might involve failure to mention events covered by a previous question because re-mentioning them seems redundant to the respondent.

Research indicates that survey context effects operate in most countries, but experiments and detailed studies have been conducted in only a small number of countries and coordinated, cross-national, experimental studies are lacking.

It is also quite plausible that context effects would not be constant cross-nationally. For example, in a study in Germany, Schwarz (1999a) demonstrated that respondents' favorability ratings of politicians shift, depending on the order in which respondents are asked about the politicians. When first asked about a discredited and widely disliked politician, respondents offer a higher rating of political leaders in good standing who immediately follow the 'bad example.' Moreover, it is likely that this effect would appear in other countries. If so, then in a cross-national survey of political leadership asking about party leaders in a predetermined order (e.g., head of government, leader of largest opposition party, leader of next largest party, etc.), the ratings of subsequent figures (e.g., opposition leaders) could easily vary artificially because of true differences in the popularity of previously mentioned leaders (e.g., head of government). This artificial variation could in turn lead to misinterpretations about the absolute popularity of opposing leadership across nations. Similarly,

Sudman, Bradburn, and Schwarz (1996) describe a context experiment in Germany that hinges on the fact that beer—not vodka—is a popular, national drink. It is entirely possible that the opposite effect would appear in Russia due to the reverse positions of the two beverages.

Conditional context effects are particularly likely to vary across cultures. As Smith (1991b) has shown, context effects are often conditional on respondent attitudes toward the preceding, context-triggering question. In the United States, asking first about generally popular government spending programs leads people to respond less strongly that their federal income tax is too high. However, this effect depends on the popularity of the spending programs. Among those in the most pro-spending group, asking the spending items first reduces by 25 percentage points the number indicating their federal income tax is too high. But for those most opposed to government spending, asking the spending items first reduces by 7 percentage points the number indicating that their taxes are too high. People with intermediate support for government spending programs had intermediate context effects. In this and similar conditional context effects, if the conditional context effects were cross-nationally similar, the net effect would be similar only if the popularity of government programs was also cross-nationally comparable.

Finally, question responses can vary by mode of administration (e.g., self-completion, in-person, or telephone). Analysis has demonstrated many mode effects. Among the most consistent is that more socially undesirable or sensitive behaviors (e.g., alcohol consumption, illegal drug use, sexual activity) are reported in self-completion modes than in interviewer-assisted modes (Tourangeau, Rips, and Rasinski 2000). Maintaining consistency of mode does not automatically solve the mode effect problem, since mode may not have a constant cross-national impact. For example, not only would showcards with words be of little use in low literacy societies, but the inappropriate use of these may artificially generate greater differences between the literate and illiterate segments of the population than a survey mode not as interactive with literacy and education.

In sum, various measurement effects can influence survey responses. In a number of cases, we know that these effects can vary across subgroups and/or countries, and in other cases such variable effects are plausible, albeit not empirically demonstrated.

This is not to suggest that response effects always—or even typically—differ among different groups and across societies. Though the body of rigorous cross-national measurement studies is small, the research has documented a number of consistent results. For example, some social desirability effects are demonstrably similar in Canada, the Netherlands, and the United States (Scherpenzeel and Saris 1997). Moreover, telephone surveys produce lower quality data in the same countries (Scherpenzeel and Saris 1997), while forbid/allow question wording variations have the same effect in both Germany and the United States (Hippler and Schwarz 1986). But variable measurement effects remain a serious concern to which researchers must be alert.

5.4   STEPS TOWARD COMPARABLE QUESTION WORDING

What can survey research do to lower—if not eliminate—hurdles to equivalence, thereby achieving valid cross-national research? First, avoid research imperialism or safari research, that is, the tendency of a research team from one culture to develop a research project and instrument, rigidly imposing both on other societies. As Van de Vijver and Leung (1997b) have observed, "Many studies have been exported from the West to non-Western countries, and some of the issues examined in these studies are of little relevance to non-Western cultures." Instead, survey research needs a collaborative, multinational approach (Szalai 1993; Van de Vijver and Leung 1997b; Jowell 1998; Schooler et al. 1998). Greenfield (1997) has pointed out, "To elucidate cultural differences in an unbiased fashion, then it is best to have a bicultural (or multicultural) team and to collaboratively develop a single instrument for all cultures before the study begins."

The ISSP adopted this approach, as Sanders (1994) noted: "One of its [the ISSP's] greatest strengths is that a country can only be incorporated in the survey if a team of researchers from that country are available . . . to ensure that the translation of the core questions can be achieved without significantly altering their meaning. The potential problem of cross-national variation in meaning is accordingly minimized."

A different example of the same joint development principle is provided by a study of AIDS/HIV in three preliterate tribes in rural Mali. The research team consisted of American health and African specialists, Mali health researchers, and local tribal informants. They worked together to design and execute a survey that was workable, linguistically compatible, and culturally appropriate for the three target populations (Schooler et al. 1998).

The joint development approach helps to culturally and linguistically decenter questions so that they are more likely to be functionally equivalent across societies and languages (see Chapter 2). This is essential since the more the societies/groups being compared differ, the harder it is to achieve equivalence. As Johnson et al. (1997) observed, "The balance of shared-to-unique contextual elements between cultures will define the degree to which the concept in question is generalizable across culture." Important dimensions to consider in the quest for cross-cultural equivalence include language, level of development, and culture (including such elements as religion, shared histories, geographic proximity). Thus, comparability problems are less likely between surveys in Norway and Sweden, which have highly similar languages, a closely shared history, the same main religion, and a comparable level of development. However, they are more apt to occur between Norway and Portugal, which differ more on each of these dimensions than the two Scandinavian countries do. It follows, then, that the challenges of designing comparable surveys in Norway and Bangladesh will be greater than establishing comparability between Norway and Portugal. The divide between preliterate, tribal societies and advanced, post-industrial nations is the widest that survey research must bridge. Not only are there linguistic and cultural differences to contend with, but researchers must grapple with the fact that the surveys themselves are different experiences to different groups of respondents. For example, in new democracies in general and developing nations in particular, the interview experience is likely to be strange and perhaps threatening. As a result, survey designers must often build in extra steps in order to lay the groundwork for a successful interview (Wilson 1958; Devereux and Hoddinott 1992). As Greenfield (1997) notes: "For participants from collectivistic cultures, the tester may need to have or establish a personal relationship with the testee outside the testing situation before a valid assessment can be done."

The second step in lowering hurdles to equivalence involves extensive pretesting and piloting. This developmental work must establish that both items and explicit scales meet acceptable technical standards (e.g., of comprehension, reliability, and validity) in each country and are comparable (or of equivalent validity) across countries. To this end, a number of useful developmental and testing procedures exist. First among these are cognitive interviews using 'think-alouds'—a protocol in which respondents verbalize their mental processing of questions—and computer-assisted concurrent evaluations (Bolton and Bronkhorst 1996; Krosnick 1999; Tourangeau, Rips, and Rasinski 2000). Another useful procedure is behavioral coding, in which the interviewer-respondent exchanges are recorded (usually on audio, but sometimes with audio-video), coded in detail, and then formally analyzed (Fowler and Cannell 1996; Krosnick 1999). Also useful is conventional pretesting (Converse and Presser 1986; Fowler 1995). Presser and Blair (1994) have conducted formal comparisons of various pretest methods and report that each has special advantages in identifying questionnaire problems. The value of such pretesting for intercultural studies is demonstrated by the use of cognitive follow-ups by Johnson et al. (1997) to determine how respondents in general—and especially respondents from different cultures—understand items.

Other survey instrument tests include concurrent ethnographic analysis, a process that cross-validates the results from surveys and ethnographic studies (Gerber 1999); exemplar analysis, which assesses scales by asking people to describe the types of events that represent the response options (e.g., 'Give an example of someone completely satisfied with his/her job, somewhat dissatisfied, etc.,' Ostrom and Gannon 1996); and the quantitative scaling of response terms described earlier (Smith 1997; Mohler, Harkness, and Smith 1998). Through such careful development, designers can select items with maximum comparability in question meaning and response scales.

Both the goal of comparability and the task of translation are aided when, at the design stage, colleagues from all countries and language groups collaborate fully and designers adhere to standard practices that facilitate translation (see Chapters 2 and 3). A number of general question-design rules have been proposed to ease translation burden and enhance the likelihood of comparability. Brislin (1986) in particular has developed 12 guidelines for producing readily translatable items. In abbreviated form they are as follows:

(1)   Use short, simple sentences of fewer than 16 words. (But items can be of more than one sentence.)

(2)   Employ the active rather than the passive voice.

(3)   Repeat nouns instead of using pronouns.

(4)   Avoid metaphors and colloquialisms.

(5)   Avoid the subjunctive.

(6)   Add sentences to provide context to key items. Reword key phrases to provide redundancy.

(7)   Avoid adverbs and prepositions telling 'where' or 'when.'

(8)   Avoid possessive forms where possible.

(9)   Use specific rather than general terms.

(10)   Avoid words indicating vagueness regarding some event or thing (e.g., probably, maybe, perhaps).

(11)   Use wording familiar to the translators.

(12)   Avoid a sentence containing two different verbs if the verbs suggest two different actions.

Others have suggested additional rules about formulating questions and designing questionnaires, though usually within a monocultural context (e.g., Sudman and Bradburn 1982; Converse and Presser 1986; Fowler 1995; Scherpenzeel and Saris 1997; van der Zouwen 2000). However, these recommendations do contain helpful direction:

First, avoid vague quantifiers (e.g., 'frequently,' 'usually,' 'regularly') since these words have highly variable understandings across both respondent and question contexts (Bradburn, cited in Miller, Slomczynski, and Schoenberg 1981).

Second, avoid items with ambiguous or dual meanings (Tanzer forthcoming). For example, Tanzer (forthcoming) noted that an anger-in item in the State-Trait Anger Expression Inventory ("I am secretly quite critical of others") could be understood both as keeping anger inside ("I keep my criticism of others to myself") and as indicating how anger is expressed ("I talk about or criticize other people behind their backs"). Americans tended to read the first meaning into the phrase. A study in South Tyrol indicated that German speakers were more likely to understand the question in the first sense, whereas Italian speakers more frequently understood the question in the second sense.

Third, ambiguity also arises from complex questions with more than one key element. Thus, so-called double-barreled questions such as "Do you support the admission of Malta and Turkey to the EU?" are particularly problematic. If a respondent favors admitting or excluding both, the suitable response is clear. But if one favors the admission of one and the exclusion of the other, there is no appropriate response (Fowler 1995; van der Zouwen 2000).

Fourth, avoid hypothetical and counter-factual questions. People rarely produce coherent thoughts on most imagined situations and may not be able to even grasp the state of affairs described in the item (Fowler 1995; van der Zouwen 2000).

Fifth, use simple terms that are widely and similarly understood across all segments of the population. When necessary, provide definitions to clarify the meanings of terms (Converse and Presser 1986; Fowler 1995; Tourangeau, Rips, and Rasinski 2000).

Sixth, use clear and precise time references (Fowler 1995). For example, "Do you fish?" might be understood to mean "Have you ever gone fishing?" or "Do you currently go fishing?" It would be better to phrase the question as follows: "Have you gone fishing during the last 12 months?" (Tourangeau, Rips, and Rasinski 2000 discuss the many difficulties in asking questions with time references—for example, telescoping and forgetting curves—and means to deal with same—such as bounding and dating aids.)

Finally, avoid the particularistic and use questions with a higher abstraction level (van Deth 1998a). As Inglehart and Carballo (1997) argued: "If we had asked questions about nation-specific issues, the cross-cultural comparability almost certainly would have broken down. In France, for example, a hot recent issue revolved around whether girls should be allowed to wear scarves over their heads in schools (a reaction against Islamic fundamentalism). This question would have had totally different meanings (or would have seemed meaningless) in many other societies. On the other hand, a question about whether religion is important in one's life is meaningful in virtually every society on earth, including those in which most people say it is not."

However, other research indicates that people, particularly the less educated, have greater difficulty with abstract than concrete questions (Converse and Presser 1986). Even within a culture, abstract and specific items on related topics can yield quite different results. For example, studies conducted in the United States show that a large majority of respondents endorse unrestricted free speech for all, but the number endorsing free speech for particular suspect groups such as communists and militarists is much lower (Sullivan, Piereson, and Marcus 1992; McClosky and Brill 1983). The danger of abstract-concrete item disparity must be weighed against the problem of developing idiographic items.

As a general rule regarding survey content, researchers should follow the idea that 'more is better.' As discussed earlier, they should use multiple indicators in order to enhance scale reliability and overcome linguistic artifacts. Second, they should consider using both emic and etic items. Suppose that one wanted cross-national data on political participation, in general, and information on cross-national patterns of contacting government officials, in particular. In the United States, relevant questions would include items pertaining to the display of bumper stickers on cars, visits to candidate Web sites, and e-mails sent to public officials. In most developing countries, however, these items would range from rarely relevant to meaningless. Likewise, an item on asking a village elder to intervene with the government on behalf of an individual might touch on a major avenue of political participation in developing societies, but have little relevance in developed nations. In such circumstances solutions might include using general items that cover country-specific activities within broader items, asking people in each nation to respond to both relevant and irrelevant items, or asking a core set of common items (e.g., voting in local and national elections, talking to friends about politics) in addition to separate lists of country-specific political behaviors. Of course, even identical actions, such as voting in the last national election, may not be equivalent. In some countries, voting is legally mandatory and is thus not a meaningful measure of voluntary, political activity. In other countries, elections are meaningless charades, so voting is not a meaningful measure of participation in democracy or exercise of political choice.

The general-items approach is perhaps minimally useful since necessary loss of detail is often a heavy price to pay, and general items may be too vague and sweeping. The relevant-irrelevant approach is reasonable, provided the number of low relevancy items remains relatively low and the items are not so irrelevant as to be nonsensical or otherwise inappropriate. The ISSP has successfully used this technique in its study of global environmental change, asking about personal car use in all countries, though a few countries had very low ownership levels.

Provided the common core is adequate for direct comparisons, this last approach can be useful. It is well illustrated by a study of obeisance to authority in the United States and Poland, for which researchers developed and used five common items plus three and four country-specific items for Poland and the United States, respectively (Miller, Slomczynski, and Schoenberg 1981). This structure permitted both direct cross-national comparisons as well as more valid measurement of the construct within countries—and presumably better measurement of how that constructs works within models. For example, if the core items and the core plus country-specific items formed reliable scales showing the same basic relationships in both models, then results would be clear and robust. The appearance of different patterns for the core and country-specific items would, of course, raise questions about cross-national validity. Another example is the development of the Chinese Personality Assessment Inventory. Research indicated that several constructs that were important parts of Chinese personality matched no dimension measured on traditional Western scales (e.g., *ren quin* or 'relationship orientation'). To complete the instrument, developers added these items (Cheung et al. 1996), viewing the adjustment as an illustration of the importance of a combined emic-etic approach to personality assessment in non-Western cultures. Prezeworski and Teune (1966) substantiated this view, stating that "the cross-national and nation specific indicators combined provide a scale for reliable and valid measurement of the same phenomenon in various countries."

Fourth, at the analysis stage, one should follow the principle laid down by Kohn (1987): "Prudence dictates that the first hypothesis one entertains is that the inconsistent findings are somehow a methodological artifact." Large and/or unanticipated cross-national differences should always be suspected as the results either of unintended variation in some aspect of measurement (as in the earlier Dutch ladder example) or as failure to achieve functional equivalence though procedures appear comparable and translations were carefully executed. (On establishing the comparability of items used in psychological scales, see Van de Vijver and Leung 1997b.)

The final step is solid documentation. Jowell (1998) observed that good documentation and "detailed methodological reports about each participating nation's procedures, methods, and success rates" are needed. However, Hermalin, Entwistle, and Myers (1985) noted that "maintenance and documentation are painstaking tasks for which little provision is made." In the case of their work associated with the World Fertility Surveys, they found that results from a number of surveys no longer exist and that "the documentation for surviving surveys is often confused and incomplete." Of course all survey phases—from sampling to data processing—must be carefully recorded. However, it is particularly useful to have available the original questionnaires used in each country so that researchers can consult them when seeking to understand results (and particularly differences in results) across countries. The ISSP follows this practice, storing copies of original instruments on CD-ROM.

In sum, the likelihood of achieving valid cross-national research is enhanced when, from the earliest stages, survey developers elicit the collaboration of experts from all involved countries, execute careful development work, cross-validate findings with multiple indicators, subject all apparent results to critical review for measurement artifacts, and clearly, precisely document all procedures.

5.5   CONCLUSION

Surveys involve both social and cognitive processes (Sudman, Bradburn, and Schwarz 1996). First, they are social encounters between respondents and researchers. (Researchers may be directly and personally represented by an interviewer or indirectly and impersonally represented via a questionnaire.) Respondents must discern the nature of the interaction facing them and determine their role in the encounter. Surveys are also cognitive tasks in which the respondent is asked to comprehend various terms and inquiries, search his/her memory for relevant information, and formulate these ideas into the proffered response options. The great challenge in cross-national survey research is that both social conventions and cognitive abilities and styles vary across societies (Fiske et al. 1998). To obtain valid, equivalent measurement across countries and cultures, one must minimize and equalize measurement error from these sources so that emerging information is valid, reliable, consistent, and substantive. To achieve this high quality, substantive information, the cross-national survey researcher, like Oliver Twist, must want 'more.' Achieving cross-national comparability is so difficult that researchers must be more careful, invest more effort in instrument development, and include more items in their surveys. But when they do this successfully, researchers will also get more from their efforts.

Chapter 6

CULTURE-SENSITIVE CONTEXT EFFECTS: A CHALLENGE FOR CROSS-CULTURAL SURVEYS

NORBERT SCHWARZ

As every survey researcher is aware, minor variations in question wording, format or order can profoundly influence respondents' reports. Since the mid 1980s, psychologists and survey methodologists have made considerable progress in understanding the underlying cognitive and communicative processes, increasingly turning the "art of asking questions" into an applied science, grounded in basic psychological research (for reviews see Sudman, Bradburn, and Schwarz 1996; Schwarz 1999b; Sirken et al. 1999; Tourangeau, Rips, and Rasinski 2000). As this work developed, it became apparent that cultural differences in human cognition and communication are likely to influence the processes underlying survey responding, potentially giving rise to cultural differences in the emergence and form of context effects. This chapter addresses this possibility and illustrates how *culture-sensitive context effects* can invite misleading conclusions about cultural differences in the actual behaviors or attitudes under study.[1]

6.1 CULTURE AND COGNITION

Psychologists have long portrayed "the mind as a machine or computer that is the same in all times and places, while only the raw materials processed by the machinery or the data in the computer vary" (Fiske et al. 1998, 918). Challenging this assumption, recent comparisons of Western and East Asian reasoning documented pervasive cultural differences across a variety of cognitive tasks (for reviews see Fiske et al. 1998; Miller 2001; Nisbett et al. 2001). Here, I focus on only two aspects, namely the impact of independent versus interdependent self-construals on individuals' knowledge about their own and others' behavior and their sensitivity to the common ground of a conversation.

In a nutshell, Western cultures foster an *independent* perspective on the self. They conceptualize the self as fundamentally distinct from others and defined in terms of internal features such as attributes, abilities, and attitudes (e.g., Markus and Kitayama 1991; Oyserman 2002; Oyserman, Coon, and Kemmelmeier 2002). In contrast, most

East Asian cultures foster an *interdependent* perspective on the self (for a review of commonalties and differences in self-construal in different East Asian cultures see Oyserman, Coon, and Kemmelmeier 2002). They conceptualize the self as fundamentally connected to others and defined primarily in terms of relationships, group memberships, and social roles. Maintaining the interdependence of the self requires pervasive attentiveness to others in the social context, a need that is further compounded by an emphasis on maintaining harmony in relationships and "fitting in." These differences in attention to the social context have important implications for the emergence of context effects in survey measurement.

First, if East Asian cultures put a premium on "fitting in," we may expect that Asians are more knowledgeable about their own and others' behavior than Westerners. After all, ensuring that one "fits in" requires that one monitors one's own and others' behavior to avoid unwanted discrepancies. If so, Asians should have more detailed representations of their own and others' behavior available in memory, attenuating the need to rely on contextual cues when asked to provide behavioral reports.

Second, East Asian cultures put a premium on maintaining harmony in social relationships and value indirect forms of communication that avoid the direct expression of personal preferences and thus require the recipient to "read between the lines." If so, we may expect Asians to be more sensitive to the conversational context of a given utterance, with important implications for question order effects in survey measurement.

6.2   FREQUENCY REPORTS ACROSS CULTURES: CULTURAL DIFFERENCES IN AUTOBIOGRAPHICAL MEMORY

As numerous studies with Western samples have demonstrated, respondents can rarely retrieve the frequency with which they engage in a behavior from memory. Instead, they rely on a variety of estimation strategies (see Sudman, Bradburn, and Schwarz 1996) to arrive at a plausible answer. One of these strategies entails the use of the frequency scale provided by the researcher: Respondents assume that the researcher has constructed a meaningful scale, with values in the middle range of the scale representing the "average" or "typical" frequency and values at the extremes of the scale corresponding to the extremes of the distribution. Given these assumptions, respondents use the range of the scale as a frame of reference in estimating their own behavioral frequency, resulting in higher frequency reports along scales with high rather than low frequency response alternatives (for a review see Schwarz 1996). The more poorly the behavior is represented in respondents' memory, the more pronounced these scale effects are (Menon, Raghubir, and Schwarz 1995).

If Asians are more knowledgeable about their own behavior, they should have less need to rely on estimation strategies, resulting in an attenuated impact of frequency scales. Importantly, this prediction should only hold for public behaviors, which can be observed and thus need to be monitored to ensure that one "fits in." In contrast, it should not hold for private, unobservable behaviors, which do not require monitoring. Hence, the frequency scale provided by the researcher should influence the reports of Asian as well as Western respondents to a similar degree when the behavior is unobservable, but should influence the reports of Asians less when the behavior is observable.

To test these predictions, Ji, Schwarz, and Nisbett (2000) asked American students at the University of Michigan and Chinese students at Beijing University to report the frequency with which they engaged in a variety of public (e.g., coming late to class) and private (e.g., having nightmares) behaviors. Respondents provided their reports either along a high or low frequency scale or in an open response format. For example, the response alternatives for "going to the library last month" read:

Low Frequency Scale	*High Frequency Scale*
( ) 0–1 times	( ) less than 10 times
( ) 2–3 times	( ) 10–11 times
( ) 4–5 times	( ) 12–13 times
( ) 6–7 times	( ) 14–15 times
( ) 8–9 times	( ) 16–17 times
( ) 10 times or more	( ) 18 times or more.

Reports on these scales can be compared by assessing the percentage of respondents who report a frequency of 10 times or more. All behaviors were selected to be of similar frequency on both campuses, based on extensive pretest data.

The top panel of Table 6.1 shows the reports of private behaviors. The open-ended reports confirm that all behaviors were of similar frequency across campuses, as desired. More important, Chinese as well as American respondents reported higher frequencies of *unobservable* behaviors when given a high rather than a low frequency scale, indicating that respondents in both cultures relied on the scales in estimating the frequency of private, unobservable behaviors. Not so, however, when the reports pertained to public, *observable* behaviors. In this case, American respondents still reported higher frequencies along the high than the low frequency scale, indicating that they needed to rely on estimation strategies for public as well as private behaviors. In contrast, no scale effect was observed for Chinese respondents, who apparently did not need to rely on estimation strategies for public behaviors.

Following the reports of their own behaviors along high or low frequency scales, all respondents were asked to estimate the frequency of other students' observable and unobservable behaviors in an open response format (e.g., "How often do you think students, on average, go to the library per month?"). The bottom panel of Table 6.1 shows the mean responses. When the behavior was *unobservable*, Chinese as well as American respondents provided higher frequency estimates when they had reported their own behavior on a high rather than a low frequency scale. The same held true for American respondents when the behavior was *observable*. The estimates provided by Chinese respondents, however, were again independent of scale range.

TABLE 6.1. Behavioral   Frequency   Reports   across   Cultures: The   Impact of Response Scales

	Open Format	Low Scale	High Scale
Reports of own behavior			
Unobservable behaviors			
American respondents	23%	13%	24%
Chinese respondents	27%	16%	27%
Observable behaviors			
American respondents	32%	22%	38%
Chinese respondents	29%	29%	30%
Reports of others' behavior			
Unobservable behaviors			
American respondents	3.9	4.8	6.1
Chinese respondents	4.6	5.4	5.0
Observable behaviors			
American respondents	6.3	7.5	7.9
Chinese respondents	7.3	7.2	7.3

NOTE: Shown is the mean percentage of respondents reporting a frequency above the comparison point of the respective frequency scales for 3 unobservable and 3 observable behaviors (top panel) and the mean frequency of others' behavior, reported in an open-response format (bottom panel). N = 246 Chinese and 255 American respondents, randomly assigned to conditions. Adapted from Ji, Schwarz, and Nisbett (2000), reprinted by permission.

In sum, these findings replicate the familiar observation that Western respondents rely on the frequency range presented in the response scale to estimate the frequency with which they, or others, engage in a given behavior (see Schwarz 1996 for a review). In contrast, Chinese respondents were only influenced by scale range when the behaviors were unobservable. This presumably reflects that members of interdependent cultures pay more attention to their public behaviors, as would be expected in a culture that puts a premium on "fitting in." This increased attention to one's own and others' behavior, in turn, results in better memory representations, which have been found to reduce the need for estimation (Menon, Raghubir, and Schwarz 1995).

What renders these findings troublesome from a survey methods point of view is the observation that cultural differences in the use of estimation strategies invite misleading conclusions about actual differences in the frequency of observable behaviors, as shown in the top panel of Table 6.1. Based on the reports provided in an open response format, we would conclude that the behaviors are of *similar* frequency in both countries. In contrast, the reports provided along low frequency scales suggest that the same behaviors are *less* frequent in the United States than in China, whereas the reports provided along high frequency scales suggest they are *more* frequent. Quite clearly, we need to understand cultural differences in the cognitive processes underlying behavioral reports to avoid substantive misinterpretations. At present, little is known about cultural differences in the content and structure of autobiographical memory (see Ji, Schwarz, and Nisbett 2000 for a discussion) and the exploration of these issues provides a promising avenue for future research at the interface of psychology and survey methodology.

6.3 QUESTION ORDER EFFECTS ACROSS CULTURES: CULTURAL DIFFERENCES IN CONVERSATIONAL CONDUCT

A large body of research indicates that survey respondents bring the tacit norms underlying the conduct of conversation in daily life to the survey interview (see Schwarz 1995, 1996 for reviews). One of these norms requires speakers to provide information that is new to the recipient, rather than to reiterate information that the recipient already has (Grice 1975). Observation of this norm requires close attention to the common ground established in a conversation and strongly influences question order effects in surveys.

For example, Schwarz, Strack, and Mai (1991) asked participants to report their marital satisfaction and their general life-satisfaction in different orders. When the life-satisfaction question preceded the marital satisfaction question, the answers correlated $r = .32$. But this correlation increased to $r = .67$ when the question order was reversed. This increase reflects that answering the marital satisfaction question first brought marriage-related information to mind, which participants then drew on in evaluating their lives in general. For a third group of participants, both questions were introduced by a lead-in designed to evoke the conversational norm of nonredundancy. This lead-in informed them, "We now have two questions about your life. The first pertains to your marital satisfaction and the second to your general life-satisfaction." Under this condition, the correlation dropped from $r = .67$ to $r = .18$. Apparently, the latter participants interpreted the general life-satisfaction question as a request for new information, pertaining to aspects of their lives they had not yet reported on. Confirming this interpretation, a condition in which the general life-satisfaction question was reworded to read, "Aside from your marriage, which you already told us about, how satisfied are you with other aspects of your life?" resulted in a nearly identical correlation of $r = .20$. As this example illustrates, a theoretical understanding of question order effects requires the consideration of cognitive as well as communicative processes.

Given that interdependent cultures often value the skill of "reading between the lines," and expect conversation partners to be sensitive to subtle cues, we may conjecture that interdependent individuals are more likely than independent individuals to attend closely to what others are directly or indirectly communicating. To explore this possibility, Haberstroh et al. (2002) conducted a conceptual replication of the above question order experiment with students in Heidelberg, Germany, and Beijing, China. Specifically, they asked respondents to report their academic satisfaction and their general life-satisfaction, either in the academic-life or the life-academic order. Table 6.2 shows the results.

In the German sample, the correlation *increased* from $r = .53$ in the general-academic order to $r = .78$ in the academic-general order, replicating the earlier pattern (Schwarz et al. 1991b). As is typically observed in Western samples, the German respondents drew on the information brought to mind by the academic question when they subsequently evaluated their general life-satisfaction.

If Chinese respondents are indeed more sensitive to the conversational context, however, this pattern should reverse. Having just reported on their academic satisfaction, these respondents may now interpret the general question as if it were worded, "Aside from your academic life, which you have just told us about, how satisfied are you with the rest of your life?" Empirically, this is just how respondents reacted; the observed correlation *decreased* from $r = .50$ in the general-academic order to $r = .36$ in the academic-general order. Thus, the Chinese respondents spontaneously behaved like Western respondents in earlier studies explicitly made aware of the conversational context through a joint lead-in. They disregarded information that they had already provided, consistent with the conversational norm of nonredundancy.

Note that the correlations shown in Table 6.2 would suggest different substantive conclusions about the role of academic satisfaction in students' overall life-satisfaction, depending on the order in which the questions were asked. When the general question precedes the academic satisfaction question, we would conclude that academic satisfaction contributes equally to students' life-satisfaction in both cultures. Yet, when the academic satisfaction question precedes the general question, we would conclude that academic satisfaction figures more prominently in the lives of German students than in the lives of Chinese, and hence contributes differentially to their respective overall life-satisfaction. A substantive explanation could easily be generated and one might propose that individual achievement plays a more important role in independent than in interdependent cultures. Most likely, however, the observed differences reflect solely that East Asian respondents are more sensitive to the conversational context than are Western respondents.

TABLE 6.2. Question Order Effects Across Cultures

	Germany	China
Question Order		
Academic-life	.78	.36
Life-academic	.53	.50

NOTE: Shown are Pearson correlations; N = 58 German and 109 Chinese respondents, randomly assigned to conditions. Adapted from Haberstroh et al. (2002, Experiment 2), reprinted by permission.

To provide a more direct test of this conversational interpretation, Haberstroh and colleagues (2002) conducted an experiment with German students and manipulated their temporary interpersonal orientation through an experimental task. Under the pretext of studying language comprehension, participants were asked to read a short paragraph about a "trip to the city" that was written in the first person singular or first person plural. Participants' task was to circle either singular pronouns (*I*, *mine*, etc.) or plural pronouns (*we*, *ours*, etc.) in the otherwise identical text. Earlier research had demonstrated that this task elicits temporary changes in interpersonal orientation that parallel the differences observed between independent and interdependent cultures (e.g., Brewer and Gardner 1996; Gardner, Gabriel, and Lee 1999). After students completed the pronoun-circling task, they reported their academic satisfaction followed by their general life-satisfaction.

When German students were induced into a temporary interdependent mind set by circling plural pronouns, their answers to the academic and general satisfaction questions correlated $r = .34$, paralleling the correlation of $r = .36$ observed in China. Yet, when they were induced into an independent mind set, their answers correlated $r = .76$, paralleling the correlation of $r = .78$ previously observed in the German sample. In other words, inducing the independent orientation that they already brought to the study had little additional effect; but inducing the interdependent orientation that normally characterizes other cultures changed the pattern of the order effect, as expected on theoretical grounds. Subsequent experiments, with different questions, provided conceptual replications of this observation (Haberstroh et al. 2002).

6.4 CONCLUSIONS

In combination, the reviewed studies highlight that cultural differences in survey responding extend far beyond the problems of adequate translation and culture-appropriate question meaning that have received the bulk of attention in survey methodology to date. Drawing on basic psychological research into interdependence and independence, we conjectured that East Asian cultures' emphasis on social sensitivity, harmony in relationships, and "fitting in" would foster closer attention to one's own public behavior as well as closer attention to the common ground of the ongoing conversation. Confirming the conjecture that interdependent individuals would have better knowledge about their own and others' behavior, Chinese respondents were less likely than American respondents to rely on the frequency scales presented by the researcher in estimating behavioral frequencies, but only when the behavior was public and hence needed to be monitored. Confirming the conjecture that interdependent individuals would be more sensitive to the common ground of the conversation, Chinese respondents, as well as German respondents who were induced into an interdependent mind set, were more likely to disregard information they had provided earlier when answering a subsequent question.

Most important from a survey methods point of view, the *differential* nature of the emerging context effects suggested misleading conclusions about cultural differences in the actual behaviors or attitudes under investigation. To avoid such misleading conclusions, we need to gain a better understanding of how cultural differences in cognition and communication influence survey responding. At present, the availability of detailed cognitive and communicative models of survey responding (see Sudman, Bradburn, and Schwarz 1996; Tourangeau, Rips, and Rasinski 2000) and the rapid progress made in understanding differences between East Asian and Western reasoning (see Fiske et al. 1998; Nisbett et al. 2001) have set the stage for a fruitful collaboration that promises to improve comparative surveys involving these cultures. In the process, we are likely to gain new insights into cultural psychology as well as survey responding, making this endeavor a promising avenue for all involved.

Note
1. Preparation of this chapter was supported through a fellowship from the Center for Advanced Study in the Behavioral Sciences, Stanford, CA, which is gratefully acknowledged.

Chapter 7

BACKGROUND VARIABLES

MICHAEL BRAUN
PETER PH. MOHLER

Demographic information about respondents and their social context is referred to variously as context variables, collateral variables, demographics, or as used here, background variables. Background variables provide the 'independent' information against which study-specific 'dependent' data are analyzed. Not surprisingly, they are among the most widely used variables in social surveys. Background variables include demographic, biographic, and face-sheet data (Scheuch 1968, 1973; Sudman and Bradburn 1982; Statistisches Bundesamt 1999). Typical background variables cover information on a respondent's ethnicity, age, gender, education, income, and voting behavior, as well as on the date and place of an interview, the size of community, or kind of neighborhood. Thus they present a diverse assortment of information on the social, economic, cultural, geographical, and biophysical settings in which respondents live and act.

At first glance, background variables look like easy items for comparative survey implementation and easy items for comparative analysis. This chapter illustrates, however, that implementing and analyzing background variables involves many of the same problems of equivalence as implementing and analyzing substantive questions.

The chapter begins with general issues of background variables followed by equivalence issues. These provide a framework for the discussion which follows of a number of background variables and harmonization. The last section makes some recommendations for comparative research.

7.1 BACKGROUND VARIABLES FOR SOCIAL SURVEYS

7.1.1 Uses of Background Variables

Background variables often serve multiple purposes in a social survey. In addition to providing general contextual/collateral information, they are used as independent variables, as socio-economic covariates of attitudes, behavior, or test scores, etc. and in all sorts of statistical models, in particular, as exogenous factors in causal analysis. They enable analysts to establish homogeneous subgroups, explain differences of scale scores due to different composition, and to identify spurious correlations and causal relationships. Background variables are also used to assess the quality of a realized sample and to decide on any corrections necessary. The distribution (of a combination) of background variables in the realized sample is compared with population characteristics from official data. Deviations from the known population distributions can be corrected by appropriate weights.

7.1.2 Levels and Sources of Background Variables

Typically background variables span a wide range of aggregation levels, ranging from individual-level data (respondent characteristics such as age, education, personal income, marital/partnership status, membership in organizations) and group-level data (for instance, household composition, household income) to macro-level data (for instance, size of town, country GNP). The accuracy of information may differ from level to level, depending on the different sources of information and the measurement errors involved. Many background variables are reported by respondents about themselves and their household (Groves 1996; Schwarz 1999c). These individual-level background variables include age, education, income, marital or partnership status, occupation, and household composition. Depending on study and sample, some of these individual-level background variables may also be taken from external sources, such as community registers (for instance, citizenship, age, or income). Macro-level data are obtained from regional and national statistics (background variables such as urbanization of respondent location, number of foreigners in respondent's community, GNP). However, it would be misplaced to assume that published data or official statistics are generally more accurate than respondents' assessments or reports of facts. Interviewers standardly record respondent gender and contact protocol information and may record information on housing (type of house, neighborhood) or assess respondent cooperation during the interview.

7.1.3 Measurement Issues

The distinction between theoretical concepts or constructs and their indicators and between indicators and concrete items as measurement devices is just as important for background variables as it is for attitudinal variables. The common epistemological understanding among survey researchers is that 'reality' can be described on the basis of theoretical (latent) constructs that are measured by indicators formulated as questions. Age, for example, is an indicator of often complex theoretical concepts such as life cycle and cohort experience (Ryder 1965).

Once indicators for a given construct are identified, robust and askable survey items have to be worded. Multiple indicators (and items) afford better measurement. For affiliation to a religious group or church, for example, membership, frequency of service attendance (if applicable) and prayer provide a better measure of the underlying construct than does just one measurement. The same applies, of course, to measurements of health, physical performance, or occupation.

Most background variables target factual or behavioral information (Kruskal 1991; Schwarz and Sudman 1994, 1996; Sudman, Bradburn, and Schwarz 1996). It is sometimes suggested that for factual or behavioral characteristics (age, visit to a doctor, marital status) respondents can automatically produce 'true' and unbiased values (Kruskal 1991, xxiv). This view assumes that survey contexts, question contexts, and question wording have no effect on the data collected for factual data of this kind. Groves indicates that this view of data collection ends in a "true value morass" (1991, 22). There is no principal difference between the measurement of facts and attitudes. The crucial issue here is the term "measurement" in contrast to "getting/collecting information" (Groves 1996) and whether we assume reality is directly accessible and assessable, or whether we assume that we always have to "measure reality with some kind of error" (cf. Kruskal 1991, xxiv–xxvii and further references there).

The view presented here is that background variables are measurements based on reports of information of different kinds and not just facts verbalized by respondents (Harkness, Mohler, and Thomas 1997). As a result, the rules and techniques relevant for identifying measurement error for attitudinal variables are relevant for background variables, too. Consequently, seemingly simple facts, such as the number of children a respondent has, call for careful question design. Information about household composition provides a good example. Respondents are often asked how many children live in their household. The number of children in a household is relevant for many different kinds of analysis. In answering, respondents need to know which children to count (as well as being able, for all practical purposes, to count them). The issue, then, is whether to define children in the survey question as biological children or as biological children plus stepchildren, adopted children, and fosterchildren, or as any children in the household. In the Philippines it would be relevant to indicate whether godchildren are to be counted or not, whereas instructions to count or omit godchildren in other contexts might result in mirth or confusion. Moreover, a precise definition of "household" is also necessary.

If the question asks about "how many children you have," in some contexts respondents might also count children no longer alive. In these contexts, provided it was culturally accepted to mention this, that, too, would need to be specified. Social surveys do not have the questionnaire time to ask background questions in the detail necessary to capture all these aspects. Instead, a list of household members is often used which records the gender and age of all people living in what the respondent defines to be her/his household. Sometimes the relationship to the selected respondent is also recorded.

Systematic errors in measuring background variables are connected to factors like social desirability (reporting higher or lower incomes), fuzzy definitions ("working for pay"), or memory lapses (forgetting additional income; see Groves 1991) and to cultural frames of reference (age, number of children, marital status, premarital sex experience). Not surprisingly, reliabilities of background variables vary markedly. In a German General Social Survey (ALLBUS) three-wave test-retest panel, agreement over the three waves was at about 90% or more for age, education, marital status, and religious affiliation, but only about 50% or lower for respondent's first occupation and father's occupation (Porst and Zeifang 1987).

7.2   EQUIVALENCE OF BACKGROUND VARIABLES

Both for constructs and the items designed to measure background variables, researchers have to identify and establish the equivalence level on which they can be measured (see Chapter 10 for levels of equivalence) and what types of error are likely to occur. In intercultural comparative research, the differences between the objective realities of different cultures and nations make numerous demands on the survey design. A central question at the outset is whether indicators can be measured in an identical way ('Ask-the-Same-Questions' (ASQ) approach) or are better measured in an equivalent, but not identical fashion in different countries (van Deth 1998a; see Chapter 19). Background variables cannot be measured by asking the same questions if the realities in the different countries make this impossible. The first step, therefore, is to establish construct equivalence for background variables across the cultures involved.

7.2.1   Construct Equivalence

Without a precise definition of the measurement properties of background variable constructs, no equivalent implementation can be attempted; proper definition is the prerequisite for constructing equivalent indicators, and they are the prerequisite for deciding on wordings. National idiosyncrasies, different social structures, and legal institutions may make it difficult to define the relevant latent constructs. Indicators chosen are then often not equivalent across nations and cultures and neither are the questions. What is compared in the end, might be 'apples' and 'oranges,' i.e. disparate things. One way out of the apple and orange trap is to move up one step in the level of abstraction: on a higher level apples and oranges are *fruit* (cf. Verba 1971). The price for gaining construct equivalence by measuring on the more abstract level ("How much fruit do you have at home?") is a loss of culture-specific information. Moreover, the level of comparability could miss what the background variable is trying to measure (van Deth 1998a). Alternatively, one could test culture-specific sets of indicators during the source questionnaire development and establish functional equivalence by proving that buying apples in one culture is functionally equivalent to buying oranges in another. Adopting this approach, the construct would be the same in each culture. However, whether the scores/values can be meaningfully compared across cultures is a different question (van de Vijver and Leung 1997b). Some researchers call this type of measurement "functionally equivalent" in order to distinguish it from so-called "identical measures" (van Deth 1998a; but see Johnson 1998a on the many different uses of "equivalence"). To sum up, background variables used in harmonized cross-cultural data sets can suffer badly from construct inequivalence, indicator inequivalence, and from various forms of item bias.

7.2.2 Harmonization of Background Variables in Multinational Surveys

Centrally designed surveys such as the European Community Household Panel (ECHP; see European Commission/Eurostat 1996, 1999) or centrally designed guidelines for the measurement of background variables such as those of the European Society for Opinion and Marketing Research (ESOMAR, Bates 1998) formulate source items and recommended wordings, i.e. they pursue an ASQ approach. In projects where the comparative study is simply added on to a national project (piggy-back), the background variables of the first project will often be used for the piggy-back survey. A number of members of the International Social Survey Program (ISSP) use this strategy to reduce the costs of the ISSP module. In such cases, asking for background variables twice would be both a burden to respondents and would increase survey costs. Abandoning the background variables for the other survey is usually out of the question; this will often be the larger study with either more background variables or other background variable needs. Changes could then affect national measurement goals and time series. The *social grade* scale recommended by ESOMAR for the measurement of socio-economic status illustrates problems connected with centrally designed 'comparable' classifications (Bates 1998). The scale is derived from the education level of the main income earner in a household (measured by school-leaving age adjusted for any later period of education) and occupation. If the main income earner is not active in the work force, occupation is replaced with a measure of the economic status of the household based on the ownership of ten consumer durables such as a radio-clock, two or more cars, and a video camera.

Almost every feature of the *social grade* instrument is likely to result in measurement error and lack of comparability across countries. Countries differ enormously, for example, in whether there is a main income earner or not; often both partners work. The *main wage-earner* might then be taken as the person who earns most. However, some couples earn similar amounts. In other contexts, neither partner may work. Using school-leaving age as an indicator for education, moreover, blurs important distinctions in national educational systems in general. In particular, the lack of differentiation beyond a school-leaving age of between 20 and 21 fails to distinguish between people with a university education and those without. In addition, the ESOMAR scale ranks occupational groups in an arbitrary hierarchy (e.g., the "employed professional" category is placed above the "general management, director or top management with responsibility for five employees or less" category). Finally, the measurement of economic status by possession of durables ignores, for example, the impact of life-style on the purchase of some of the items included in the instrument; cross-national differences in personal and household accoutrements, as well as a number of issues related to the price of items which could fall under each rubric (designer or supermarket clock) and the age and origin of the items owned.

7.2.3 Choosing Background Variables

The choice of background variables to be collected in a study depends on the study budget, the research topic, and designer/researcher preferences as well as on the burden the questions place on respondents. Sudman and Bradburn (1982), reporting discussions within the Social Science Research Council (SSRC), suggest that eight classes of demographic information using about fifty items are adequate for any in-depth analysis. However, the number of background variables actually used in many published analyses is much lower, even in monocultural analyses. Two national reviews (Koch 1993; Smith and Heaney 1995) indicate that only three background variables are widely used in general social surveys: age, gender and education (A.G.E.; see Harkness, Mohler, and Thomas 1997). Other sources add occupation, place of birth, and type of dwelling (Stacey 1969). Our own unpublished analysis of the five volumes on Beliefs in Government indicate similar results for comparative studies (Kaase, Newton, and Scarborough 1995).

Background variables are not only a fielding cost factor, harmonizing background variables in multinational studies is a labor-intensive and costly process (European Commission/EUROSTAT 2000). In limited study budgets, background variable fielding time may compete with fielding time for substantive items. If past usage is indicative of what researchers need, one conclusion could be that it is better to include a small number of well-measured background variables than a larger number of roughly measured background variables. However, taking this approach in multi-purpose studies would create problems. Multipurpose studies, as their name suggests, are intended to provide data for many researchers who have very different research questions in mind and who will need a number of different background variables.

7.3   DESIGNING BASIC BACKGROUND VARIABLES FOR CROSS-CULTURAL SURVEYS — A.G.E AND BEYOND

In this section we review a number of important background variables with an eye to general problems of design as well as specific problems for comparative research.

7.3.1   Age

Age is a multidimensional variable at even the national level of measurement, since it relates to aspects of life cycle and cohort. In comparative analysis, the effects of life cycle and cohort can vary greatly across countries. Consider, for example, comparing the following types of societies: on the one hand, a developing society where people marry early and begin having children early. Life expectancy is low (life-cycle-effect components) and the economy is consistently weak. The country has not experienced war or major disasters in recent decades (cohort-effect components). Consider, in contrast, a country (or countries) where marriage and fertility rates are low and perhaps still on the decline, and both marriage and births occur increasingly late while life expectancy is high. Cohorts in Western European countries fitting this second description also differ with regard to their experience of periods of war and peace, and periods of economic depression and growth. Against such backgrounds, it is evident that the 'meaning' of age will differ importantly, even between relatively similar societal contexts (Scheuch 1968).

On paper, an age background variable looks fairly straightforward to collect. A standard format in Western social research measures age by asking for a person's birthdate. In order to facilitate cohort analysis, respondents are asked to provide month and year of birth (but not the day, for data confidentiality reasons). This format assumes a calendar frame of reference and takes into account that it is easier to remember a fixed event than a variable age in years. However, it works less well in cultures not accustomed to locating such events in a calendar frame of reference and where people perhaps do not have copies of birth certificates in a neat folder at home (i.e., outside the birth certificates belt). If respondents locate their births in terms of local events (two years after the great flood) or in relation to other measures (four winters after my sister was born) individual computations will be needed. This is not an unusual procedure. In many contexts mothers of large families may begin working out the age of one child relative to the age of other children.

7.3.2 Gender

In sociological terms, gender is not strictly equatable with biological sex. Gender is socially constructed on the basis of biological sex (which is comparable) as a visible distinguishing criterion (Hofstede 1998). Gender differences vary enormously by country. In many countries, the life opportunities and rights of both genders differ, not just in degree, but in kind. Depending on the focus of a comparative study this might create confusion, for example, if differences in women's political consciousness are to be analyzed. Data from the ISSP 1985 module show that 43% of female respondents in Italy say that women with equivalent education and experience earn less than men do. In Germany 88% of women stated this. However, the figures in Germany are more likely to reflect a culture-specific heightened awareness of inequality than actual differences in pay between sexes for the two countries at that time.

7.3.3 Education

Education is a multidimensional and multifaceted concept; it is a certificate of formal education and has effects for labor market opportunities, it measures exposure to specific types of socialization, and can be a measure of social class (Braun and Müller 1997). Educational systems also differ markedly across countries. Decisions about what levels of education to compare across countries — which categories of an educational classification can be taken as roughly equivalent — depend on which dimension of education is of relevance to the analysts. Moreover, educational systems change fairly rapidly. Thus items used to measure educational levels ideally need to cover both national and international frameworks and to cover changes over time within countries (see Section 7.4).

7.3.4 Employment Status and Occupation

Instruments to measure employment status often distinguish between *full-time* and *part-time* work. Countries (and firms within countries) differ considerably in how many hours actually count as full-time. In addition, societies differ in what they consider one or the other category to consist of. Some distinguish consistently and linguistically between part-time work which is half-time and other part-time. In societies where people regularly have two or more jobs because everyone needs to find as much work as possible, these superficially neat distinctions may disintegrate completely.

The social and legal implications of having one or the other form of employment can be very different across cultures. In Sweden, for example, working part-time as an employee just means working fewer hours than full-time colleagues for a correspondingly lower salary. In other countries, such as the United States, social benefits related to employment differ for part-timers and full-timers.

In addition, the definition of what counts as *unemployment* varies considerably across countries, as do the social and psychological consequences of unemployment (Maier 1991). Official definitions may also not coincide with people's subjective perception of their work status. Official definitions may also change according to political dictates. As a general rule, the fact that someone is not employed does not automatically mean that he/she is unemployed. To count as unemployed and qualify for any benefits other criteria have to be met; for example, people have to be of working age, have to be fit for work and seeking work, but may also need to have already been part of the work force. Depending on the system, young people looking for their first job may or may not qualify.

A comparable measurement of occupation is difficult, too. Is a *farmer* in the United States comparable to a farmer in a developing country? The first may well run a farm as a business to make profit, whereas farmers in rural parts of a developing country may use their farm product to feed their family. In addition, the relatively low status attributed to farmers in status classifications will differ in goodness of fit within and across countries, depending on the farm type, size, and location. Similarly, the relatively low status and salary grade-school teachers have in the United States stands in contrast to the standing and salary of school teachers in Germany.

Such differences raise basic questions for measurement and comparability, even at the relatively simple level of background variables. How do we decide which groups are comparable? Does it make sense to compare blue-collar workers in developed and developing countries, or is it more useful to compare blue-collar workers in the developed countries with farm laborers in developing countries (Scheuch 1968)? It is often difficult to decide on the correct explanation for differences on some dependent variable between what seem to be comparable groups in different countries. In essence, we have to decide whether comparable groups in the different countries do indeed differ on the dependent variable, or whether the groups themselves are different from each other.

The International Standard Classification of Occupations (ISCO) is a large-scale project aimed at harmonizing occupational variables. The main criterion of classification in ISCO88 (International Labour Office 1990) is based on skills required for an occupation. Glover (1996) illustrates how national peculiarities with regard to the occupational system cannot be properly reflected in a comparative schema of this sort.

A further complication arises from the fact that occupation is used to compute other indicators. Thus, for example, status scales are often based on occupation plus other information. Occupational status scales take ISCO classifications or similar schemes as their basis and use specific recoding instructions to condense the information from these and combine it with other information such as respondent age, education, income, employment status, and prestige ratings of occupations. As a result, these status scales are affected by all the measurement issues outlined above. Well-known scales include the Treiman prestige scale (Treiman 1977), socio-economic status (Duncan 1961; Ganzeboom, De Graaf, and Treiman 1992) and the so-called EGP class categories (Erikson, Goldthorpe, and Portocarero 1979; Erikson and Goldthorpe 1992). Ganzeboom and Treiman (1996) discuss these scales and their computation on the basis of ISCO88 codes.

The standard format for collecting information on occupation is complex and is spread over several questions. The questions dealing explicitly with occupation are open questions. In order for respondent answers to be assigned detailed ISCO codes, interviewers have to ensure that respondents provide enough ISCO-compatible information. For instance, if respondents answer the question "What kind of work do you do? That is, what is your job called?" with "teacher," this cannot be coded in a valid ISCO category. They need to provide information more like "teacher at a state high school." In interviews respondents can be prompted to provide more detail. Asking occupation background variables intended for ISCO coding in self-completion formats poses further problems. The answers to the open questions are either coded by hand or by using special computer programs (Geis and Hoffmeyer-Zlotnik 2000). Alternatively, a closed question can be constructed. It might present, for example, the categories of the EGP scheme directly and add some examples to help respondents to identify their own status position on the scale. This procedure, however, will lead to a loss in precision. Whether this is acceptable depends on how important the variable occupation is in a given project.

7.3.5 Income

Income has basically two different aspects: the absolute amount of income someone has determines their standard of living and life opportunities, while the relative amount of income compared to others' income defines their position in the system of inequality. These two aspects have to be kept analytically separate.

Income is often measured in monetary units. Obviously, using exchange rates to establish comparable incomes is not the route to follow, since this ignores the value of a unit within a culture. Indeed, with a large part of Europe now using the Euro, care will be required across countries using the same currency. The purchasing power of the Euro will continue to differ across countries for the foreseeable future. Instead, relative measures should be established which also reduce effects related to using different measures (wordings) to ask about income. Country formats differ, for example, in whether they ask for gross or net, yearly, monthly, or weekly, or ask for respondent's income and/or household income. Comparative research also needs to take into account that sources of income differ across countries, as does the importance of different kinds of income (Hoffmeyer-Zlotnik and Warner 1998; Smeeding, Ward, and Castles 2000). Assets may also not be perceived as income but as an essential part of one's livelihood. In developing countries, home-grown produce and owning where you live may be more important than the transfer income central to Western countries.

Needless to say, most income measures for general purposes are rough. Many ask respondents to place themselves in income brackets and may focus (inadvertently or not) on salary or pension-earned income. More exact measures of income are the area of special studies such as household income panels. General social surveys can work well with meaningful proxies.

7.3.6 Marital and Partner Status, Family, and Household Composition

Marital status, at least in Western countries, is usually measured by asking respondents whether they are "married," "divorced," "widowed," or "never married," irrespective of how relevant these categories are for the legal complexity or social cohabitation patterns of an individual country context and what counts as *married* etc. across cultures. Sometimes, the "married" category is further differentiated in terms of whether a respondent lives together with his or her spouse or whether they are separated. Comparative research needs to recognize the considerable intercountry variability of the legal implications of belonging in one or the other categories of the marital status variable. In some countries, divorce might have dramatic financial consequences in the form of alimony payments. In others, maintenance payments are either not legally required or not properly enforced. In the former German Democratic Republic, for example, divorce had few financial consequences, because the vast majority of women earned their own living and extensive childcare was provided by the state.

Marriage in some cultures might be the only route to securing accommodation or access to a sexual partner, while in others marriage may only become properly salient if children are planned. Moreover, countries not only differ in terms of the consequences of belonging to different marital status categories, but also in terms of who can claim a given status. In many countries, gays are not permitted to marry, and in some countries people are not permitted to divorce and then remarry. As a consequence, the match between respondents' legal status and their actual living arrangements varies considerably across countries.

Wording can complicate measurement unexpectedly. If, in English, the term *single* is used for the legal category "never married," neither the legal status nor the actual cohabitation arrangements of an individual can be ascertained for certain. Single is used in everyday language for a number of states, including "currently not married" and "currently not attached." Similar problems with legal terminology and everyday language interpretations by respondents can be noted for other languages and concepts.

In contexts where both separation and divorce are possible, it is relevant to distinguish between the two. In some contexts, however, divorce is not possible, but legal separation might be (e.g., Ireland, Chile). A suitable background variable question can then be formulated for each context but comparing the data across contexts raises questions. Should *separated* people in Ireland be compared with the separated or with the divorced in other countries? And do the separated in either context have anything in common with respondents who break up with partners to whom they have never been married in countries where consensual union is the rule? What measurement solutions are available? In countries where the relationship between actual living arrangements and legal status is relatively weak (for instance, Sweden), information on the actual living arrangements is essential. This applies in particular if marital status is used as an indicator for related factors, such as social support.

In addition to marital status, most surveys collect information about other household members. This information might be useful to identify household types, but obviously cannot be used satisfactorily to determine respondents' family relationships or interpersonal networks. In addition, differences in how 'household' is defined (and understood by respondents) obviously affect the comparability of measurements of household income.

7.3.7   Party Preference, Vote, Left–Right Scale

Background variables on voting and party preferences are collected as country-specific measures which list, for example, the national parties and ask which party a respondent supports or voted for. A direct comparison of data across countries is thus not possible. To do so, the country-specific data have to be classified according to an underlying dimension or construct. In many cases, party preference is measured to provide insight into political leanings. It can be used to locate respondents on dimensions, such as left to right, liberal to conservative, democratic to totalitarian, and so forth.

At the same time, such procedures assume that respondents' positions can be located on one dimension. Unfortunately, party systems may be far from one-dimensional. The emergence of socially liberal, but fiscally conservative parties and leaders in Western countries and of parties that simultaneously emphasize economic and civic liberties are only two indications of developments which may undermine the traditional European left–right distinction for analysis (Clark 1998; Clark and Inglehart 1998).

7.4   HARMONIZATION OF NATIONAL MEASURES

At the analysis stage there are two basic ways to deal with problems of noncomparability of national background variables (Glover 1996). One strategy is to acknowledge the differences and to analyze each national data set separately, i.e., to compare models with different background variables. However, interpreting the statistical findings comparatively calls for detailed knowledge of the different national systems. Moreover, the equivalence of findings cannot be statistically tested; one has to rely on face validity and plausibility. The second approach is to harmonize background variables. Harmonization means to construct comparable background variables out of country-specific measures that cannot be directly compared. In essence, the technique of harmonization requires a common level of generality to be identified valid for all cultures/countries involved (Verba 1971). A more negative view is that harmonized variables will always "be at the level of the lowest common denominator" (Glover 1996, 35) and therefore result in a loss of information and discriminatory power. However, this holds for all kinds of measures involving more than two units: the higher the number of different units, the higher is the level of abstraction (town surveys harmonize across town quarters, statewide surveys harmonize across counties, etc.).

Education as a background variable is used here to illustrate conceptual and methodological problems and to propose some solutions. As said above, education is a multidimensional concept. Different characteristics related to education might be relevant for different dependent variables, including skills, knowledge, value socialization, or the acquisition of credentials (Braun and Müller 1997). In addition, several measures of education are available that are suited for specific dependent variables and for specific countries.

The discussion that has taken place in the Methodology Group of the International Social Survey Programme indicates how difficult arriving at solutions for academic projects may be. Many of the participating countries add the ISSP module as a drop-off to their ongoing national surveys. As most of these national surveys are older than the ISSP, they often have national traditions for the collection of background variables that are, partly for good reason, quite resistant to change (Braun 1994). The first classification scheme used in the ISSP (Zentralarchiv 1996) harmonizes national educational qualifications of respondents in seven categories:

(1)   None
(2)   Incomplete primary
(3)   Primary completed
(4)   Incomplete secondary
(5)   Secondary completed
(6)   University incomplete
(7)   University degree completed.

A harmonization initiative was started by the ISSP in the mid-1990s to find out how different countries handle the scheme and to discover problems each country had with the seven-code scheme. It transpired that countries interpreted categories differently, especially the two "primary" and the three "incomplete" categories. This affected their coding of nation-specific education categories in the ISSP scale. In addition, it became clear that the omission of vocational training from the scheme created problems for comparability.

Category 2 and 3 (primary): If primary is understood as the first cycle of mandatory schooling (as in British English), then very few respondents in many Western industrialized countries will end in categories 1 through 4. At least half of the ISSP member countries, however, take "primary completed" to mean *having the lowest formal qualification attainable.*

Categories 4 and 6 (incomplete): these are left unused in a number of countries, because they cannot be easily derived from national measures. However, even if countries use them, incomplete qualifications or schooling have different values across cultures. Having some years of college is regarded as an improvement over a high school diploma in the United States, while some years of university can be seen as having failed to stay the course in Germany.

The scheme makes no provision for intermediary qualification categories which lie between other qualifications but are not *incomplete* as such. Some ISSP countries solve this problem by locating 'complete' levels below those in the scheme within the nearest 'incomplete' levels. As a result, what different countries allocate to one and the same category is not comparable. In addition, the scheme does not make provision for vocational education, although this is an important component of some formal education systems. As a result, some countries use the incomplete categories to accommodate information on vocational education while other countries explicitly exclude vocational training from their national measurement.

After some debate an alternative scheme was developed to improve comparability:

(1)   No formal qualification
(2)   Lowest formal qualification attainable
(3)   Qualifications above the lowest, but below the usual entry requirement for universities
(4)   Entry requirement for universities
(5)   Qualifications above the higher secondary level, but below a completed full university degree
(6)   University degree completed.

The scheme alleviates some of the problems just mentioned but still poses problems for countries in which there is no obligatory formal qualification (certificate) at the end of compulsory schooling. In this case, people can complete schooling, but leave the educational system without a certificate, as used to be possible in Britain. An additional problem arises if entrance to university is not tied to a specific formal qualification. In New Zealand, for example, every citizen over 20 is entitled to attend a university, irrespective of educational qualifications. Both versions of the ISSP scheme end measurement at university level, thereby excluding the 'life-long learning' which has become a reality of working life in many countries.

Measures are available in social stratification research to distinguish between secondary education with and without a vocational component. The CASMIN scale, for example, takes vocational qualifications into account (König, Lüttinger, and Müller 1987; Brauns and Steinmann 1999) and is useful for explaining a variety of dependent variables (Braun and Müller 1997). However, it calls for information which is not available for most countries in almost all international surveys and collecting it would require additional questions. The CASMIN scheme is also only available for a small number of Western European countries.

The ISSP solution to date has been to collect two independent measures of education: the highest qualification achieved and years of schooling. Together these can be used to construct different sets of indicators suited for a variety of analyses. Since the full national codes are also provided in the merged data sets, analyses calling for more specific detail are also possible.

7.5   CONCLUSION

Background variables provide the backbone of statistical analyses in national surveys. They contain information necessary to define homogeneous subgroups, to establish causal relations between attitudes and societal facts, and to define differences between scores on scales. In short, they allow us to define contexts in which respondents' opinions, attitudes, and behavior are socio-economically embedded.

The definition and design of background variables for comparative projects is an integral part of the overall survey. Functional equivalence and levels of generality are basic issues in designing appropriate measures for background variables. As with other indicators, it is essential to define the latent construct to be measured by background questions.

Comparative research has not yet formulated an agreed standard set of background variable concepts or definitions. Corresponding items—common or otherwise—are thus obviously also not yet available. However, projects like the ISSP set standards that ease the burden for secondary analysts. At the same time, the solutions offered are not always compatible with the analytical requirements. Researchers should be always given the opportunity to assess the validity and reliability of background variables by inspecting national codes and recoding categories, if necessary, to their requirements. Thus any move to include only 'harmonized' background variables in published data sets must be avoided. Harmonization is only possible with a specific underlying dimension and a set of research questions in mind, and the quality of any harmonization can only be judged by its fit to the set of research questions which it should help answer (see Chapter 19). Researchers can also take advantage of relative, rather than direct, comparisons (such as gender-by-cohort income measures) because scores of relative measures are often directly comparable across groups, nations, or cultures.

Chapter 8

SAMPLING AND ESTIMATION

SABINE HÄDER
SIEGFRIED GABLER

8.1 INTRODUCTION: SAMPLING AS THE KEY TO GOOD RESEARCH

Sampling is a "highly technical aspect of survey research" (Backstrom and Hursh-César 1981, 52) that has significant implications for the quality of the data being collected. This observation applies particularly to comparative research projects where variation in sampling design is more or less unavoidable. It is essential that this be kept in mind at the planning stage and that the designs selected for a comparative project allow cross-sample data comparison and analysis. From this perspective, as well as being highly technical, sampling can also be described as a "scientific art."

The theory and practice of sampling for multipopulation studies remains a somewhat underdeveloped aspect of social research. The limited availability of properly executed and well-documented multipopulation studies leaves a vacuum that presents problems for comparative research. Inconsistencies between the theory of sampling and the sampling designs actually used persist, for example when random and quota samples are combined in merged data sets. Little has changed since Bulmer observed: "Conditions may vary considerably between different parts of the world but in most cases sampling will involve departure from well-codified textbook norms" (Bulmer 1983, 92).

The view taken here is that such departures, which are likely in cross-cultural studies, are acceptable departures provided probability samples and only probability samples are used. As Kish points out with reference to comparative research: "Sample designs may be chosen flexibly and there is no need for similarity of sample designs. Flexibility of choice is particularly advisable for multinational comparisons, because the sampling resources differ greatly between countries. All this flexibility assumes probability selection methods: known probabilities of selection for all population elements" (Kish 1994, 173). Only when probability samples alone are combined is there an adequate theoretical basis for inference from samples to the whole population (e.g., Europe), or to parts of it (individual European countries). Thus an optimal sampling design for cross-cultural surveys would seem to consist of the best practices available in each participating country, provided that only probability methods are used. This is the procedure followed by the Demographic and Health Surveys (http://www.measuredhs.com) and the European Social Survey (www.europeansocialsurvey.com).

In what follows, we focus on the problem of sampling designs for comparative research with the intention of establishing a more solid foundation for future studies. The purpose is to offer recommendations and point out some of the potential pitfalls involved. First, a theoretical basis is set out and methods of sampling and estimation are explained. We discuss the need for careful planning of sampling in cross-cultural surveys and outline the practical considerations involved. We then look at designs of some international studies, such as the European Social Survey and the International Social Survey Programme. Finally, since secondary data analysis depends on the proper documentation of the primary research, we outline a standard set of criteria for documenting the details of sampling design and implementation.

8.2   THEORETICAL BACKGROUND

Cost and time considerations make it impractical to gather information on every element in a population. The solution to this problem is to take a sample of a limited number of elements and to infer the parameters of the population from the sample. Representative sampling is a method first devised by Kiaer, a Norwegian, in 1895. Sampling theory was subsequently developed as a separate branch within statistics and was given a unified theoretical grounding by Godambe (1955).

The ideal sample resembles the structure and characteristics of the whole population. In other words, it is a miniature of the whole. However, in practice, due to total survey error (Andersen, Kasper, and Frankel 1979), samples only approximate this ideal standard. The sources of error lie in areas such as the design, noncoverage, and nonresponse, some aspects of which are discussed in the following section and in other chapters.

In cross-cultural surveys, independent samples are normally taken from each population under study. The parameters of interest are estimated taking into account the underlying sampling designs. The estimates from the different populations are compared and then combined to produce estimates for the overall population.

8.2.1   Sampling Frames

A sampling frame is an enumerated list of units. These units may themselves be the target elements or they may be units that include one or more target elements. In other words, it is essential that one can make a link between the target elements and the listing units. For example, suppose the population to be surveyed is that of the residents of a particular state, excluding those who are non-nationals. If the only available complete listing of the population is of households, without reference to nationality, this list necessarily becomes the sampling frame. However, only household members who are nationals comprise the population from which the actual sample will be drawn.

As in this case, problems sometimes arise where available sampling frames are not perfect in terms of the survey requirements. They may contain listings of units that are not relevant to the target population, and so provide overcoverage; or relevant units may be missing, that is, undercoverage. A common example of an imperfect frame is the ubiquitous telephone book, a section of which is devoted to private households. Some households have multiple entries to accommodate differing surnames within the household, while others are unlisted, either by choice or in the absence of a telephone. The use of the telephone book as a sampling frame can therefore be fraught with difficulties. If not taken into consideration, these can yield a biased sample.

The degree of bias can differ from country to country and across time. For instance, in 1992, the publication of household telephone numbers in the telephone book became optional in Germany. Between 1992 and 1998, the percentage of households listed dropped from about 97% to about 72%. Moreover, significant socio-demographic differences were observed between those who could be reached via the telephone book and those who could not. Thus by 1998, German telephone books no longer provided an adequate sampling frame for population surveys. When Switzerland made listing optional in 1998, 95% of private households in Switzerland were actually listed and the telephone book offered a reliable sampling frame. By the year 2000, however, the proportion of private households listed in Switzerland had fallen to 87.75% (Meisen 2001). These contrasting experiences draw attention to the importance of regularly reassessing the viability of traditional sampling frames both over time and across political borders.

Another potential problem in this context is that listings of persons or households are frequently out of date. This situation is compounded by different updating schedules from country to country. For example, in Denmark registers of residents are updated weekly, while in the Czech Republic they are updated every three months. Failure to take these differences into consideration in comparative research would result in an under-representation of the more mobile sections of the population.

It is apparent, therefore, that for cross-cultural surveys the proper definition of the population to be studied is of paramount importance if the resulting sample is to be scientifically reliable. Definitions need to be spelled out clearly and unambiguously (e.g., all persons living in private households aged 18 years and older) and should be applied to all participating countries. Each person who meets the definition should have a nonzero chance of being selected for the sample. This criterion lies at the core of successful sampling. It follows that the more comprehensive the sampling frame is, the better the sample will be. Variations in sampling frames from one country to another should be pinpointed, documented, and kept in mind at the analysis stage of the research (see Särndal, Swensson, and Wretman 1992; Lessler and Kalsbeek 1992 for overviews of frame types).

8.2.2 Probability Samples

There are several different ways to draw samples from a population. A probability sample is achieved by using a probabilistic selection method that assigns each population unit a probability of being selected. The probability distribution on the whole set of samples is called *sampling* design. Simple random sampling, in which each sample of fixed size has the same selection probability, is the most basic sampling design. It is a useful method when drawing samples from population lists that merely enumerate the units; for example, blocks of telephone numbers within one area. When a population list provides further information about each unit, however, a stratified sample may be used. In this case, probability samples are selected independently from each stratum. The main purpose of stratified sampling is the reduction of the sampling variance relative to that given by a simple random sample (Kish 1965). It can be used, for instance, in a study of a whole population for which there is a central register of inhabitants with information on each individual. It can also be used for telephone samples drawn from different areas, because the regional information that can be inferred from the area codes can be taken into consideration.

Multistage sampling is also commonly used, especially when suitable sampling frames are not available. Most large population surveys are based on this method, including ISSP surveys, the European Social Survey and Labour Force Surveys. In multistage sampling, as the name suggests, sampling is carried out in stages. The first stage sampling units are called primary sampling units, the second stage units are called secondary sampling units, and so forth. This design frequently makes use of cluster sampling. For instance, if at the second stage all household members are included in a survey, the result is a cluster sample in which the household is the cluster.

> "The motivation for multi-stage sampling is . . . to reduce sampling error for a given level of expenditure. Although clustering per se tends to increase sampling variance, the reduction in field work costs permits a larger sample size, thus reducing sampling variance. If the latter effect outweighs the former . . . then the overall effect is a reduction in sampling error for a fixed survey budget. Thus, in summary, techniques like stratification and multi-stage selection are not used on . . . social surveys as a 'second best' because researchers cannot afford simple random samples. Rather, the designers of survey samples utilize the full range of sampling techniques at their disposal to produce the design that is estimated to deliver maximum accuracy (minimum total survey error) for a fixed cost (or alternatively, to deliver a prescribed accuracy for minimum cost)." (Lynn 1998)

As said, the sampling design for each participating country in a cross-national study does not need to be identical. The optimal solution may even be that one country uses a simple random sample, another a stratified, and a third a multistage design. Various factors affect the choice of design from country to country, such as available sampling frames, previous experience, and, of course, the costs involved. What is important, however, is that the first-order inclusion probabilities are known and documented for each sample unit in each country. This was a clearly stated specification in the International Adult Literacy Survey: ". . . each participating country should use a high quality representative sample of individuals. 'High quality' and 'representative' should imply probability sampling, i.e., sample selection probabilities should be known and every member of the target population should have a nonzero probability of being included in the sample" (Lyberg 2000, 20). The European Social Survey lays down similar specifications. However, strict probability sampling is still not universally accepted for all cross-cultural studies, and it is appropriate to say something here about quota sampling.

8.2.3 Quota Samples

Quota sampling is widely used, especially in market research (cf. Deville 1991). It is the nonprobability equivalent of stratified random sampling and was used, for example, in some countries in the multination World Values Survey 1996. In quota sampling, the population is divided into different subpopulations or strata, from each of which the interviewer selects a fixed number of respondents. The aim is to get a representative sample with respect to the variables or characteristics which define the subpopulations, such as gender or age. Since it is the interviewer who actually makes the selection of the sample units within the subpopulations (in contrast to stratified random sampling), the influence of extraneous variables in making this selection cannot be ruled out. This aspect of quota sampling may give rise to varying degrees of bias. Stocks (1999) aptly describes the potential pitfalls:

> "While a quota sample would be more likely to resemble the parent population than a comparable convenience sample, drawing inferences about a parent population would be inappropriate for a couple of reasons. The first of these is that the researcher may choose the incorrect variable on which to stratify. For example, a researcher wishing to investigate political attitudes on race might stratify on racial/ethnic identification. However, the variable that was most predictive of political attitude might be something else entirely (e.g., economic status). Another difficulty would involve bias within the categories. Assume again that the researcher chose race/ethnic identification categories. Interviewers are sent out to collect data, but they stay away from 'bad' neighborhoods, conducting their interviews in more affluent neighborhoods. This bases all categories in favor of more affluent individuals, thus biasing the sample. As with all nonprobability samples, there is no possibility of evaluating sampling error."
> (Stocks 1999, http://www.msu.edu/user/sswwebed/)

Despite its disadvantages, quota sampling appears attractive for market research since it is both cheaper and less time-consuming than probability sampling. However, it lacks any theoretical basis for statistical inference from sample to population and is certainly not suitable for cross-national comparisons.

8.2.4   Estimation

Let us refer to the characteristics of a population that are of primary interest as y-values and to the secondary or auxiliary characteristics related to these primary characteristics as z-values. A population parameter is a function of the unknown y-values. Examples include the population mean, the population variance, or the population correlation coefficient if several variables are of interest. An estimator for a particular parameter is calculated using only the y-values that are known from the sample. However, if the auxiliary values for the population are known, this information can also be taken into account. For example, in random sampling the sample mean of the y-values provides a good estimator of the population mean. However, if z-values are known for all units of the population and if the relationship between the z- and y-values is almost homogeneously linear, the ratio estimator is better than the sample mean. A strategy consists of a sampling design and an estimator. Simple random sampling combined with the sample mean is called *standard strategy*.

An estimator may be said to be unbiased if its design expectation is equal to the population parameter. Under simple random sampling, the sample mean is unbiased for the population mean. However, the ratio estimator is unbiased only asymptotically. The well-known Horvitz–Thompson estimator is always design-unbiased; the stratified estimator is an example of this type.

In cross-cultural surveys, it is not usually necessary to develop new types of estimators; population parameters in the different nations or cultures can be estimated from their respective sample statistics. This is similar to the use of a stratified estimator which combines the estimators of each stratum using suitable weights.

8.2.5   Design Effects and Confidence Intervals

Calculation of design effects provides a way of comparing and evaluating different sampling designs employed in different countries. Its use shows how much better the preferred strategy is compared to the standard strategy using a sample of the same size. If a stratified design is used and the y-values are homogeneous within the strata, then the design effect is less than 1, assuming that the final sample sizes are equal. Cluster sampling usually provides a design effect greater than 1. An example for considering design effects in a multipopulation survey is provided by Verma and Le (1995). The effective sample size is the sample size necessary to get the same precision as could be obtained using the standard strategy with fixed sample size (see example in 8.4). In multipopulation surveys, sample size should be determined on the basis of achieving the same effective sample size across all countries; this is the approach adopted, for example, in the European Social Survey. In order to be able to compute the design effects in such cases, the estimator must be known for each country and the various probabilistic aspects of the sampling designs need to be clearly documented. Quota sampling is thus not an adequate method for such projects.

Confidence intervals are computed in order to assess the quality of the chosen estimator. The interval includes the unknown population parameter with a specified degree of probability or confidence (usually $p = 0.95$). Since the confidence limits depend on the unknown variance of the estimator, it must be estimated by a consistent variance estimator, assuming a central limit theorem. In practice, the variance estimator used is the estimator based on simple random sampling multiplied by the design effect. If the variance estimator is not a linear function, an approximate design-unbiased estimator of the variance is derived by estimating the design-variance of the linearization. Since in complex sample surveys only the inclusion probabilities of the primary sampling units and cluster membership are known, replication techniques are used to obtain variance estimates, for example, the random group method (see Wolter 1985). In addition, several jackknife and bootstrap methods have been developed and applied in order to estimate design-variances of estimators. These are resampling methods, that is, the sample in hand is used to construct artificial samples that allow one to assess the variability of the underlying sampling design (see Rao and Wu 1988). The *Data Quality in Complex Survey within the new European Information Society* (DACSEIS) project uses this approach. It describes its goals as follows:

> "The European statistics faces the difficult task of creating a harmonized and reliable socio-economic database for the New Economy in a united Europe. The definitions used in the Member States of the EU need to be standardized and the quality of the data gained from complex surveys need to be made more homogeneous. The core of the problem is to obtain practical and usable methods for variance estimation in complex multipurpose sampling, which enable the user to effectively and reliably estimate variances with comparable standards for the relevant national surveys by choosing from a list of criteria. This leads to a harmonized and standardized European quality management system in statistics. ... Errors in data may have unfortunate consequences in economic and social applications if they remain unknown. Therefore the aim of DACSEIS is to advance the provision of reliable information on data quality to accompany statistics databases. Best practice recommendations will be prepared for users as well as a discussion of the possibilities for harmonizing sampling methods." (http://www.dacseis.de)

Research institutes from six European countries are involved in this study which runs until February 2004. Interesting and valuable results are anticipated.

8.2.6 Weighting

There are at least two reasons for weighting data. Weighting may be undertaken because of certain features of the design and it may also be undertaken to reduce bias arising, for example, from nonresponse. The Horvitz-Thompson estimator is a well-known estimator that includes weights. The sampling weights are the inverse of the inclusion probabilities. This implies that the estimator is always design-unbiased. In stratified sampling, the weights are proportional to the ratio of stratum size and of sample size within the stratum. Ignoring weighting can lead to unacceptable biased estimators. If the probability of response of a unit is known or can be estimated, the weights to adjust for nonresponse are the inverse of the product of the inclusion probability and the response probability. Sampling weights are often interpreted as the number of population units represented by each unit of the sample.

Weighting-class methods are an extension of this approach. In this instance, the sample is divided into weighting classes on the basis of known auxiliary variables. It is assumed that respondents and nonrespondents belonging to the same weighting class respond in similar ways. This implies that the y-value is a class value and is independent of the response value.

Weighting adjustments are usually used for unit nonresponse, not for item nonresponse. If there are missing values on certain items for particular respondents, imputation methods can be used to assign plausible values, with a view to producing a clean data set. An overview of different imputation methods is provided in Särndal, Swensson, and Wretman (1992) and Lohr (1999).

8.2.7 Summary

Since there are no specific sampling designs for cross-cultural survey research, the recommended procedure is to use the best sampling designs from monopopulation studies. These include stratified and clustered (multistage) samples from central registers of individuals (e.g., for studies with face-to-face interviews); random digit dialing samples (e.g., for telephone surveys) and simple random samples from electoral rolls (e.g., for mail surveys).

Other design options which may be appropriate for a given context can depend on factors such as population density, level of literacy, and telephone density. As far as sampling itself is concerned, data collection methods are insignificant. What is important is that the sampling design is logically appropriate for the topic under study (Backstrom and Hursh-César 1981, 53). Thus, for example, if reliable inferences are to be drawn from the sample to the population, it is essential that a probability sampling scheme is used. In cross-cultural surveys, it is quite likely that the (probability) samples involved will be selected in quite different ways. The important issue here is that the estimators used must be adequate for the sampling designs.

8.3   RECOMMENDED SAMPLE PLANNING

The following is an overview of some of the considerations to be kept in mind in the case of a multipopulation study.

8.3.1 Selection of Participating Countries

The first step is to decide which nations or cultures are to be included in the study. Van de Vijver and Leung (1997b) suggest three alternative ways to proceed at this stage. The first is convenience sampling, which selects cultures purely on the basis of convenience. This can often mean that there is in fact no linkage between the goal of the study and the countries selected and the population parameters being explored. This sampling procedure is often used for psychological differences studies. The second way to select nations or cultural groups is by systematic sampling, which has a firmer theoretical basis: "These studies usually fall into the categories of theory-driven or generalizability studies. Cultures are selected in this procedure because they represent different values on a theoretical continuum. . . . To maximize the effectiveness of systematic sampling, effort should be made to select cultures that are far apart on the theoretical dimension on which they vary . . . This strategy will maximize the possibility of detecting cultural differences if they truly exist." (Van de Vijver and Leung 1997b, 27f.; note that 'systematic sampling' as used by Van de Vijver and Leung is distinct from systematic random sampling.) The third and final approach is the random sampling of a large number of cultures: "This strategy is preferable for generalizability studies, in which a universal structure or a pan-cultural theory is evaluated. Obviously, it is almost impossible to collect a truly random sample of cultures because of constraints of time and resources" (Van de Vijver and Leung 1997b, 28f.).

The authors recommend that if the objective of the study is to look for differences, one should start with cultures that are more similar. In contrast, if the goal is to look for universal patterns, one should include cultures that are as different as possible. However, as already mentioned, the choice between these alternatives may often be determined by extraneous factors such as available funding and expertise, or by political, cultural and/or methodological restrictions. It is apparent, therefore, that both practical and theoretical considerations need to be considered in making decisions regarding design.

8.3.2 Types of Multipopulation Studies and Their Design

Kish (1994, 167 ff.) provides a useful outline of a number of alternative multipopulation survey designs, distinguishing five in all. These can be summarized as follows:

(1)   Periodic surveys (panels, or overlapping, rotating, or distinct samples, e.g., analysis of data with periodic General Social Surveys to examine changes over time);

(2)   Comparisons of distinct domains from the same survey (e.g., a comparison of the attitudes of western Germans and eastern Germans with data from the German General Social Survey);

(3)   Multinational comparisons (e.g., the Eurobarometer studies, the Demographic and Health Surveys, or the ISSP);
(4)   Combinations and cumulations of separate samples (e.g., the cumulative 1980–1998 German ALLBUS for analysis of the social structure in Germany);
(5)   Controlled observations of diverse types (as widely used in epidemiology and education).

Irrespective of the choice of participating countries and of the overall purpose of the study, multipopulation studies tend to be similar in all major aspects of design. Kish (1994) identifies the following essential considerations when planning such a survey:

(1)   Definition of concepts, variables, and populations should be synchronized;
(2)   Measurements and data collection methods should be standardized and made as similar as possible across all comparisons;
(3)   Substantive analysis should be closely related to survey concepts and variables, but not to the sampling design;
(4)   Weighting will in some aspects be related to sampling design, but also to analysis;
(5)   Statistical analysis should be closely related to the sampling design. It includes the choice of the statistics to be compared, also the computation and presentation of sampling errors;
(6)   Sampling designs should be chosen flexibly, depending on the available resources such as sampling frames and the experience of the participating institutes. However, the selection method has to be a probability method;
(7)   In calculating sample sizes, design effects should be considered and effective sample sizes should be computed.

In dealing with international comparability of business surveys, Laaksonen (1995, 861) raises two fundamental concerns: (1) how to make the comparisons "qualitatively" (in particular in terms of the points made by Kish) and (2) how to organize the work at an international level. He expands Kish's list to ten items and introduces the interstatistician, whose role it is to make statistically-based international comparisons. When designing a multinational survey, special attention has to be paid to coverage problems in general and, in particular, to coverage differences between countries. These are the most difficult problems likely to arise in striving for good sampling frames and good sampling designs.

One example of a well-planned multipopulation study is the European Social Survey. The design specifications spell out clearly what is required. The definitions of the target populations are the same in each country participating in the first round.

"The survey will be representative of all persons aged 15 and over (no upper age limit) resident within private households in each country, regardless of their nationality, citizenship, language or legal status. [Please note that questionnaires are to be available in

all languages spoken as a first language by 5% or more of the population and interviewers must be available to administer them. For speakers of certain minority languages, however, it may be possible to use or adapt the questionnaire from another participating country]"
(*European Social Survey: specifications for participating countries*, 4; http://www.esf.org)

Strict probability methods are specified for each stage, so that the relative selection probability is known for each sample member. These probabilities are to be recorded in the data set. Quota sampling is not permitted at any stage, nor is substitution for nonresponding households or individuals, whether these are "refusals" or "noncontacts."

According to Kish (1994), the precise sampling strategies will naturally vary from nation to nation, depending on their existing sampling sources, access, and convenience. However, it is essential to ensure that every member of the relevant population in the country in question has a known, nonzero probability of selection. "For some countries, a list of addresses is the obvious and convenient starting-point for a good probability sample. But in others, these lists are either not available or, for one reason or another, not the best starting points. For instance, in countries with population registers, they are an obvious choice. Other countries may well have suitable alternatives."

The European Social Survey also specifies a minimum effective sample size.

"The minimum number of actual interviews to be achieved is 2,000 (except in countries whose total population is less then 2 million, when the minimum number is 1,000). Irrespective of the actual number of interviews, however, the minimum 'effective achieved sample size' should be 1,500, after discounting for design effects (see Appendix 1), or 800 in countries with populations of under 2 million. Thus, each country should determine the appropriate size of its initial issued sample by taking into account the realistic estimated impact of clustering, eligibility rates (where appropriate) and response rate on the effective sample size." (http://www.esf.org)

While further details of the sampling principles used in this study remain to be developed, it provides an exemplary model for the planning of future multipopulation studies.

8.3.3 Practical Problems

While many textbooks deal with the theoretical issues of sampling, (e.g., Yates 1949; Hansen, Hurwitz, and Madow 1953; Kish 1965), the statistician has also to face many practical problems. We turn now to some of the main problems in this regard.

The ideal way to draw a sample from the population of a country is by means of a (stratified) selection of individuals from a complete population register. Unfortunately for sampling designers, not many countries have central registers of the population. Even when a register does exist, it may not be complete or up-to-date

or is not available for public use. In some cases, access to registers is fast and
economical, in others drawing samples from central registers is both costly and time-
consuming. As a result, even when registers are available, other probability sampling
designs have been developed. For example, members of the Association of German
Market and Social Research frequently use random route or preselected address
procedures. The interviewer either chooses a household according to specified rules
(designed to ensure randomness) or is given the exact address of a household that has
already been selected randomly. Then an individual within that household is
randomly selected according to a specific set of rules (such as the Kish grid or the
'Last Birthday' Method).

TABLE 8.1. Telephone Connections per 100 Inhabitants in 1999

Country	Connections per 100 inhabitants
Cambodia	0.24
Indonesia	2.90
Thailand	8.42
Laos	0.60.

SOURCE: http://www.aseansec.org/economic/communication.htm

Another type of design frequently used is pure Random-Digit-Dialing and similar
designs (e.g., the Mitofsky–Waksberg Design). However, for these designs to be
useful a minimum 90% telephone density is required; Table 8.1 indicates how low
telephone availability actually is in some countries.

There can also be problems in the case of face-to-face surveys in developing
countries, as population registers, if available, are often incomplete and/or out of
date. In the absence of census data with home addresses or other comprehensive
directories of 'who is where', alternative approaches are necessary.

Often the only frame available is a list of villages, of uncertain completeness. This
leads directly to multistage designs, where in the first stage villages have to be
selected. If there is no further information on the number of residents living in the
villages, the sample has to be drawn with equal probabilities (Zarkovich 1983, 102).
In the next stages households and persons are selected. Some countries, such as India
and Pakistan, have land records which may be used to select households. However,
these are sometimes unreliable because of deliberate distortions related to their use
for tax purposes.

If there are no viable sampling frames available, a sample can be designed on the
basis of area mapping. This method uses a local map of the targeted residential area
and divides it into a matrix of grid squares. Each square is numbered in a 'serpentine'
order and then every nth grid square is selected from a random starting square. The
skip interval is determined by the sampling fraction (Bulmer 1983, 94). If maps are
not available, photographs could be used instead.

Special problems arise in countries with nomadic populations. Despite their mobility, these peoples are integral to the population under study and each individual should have a nonzero chance of inclusion in the sample. Kish (1995) outlines six methods of selecting nomads. One method is to use the local hierarchical social structure and to start by making contact with the tribal leaders. This opens a contact route to the member households. As is apparent, this is a multistage design approach. Other possible approaches in this situation are the 'Waterpoint Method' and the 'Camp Approach'. Kish does not recommend any one method as theoretically preferable to the others. However, in practical terms one or the other method may be the best option and pretesting should be carried out to establish which method is most appropriate in particular circumstances. Omission and duplication are perhaps the main pitfalls to be aware of here.

8.4   EXAMPLES OF SAMPLING IN CROSS-CULTURAL RESEARCH

The World Fertility Survey (WFS) and its successor, the Demographic and Health Surveys (DHS), are well-known and highly regarded examples of cross-cultural studies. Bulmer (1983, 17) describes the WFS as follows: "This major international survey, begun in 1972, is designed to enable developing countries in particular to carry out internationally representative, internationally comparable and scientifically conducted surveys of fertility behavior. The WFS is administered by the International Statistical Institute with funds from United Nations agencies."

In the WFS, the same sampling design was used in more than forty participant countries. The first step in the design was the creation of an area sampling frame and the listing of dwellings and/or households in the selected areas. Fertility data was then collected from a large sample drawn from this list. Lists of household members were then compiled to identify women eligible for the individual interview. The interviewees were women of childbearing age.

The results obtained by the WFS were evaluated by Chidambaram, Cleland, and Verma (1980). The authors found that the coverage of the vital events pertaining to individuals was good, while the coverage of other characteristics was partly good and concluded that the WFS data set provided a valuable database on demographic behavior that would otherwise not exist.

Kish assesses the WFS as follows: "The surveys . . . demonstrated that large-scale probability sample surveys are feasible almost everywhere; that sampling frames and resources can be found, including households, and used as sampling units; that local technicians can execute complex tasks directed by a centralized international staff; and that probability selection and 'measurability' of sampling errors can be imposed" (Kish 1994, 179). This optimistic evaluation should serve to encourage the use of probability samples for major international studies.

Another important ongoing cross-cultural survey that uses probability sampling is the Eurobarometer series, which is conducted on behalf of the European Commission at least biannually throughout the European Union. It has provided regular monitoring of social and political attitudes in the member countries of EC/EU since the early seventies. Eurobarometer documentation provides the following account of the sampling procedures used:

> In all EC/EU Member Countries samples were initially drawn among the national population, aged 15 and over. Starting with EB 41.1 the target population is the population of any nationality of an EU member country, aged 15 years and over, resident in any of the member states. The regular sample size in standard Eurobarometer surveys is 1000 respondents per country/with some exceptions.
>
> The sampling is based on a random selection of sampling points according to the distribution of the national, resident population in terms of metropolitan, urban and rural areas, that is, proportional to population size (for a total coverage of the country) and to population density. These primary sampling units (PSU) are selected from each of the administrative regions in every country.
>
> Since autumn 1989 (EB32) the basic sampling design in all Member States is a multistage, random (probability) one. In the second stage, a cluster of addresses is selected from each sampled PSU. Addresses are chosen systematically using standard random route procedures, beginning with an initial address selected at random. In each household, a respondent is selected by a random procedure, such as the first birthday method. Up to two recalls are made to obtain an interview with the selected respondent. No more than one interview is conducted in each household. In previous Eurobarometer surveys, different sample methods were used which varied between countries. Until Eurobarometer 31A in Denmark, Luxembourg, and the Netherlands a random selection from the population or electoral lists (of individuals or households) was used, in Belgium, France, Italy, United Kingdom, and Ireland quota sampling established by sex, age and occupation on the basis of census data, in Greece, Spain and Portugal a random route procedure (combining the two precedent ones). Germany used quota sampling established by sex, age and occupation on the basis of census data until EB 23 and random route starting with EB 24.(http://www.za.uni-koeln.de/data/en/eurobarometer/index.htm).

The above illustrates clearly the increasingly sophisticated nature of sampling procedures for cross-national surveys. However, while theory and associated designs have tended to converge in participating countries, difficulties continue to arise in terms of practical application. The result is a variation in data quality. This was particularly the case with the data of the Eurobarometer 41.0 of 1994 (sample sizes in brackets), a point that can be illustrated with data from Belgium (1087), France (1034), Spain (1003), East Germany (1058) and West Germany (1064). Häder and Gabler (1997) compared the distributions of selected demographic variables (age, gender, household size) with the reference data from national statistical yearbooks. Considerable differences were evident in almost all background variables and countries, suggesting that biases in other variables due to nonresponse or other errors could be expected as well.

This problem is usually dealt with by assuming a fairly high correlation between demographic variables and substantive variables. In light of this, an adjustment is made to some of the demographic variables in terms of their known distribution in the target population. This approach was pursued in this instance. Having adjusted the variables by weighting with an Iterative Proportional Fitting Approach, the efficiencies and effective sample sizes were computed. The results are shown in Table 8.2.

As the table shows, the (effective) sample size is lower than 1,000 interviews in three of the five countries when only adjustment weighting is applied. If design weighting is also taken into account, the resulting design effects would be even higher, while the effective sample sizes would be lower. This would mean the target of 1,000 interviews would not be met in most of the countries concerned.

TABLE 8.2. Efficiencies and Effective Sample Sizes for Different Nations in Eurobarometer 41.0 (1994)

Nation	Efficiency	Effective sample sizes
Belgium	92.8	1009
France	92.6	957
Spain	93.9	942
East Germany	92.5	979
West Germany	94.2	1002

Another example of the use of different designs in different countries is provided by the International Social Survey Programme. This ongoing program of cross-national collaboration began in 1985 with only four member countries and has since expanded to almost forty countries worldwide.

While probability sampling is obligatory in the ISSP, there are considerable differences in the actual designs used. For instance, no two countries that fielded the 1997 ISSP module used the same design. Norway used a simple random sample drawn from the central register of persons (between 16–79 years); Bulgaria used a two-stage cluster sample selecting households in the first stage and then persons 18 years and older using a Kish grid; and Slovenia used a systematic multistage sample of adults 18 years and older, drawn from the central register of inhabitants with substitution for nonrespondents. Several countries did not use probability sampling. Italy described its sample, for example, as "probability with quotas."

These examples from only four of what were then about thirty countries indicate the range of differences and resulting difficulties that can arise in multipopulation surveys. The variations can be summed up as follows. Definitions of the populations to be studied differ, for example, in the case of different age specifications. On another dimension of design, some countries use self-weighting samples of households, while others select individuals on the basis of equal probability. Stratification, clustering, or simple random sampling may be used. Nonresponding units may be replaced or not. In the case of Italy mentioned above, random and quota sampling were mixed, the result being uncertainty regarding the inclusion probability.

Different sampling designs, however, do not invalidate cross-cultural comparisons if the sampling procedure is probabilistic and the correct estimates are used. The Italian 1997 ISSP sample did not meet these criteria. Other studies also fail to meet essential criteria. For example, sixteen population studies of sexual behavior and attitudes to HIV/AIDS were carried out between 1989 and 1993 in eleven European countries (Hubert, Bajos, and Sandfort 1998). About 80,000 people were interviewed. The sampling designs covered a broad range: simple random sampling of national registers (Spain); stratified random sampling of households from the telephone directory followed by the 'Next Birthday' Method (France); two-stage clustered sampling (Athens); random digit dialing (Scotland) and quota sampling (Germany). The resulting data base encompassed virtually all possible designs. That, in itself, does not present a problem but errors occurred in the choice of suitable estimators in the ensuing publications. For example, design weighting to adjust for unequal inclusion probabilities in self-weighting household samples was completely neglected. In some countries the data analyzed were controlled for age and gender, in others not. Furthermore, the formula given for the computation of confidence intervals (except for the French sample) is valid only for simple random samples (Hubert, Bajos, and Sandfort 1998). As most countries used multistage complex designs, the appropriate design effect would need to be considered when computing confidence intervals. Finally, mixing probability sampling and quota sampling in the comparative context (here, as in the World Values Survey) is not good practice. For useful descriptions of additional sampling designs for cross-national surveys, see Szalai 1972; Levy and Lemeshow 1991; Verma 1992; Hayashi and Scheuch 1996; de Heer 2000; see, too, the Web sites in the Appendix.

8.5   RECOMMENDED DOCUMENTATION

Detailed descriptions of the sampling designs are needed for those conducting secondary data analysis. At the very least, the following questions should be answered and documented for all samples (see also guidelines provided by survey research associations such as AAPOR, ESOMAR, and WAPOR).

What is the target population of the study? This includes, for example, specifications of the lower and upper age limit, of the exclusion or inclusion of non-national residents, and of the exclusion or inclusion of homeless and institutional populations.

Which sample frames were used? The quality of the sample frame(s) used for the survey with respect to up-to-dateness, overcoverage, undercoverage, etc., should be described for all countries.

Which sampling designs were used? If lists are used as frames, it should be made clear how the elements of the sample were drawn from this list. Inclusion probabilities should be declared. If a design without using a list was applied, all stages have to be described in detail. Inclusion probabilities should also be calculated.

What are the sample sizes in the different countries? Information is needed on whether effective sample sizes were computed, and if they were, how this was done. If population subgroups were oversampled, this also should be documented.

What kind of weighting is necessary or recommendable? If weights are already part of the data set, documentation has to be provided on how they were computed (formulae), and a clear distinction made between design and adjustment weights. If no weights are available with the data set, the information necessary to compute design weights must be provided.

Failure to meet these criteria adequately for documentation can cause enormous problems for the assessment of the quality of the data. As Lyberg noted on the International Adult Literacy Survey: "The sampling plans are difficult to assess due to lack of detail and lack of information on the quality of master samples, replacement procedures, and nonresponse treatment" (Lyberg 2000, 20).

8.6   CONCLUSIONS

In the foreword to the DHS Analytic Report No. 3 (1997), Kish makes the following observation:

> "Sample surveys of entire nations have become common all over the world during the past half century. These national surveys lead naturally to multinational comparisons. But the deliberate *design* of valid and efficient multinational surveys is new and on the increase. New survey methods have become widespread and international financial and technical support has created effective demand for multinational designs for valid international comparisons. The improved technical bases in national statistical offices and research institutes have become capable of implementing the complex task of coordinated research. However, to be valid, multinational surveys have to be based on probability sampling designs of comparable national populations, and the measurements (responses) should be well controlled for comparability."

In fact, over the last few years a trend can be noted in this direction in sampling for multipopulation surveys. Random sampling, documentation of inclusion probabilities, and the calculation of design effects are increasingly regarded as the accepted standard procedures. When these criteria are adhered to, comparisons between estimates of different countries can be made in a theoretically appropriate manner.

Large international studies, such as the Demographic and Health Surveys, the European Social Survey, the ISSP, and the DACSEIS project, can be expected to provide further valuable experience that will serve to strengthen the methodological base for sampling in comparative research.

APPENDIX

The following provide useful descriptions of additional sampling designs for cross-national surveys:

- http://www.microcase.com/archive/wvs.html
- http://wvs.isr.umich.edu
- Demographic and Health Surveys, e.g.,
  http://www.census.gov.ph/data/technotes/notendhs.html
- http://www.measuredhs.com
- http://dpls.dacc.wisc.edu/apdu/lis_country.html (Luxembourg Income Study)
- http://www-rcade.dur.ac.uk/ (European Community Household Panel Survey)
- http://europa.eu.int/en/comm/eurostat/ (Eurostat)
- http://opr.princeton.edu/archive/wfs/ (World Fertility Surveys)
  http://www.soc.uiuc.edu/Soc280/ISJP2/methodology_of_the_1996_isjp_sur.htm
  (International Social Justice Project)
- http://www.microcase.com/archive/wvs.html; http://wvs.isr.umich.edu/ (World Values Survey)

PART THREE

ERROR AND COMPARATIVE SURVEYS

Chapter 9

ERRORS IN COMPARATIVE SURVEY RESEARCH: AN OVERVIEW

MICHAEL BRAUN

9.1   INTRODUCTION

The quality of social research depends on the quality of measurement. If social phenomena are not measured properly, the wrong conclusions may be drawn about relationships between them. Empirical evidence could thus lead to hypotheses being accepted or rejected not because they are actually true or false, but because measurement was biased. In comparative research, the quality of conclusions drawn depends on the quality of each of the separate national studies. If any of these are flawed, both similarities and differences between countries can be methodological artifacts. Surface characteristics of surveys (e.g., as recorded in study description profiles) tell us something about data collection modes, sample design, and response outcomes, and so on, across the countries involved. Unfortunately, these are not reliable indicators of whether the conclusions that the data point to are, in fact, correct. For example, if a survey is conducted by telephone, important segments of the population may be excluded, not just in countries with a low telephone density. In addition, response effects related to telephone interviews have to be taken into account and these, interacting with other factors, may differ across countries. Any superficial similarity of procedures used or the outcomes realized in different countries does not necessarily guarantee comparability. Identical response rates across countries can mask enormous differences between realized samples, depending on whether they result more from noncontact or from refusals. The effects of identical procedures may differ from country to country.

Errors in surveys may be purely random or systematic. The presence of random errors does not usually prevent researchers from obtaining good estimates of what they are interested in. Systematic errors, on the other hand, lead to biased estimates that are either larger or smaller than the 'true' population values. Groves (1989) distinguishes four sources of survey error: *sampling error*, *coverage error*, *nonresponse error* and *measurement error*. Sampling error results from using a sample instead of surveying the entire population. A coverage error results from the failure to give all the units of the population on which information should be gathered a nonzero chance to be included in the sampling frame. Nonresponse error results from the fact that not all the units of the gross sample actually participate in the survey. Finally, different processes related to the instrument, the interviewer, the respondent, and to the data collection mode contribute to measurement error. Each of these types of error can affect the collection of any information in surveys, from demographic information and behavioral reports to attitudinal data, and quiz or test performance.

Discussions of quality and error in social science survey research are usually conducted within this framework of error. Survey-based comparative research in psychology is more familiar with a framework that turns on different kinds of *bias*. The two frameworks differ both in terminology and in focus, as do the kind of survey work conducted in the different disciplines and their instrument design practices. Nonetheless, the three distinctions made in the psychology framework between *construct bias*, *item bias* and *method bias* (further divided into *sample bias*, *instrument bias* and *administration bias*) correspond to distinctions made in the Groves typology (cf. Van de Vijver 1998, Van de Vijver and Leung 1997b, see Chapter 10). At the same time, the terms and types of error covered by each reflect that each discipline has different emphases. Distinctions important for social surveys of the general population are sometimes less important for psychological studies. Thus sample bias corresponds to both coverage error and nonresponse error in the Groves scheme. The Groves framework makes fewer distinctions between sources of error related to instruments. For example, most of the various kinds of bias discussed in detail in Chapter 10 are subsumed under the heading "measurement error" in the Groves scheme.

9.2   SAMPLING ERROR

Sampling error results from using a sample instead of surveying the entire population. If the sample is drawn from the population using a random procedure, it can be exactly computed. An entire branch of statistics is devoted to dealing with analyzing sampling errors for different sampling designs. Using different sampling designs in different countries does not present a problem for cross-cultural research, as long as all of them are based on the idea of random sampling (see Chapter 8).

However, the computations of sampling statisticians hold only if the other three error types can safely be ignored. This is virtually never the case in the social sciences. Although research is ongoing on how to quantify 'nonsampling error' (Groves 1999), the size of the error involved cannot be exactly quantified and may depend strongly on the research topic.

9.3   COVERAGE ERROR

Ideally, a sample should consist of a miniature of the population sampled. Each unit of the population should have a nonzero chance of being included in the sample. If complete lists of the members of the population are not available, some members cannot be considered for inclusion in the sample and coverage error inevitably results. In comparative research, the population has to be defined in an identical way in each of the countries. If the definitions are different—either to take practical difficulties in drawing the sample into account (such as the lack of complete lists) or for any other reason—then a coverage error results. The same holds if a (gross) sample cannot be drawn to meet what were originally identical definitions of the populations across the project (see Chapter 8).

Within countries and across countries, survey modes differ in their susceptibility to coverage error (Lyberg and Kasprzyk 1991). This is most obvious with telephone surveys. Telephone surveys require a high telephone density. In addition, there are technical difficulties in drawing a sample when not all telephone numbers are listed in a directory. Mail surveys also require a complete list of the members of the target population for sampling. If such a list is not available, telephone books might be consulted instead. However, if the addresses are not listed for all the entries in the telephone book, the coverage problems are even more severe.

Coverage error is also relevant for face-to-face surveys of the general population. In many cases, complete and up-to-date lists of the entire population of a country are not available. Thus, although studies are treated as representative of an entire population, samples may be arrived at through random walk procedures. As a result, in practice the population used is often redefined as "persons living in private households." Depending on the survey topic, restricting the population to those living in private accommodation can result in serious bias. For surveys on poverty, crime and health, excluding people such as the homeless or the inmates of jails and asylums can be fatal (Schnell 1991). Across countries, the segments of the population not living in private accommodation (and their numbers) differ. In addition, the percentage of people in jails, hospitals, or asylums depends on the age structure of the population or the level of the health-care system, among others.

The impact of excluding specific segments of the national population from sampling frames also differs across countries. If, for some reason, the rural population is excluded, this will have little effect on findings in highly urbanized societies such as Monaco or the Netherlands, provided the topic does not make it important to include the rural population. However, in countries such as Poland, the rural population makes up an important segment of the entire population. Excluding them would bias the research findings for the country. Moreover, urban-rural differences are more pronounced in less affluent countries. As a result, it becomes important to control for GNP. Independent of mode, some countries restrict their target population to people with voting rights or to specific nationalities; many Israeli surveys, for example, exclude the Arab citizens of Israel although they have the right to vote. The reasons for exclusion differ from context to context; there may be a substantive preference for including only people with full citizenship rights, or problems may be met in drawing a sample to include everyone if information on voting districts is available, but none is available on residential or political districts. Alternatively, if everyone is included in the frame, then too many languages may be involved for the survey budget. The effect coverage errors have on results will depend of the survey topic. For comparative studies on voting behavior a restriction to people with full citizenship rights and the right to vote is a restriction to the relevant population. However, for research on social stratification, the consequences of excluding those without voting rights depend, among other things, on how large the numbers of people are in the countries under comparison and on whether their socio-economic position in society is markedly different from that of the population with voting rights.

Sampling and coverage error occur independent of the behavior of sampled units. However, both nonresponse and measurement error are affected by the behavior of sampled individuals; indeed motivation and cognitive factors may play a paramount role. Two theoretical approaches have been used to try to account for the underlying processes: the social-cognition approach and the rational choice approach. The social-cognition approach has developed over the last three decades in close cooperation between cognitive psychologists and survey researches (Jobe and Mingay 1991) and focuses on cognitive and communicative aspects. The rational choice approach to respondent behavior starts from the assumption that actors try to maximize their subjective utility in whatever they do and emphasizes the importance of motivational variables in choice (Esser 1986, 1993). The social-cognition approach has mostly elaborated on the interpretation of survey questions, among others. The rational choice approach is prominent in explanations of participation in surveys and of effects of social desirability. Both approaches take cognitive, communicative, and motivational variables into account. The discussion that follows borrows from both of them.

9.4   NONRESPONSE ERROR

Nonresponse error results when not all members of the (gross) sample participate in a survey and are thus not included in the net sample. There are two main reasons for nonresponse: noncontact and refusal. Noncontact refers to the inability of the interviewer to establish contact with the potential respondent; refusal refers to the decision on the part of selected respondents not to participate in a study. Other sources of nonresponse include inability to participate due to bad health or language difficulties (see Chapter 11). The consequences of nonresponse error depend on the research topic and the degree of nonresponse; they can differ greatly for noncontact, refusals, and other reasons of nonresponse. Nonresponse matters as soon as the composition of respondents and nonrespondents is dissimilar with regard to the variables under examination. The most important issue is the extent to which observed differences between countries are due to differential nonresponse.

To cope with refusals and to increase motivation to participate in a survey, costs to the potential respondent have to be reduced by the researcher (Groves and Cooper 1998). Both material costs to potential respondents, such as time foregone by participation, and subjective costs, such as concern with confidentiality and privacy, vary dramatically across cultures. The same applies to the likelihood of misinterpreting a survey situation as a marketing activity. In the United States, interviewer and respondents are frequently matched on race, but not on gender. In Arab countries, however, it will be difficult, if not impossible, for a male interviewer to get an interview from a woman, and for a female interviewer to get an interview from a man.

Even if people whose native language is not the national language are included in the sampling frame, budget restrictions may mean that the survey instrument is only available in the main language of a country. Those not proficient in this language will then be excluded from the survey. As a consequence, in countries with immigrant groups of different linguistic backgrounds, for instance, Turks in Germany or recent African immigrants in the United States, these segments of the population will be excluded at this stage. Acculturation research shows that language proficiency varies with acculturation. Thus those poorly integrated would be excluded, in Germany or the United States, because of language barriers and the research findings could then point to a greater level of integration than actually exists. This need not be a problem everywhere. Immigrants from former colonies to Britain or France, for example, could have fewer language problems. Comparative studies on the integration of ethnic minorities would be seriously affected as a result.

9.5   MEASUREMENT ERROR

The quality and comparability of instruments is just as crucial as the quality of sample design and realized samples. The instrument, the interviewer, the respondent, and the survey mode are recognized sources of measurement error (Biemer et al. 1991). The interactions between these four sources of measurement error are of particular importance.

With regard to the *instrument*, an overly complicated design of the entire questionnaire, the wording of individual questions, question ordering, and the response alternatives offered may contribute to measurement error. Countries differ in their survey tradition and this is relevant for the standards of survey research in general and the training and experience of interviewers in particular. The degree to which respondents are familiar with formalized communication with strangers, complicated questionnaires or test situations is also important. Questions that are clear and unambiguous for respondents in one country need not be clear for another culture, even when these are correctly translated. Difficulties may have much more to do with culturally anchored perception (see Chapters 4, 5, and 6).

Interviewers introduce error by deviating from the question texts or by failing to record respondents' answers properly. More importantly, sometimes, willingly or unwillingly, they fail to comply with their professional role as a friendly but neutral observer of respondents' reactions. They may, for example, give respondents cues that allow them to infer whether their responses 'please' the interviewer or not. Poor 'matching' can also affect the interaction and responses (Davis 1997), as for example when black people interview white people on racial issues.

Respondent characteristics in the sense of dispositions for social desirability and acquiescence are further sources of measurement error. These behavioral tendencies are both personality traits and activated by situational characteristics. Having agreed to participate in a survey, cognitive and motivational factors determine the role respondents assume during the process of data collection. Either they adopt a cooperative role and try to provide the response required for a given question, or they adopt a conformist role in which they provide socially desirable answers (Esser 1993). Systematic error will result only if a) an attitude exists and is important to the respondent, but b) the incentive to give a socially desirable answer is also high, and c) the situation is sufficiently structured, and d) the socially desirable response differs from the response compatible with the attitude actually held (Esser 1993). Whether these conditions are met depends on a number of factors such as the clarity of the question, the accessibility of the information to be provided, the degree of privacy of disclosure, as well as the situational tendency to answer in a socially desirable fashion, prompted by shared disclosure and individual susceptibility to social desirability as a personal trait.

All the above mentioned variables vary across countries. There is evidence that individuals from collectivist cultures are more subject to social desirability. In addition to desirability as a trait (Edwards 1957, Marlowe and Crowne 1960), trait desirability, that is the perception of what is regarded as desirable, has to be taken into account (Stocké 2001). Intercultural differences are to be expected with regard to the ability to perceive accurately what is desired in a society, by a social group or by the interviewer. People from collectivist countries will be better at guessing what is regarded as desirable, as they have learned to observe the conduct of others in order to find out how a socially acceptable response from their part should look (see Chapter 6). Moreover, what is regarded as desirable differs from country to country. Thus in countries where sexual activities of women are subject to normative restrictions, questions on the frequency of sexual intercourse in general, and on extramarital intercourse in particular, will not only be regarded as sensitive and potentially embarrassing. People will also have a clear understanding what is regarded as desirable in their society and if such questions are answered at all, the actual frequency is likely to be underreported.

One consequence of differences across *modes* relates to tendencies to answer in a socially desirable way. In particular, desirability as a trait should only have an effect if there are consequences connected to a response. These are less to be expected in 'low-cost' situations that are typical for private disclosure modes which provide a high degree of anonymity. It comes as no surprise, for example, that Tourangeau and Smith (1996) found that in the United States the difference between men and women in the number of sex partners reported was lowest in a self-completion mode. Country differences in the gender gap measured by face-to-face surveys might thus be methodological artifacts due to the effect of social desirability.

Chapter 10

BIAS AND EQUIVALENCE: CROSS-CULTURAL PERSPECTIVES

FONS J. R. VAN DE VIJVER

Most cross-cultural research is undertaken to compare countries, cultures or groups on some characteristic. When these characteristics are physical attributes, the relationship between measurement and objects/constructs to be compared is simple. Suppose that a manufacturer of blue jeans measures the leg length of a carefully sampled group of Koreans and Norwegians and that the mean leg length of the Norwegian group is longer. It is likely that the difference in the mean is attributed to the populations from which the samples were drawn. One inference from this result is obvious: one can expect different sales for the various sizes in the two countries. The differences in observed scores (average length in sample) are readily attributed to differences in the distributions of the underlying concept (distribution of length in the population).

Now, suppose that a questionnaire testing conformity was administered to the same subjects and that the Norwegians again showed a higher mean score. The inference from the mean score to the underlying construct, conformity, is less obvious, since the relationship between measurement and construct might be disturbed by various sources. Thus various alternative inferences could be drawn from the observed differences in mean scores of conformity. For example, the samples may show differences in social desirability. Such problems of alternative interpretations are characteristic for studies in which the relationship between the measurement operations and the underlying construct is not directly causal but probabilistic. The operations as specified in the measure are only samples (frequently a small sample) of the universe of all items relevant for the empirical construct. Measures with a probabilistic relationship between measure and construct abound in survey research. Many methodological issues discussed in this book arise in probabilistic models.

10.1 TWO TYPES OF STUDIES: STRUCTURE- AND LEVEL-ORIENTED

10.1.1 Structure-Oriented Studies

Two types of studies are common in cross-cultural research (Van de Vijver and Leung 1997b). The first, called *structure-oriented*, addresses the question as to whether an instrument measures the same construct across cultures. Is, for example, *national identity* a construct that conveys a similar meaning in Chile, Italy, and Tanzania? Has *religiosity* an equivalent meaning across the globe? Both universal and culture-specific patterns have been found. Schwartz (1992) reported evidence for the universality of the structure of human attitudes across 40 countries; two dimensions, labeled *openness to change* (vs. *conservatism*) and *self-enhancement/power* (vs. *self-transcendence*), sufficed to describe the patterning of human values. Similarly, *personality* (Eysenck's *psychoticism, extraversion,* and *neuroticism* factors) and *intelligence* (Jensen 1980) have been found to show universal aspects. Culture-specific patterning, on the other hand, has been found for *depression.* Tanaka-Matsumi and Marsella (1976) asked Japanese and Americans to generate words associated with *depression.* The Japanese group referred more often to mood states, whereas the Americans gave more somatic responses. The somatic responses defined the common aspects whereas the mood states were more culture specific. *Locus of control* (defined as a tendency to ascribe events to either internal or external sources, such as a high examination grade as the result of hard work or sheer luck, respectively) has also been found to show huge cross-cultural variation (Dyal 1984).

10.1.2 Level-Oriented Studies

The second type is *level oriented.* Such studies address cross-cultural differences in average scores. For instance, are Chileans, Italians, and Tanzanians equally *nationalistic*? A comparison of levels assumes that the measures show structural equivalence and, in addition, that scores can be directly compared across cultures. Because of these strict requirements, comparisons of levels have to be verified. Their validity cannot be merely assumed. Verification procedures require a classification of factors that can jeopardize the validity of inter-group comparisons. These factors are known as *bias.*

10.2   BIAS

Bias refers to the presence of nuisance factors that challenge the comparability of scores across cultural groups. If scores are biased, their psychological meaning is culture/group dependent and group differences in assessment outcome are to be accounted for, at least to some extent, by auxiliary psychological constructs or measurement artifacts.

The occurrence of bias has a bearing on the comparability of scores across cultures. The measurement implications of bias for comparability are addressed in the concept of *equivalence* (see Johnson 1998a for a review). Equivalence refers to the comparability of test scores obtained in different cultural groups. Obviously, bias and equivalence are related; it is sometimes argued that they are mirror concepts. Bias, in this view, is synonymous to nonequivalence; conversely, equivalence refers to the absence of bias. This is not the view adopted here because in the presentation of cross-cultural research methodology it is instructive to disentangle sources of bias and their implications for score comparability.

Bias and equivalence are not inherent characteristics of an instrument, but arise in the application of an instrument in at least two cultural groups and the ensuing comparison of scores, patterns or item values. Decisions on the presence or absence of equivalence should be empirically based. The need for such validation and verification should not be interpreted as blind empiricism and the impossibility of implementing preventive measures in a study to minimize bias and maximize equivalence. On the contrary, not all instruments are equally susceptible to bias. For example, structured test administrations are less prone to bias influences. Analogously, comparisons of closely related groups will be less susceptible to bias than comparisons of groups with a widely different cultural background.

10.2.1 Sources of Bias: Construct, Method, and Item

In order to detect and/or prevent bias, we need to recognize what can lead to bias. Table 10.1 provides an overview of sources of bias, based on a classification by Van de Vijver and Tanzer (1997; cf. Van de Vijver and Poortinga 1997). Sources of bias are numerous, thus the overview is necessarily tentative.

Construct bias occurs when the construct measured is not identical across groups. Construct bias precludes the cross-cultural measurement of a construct with the same/identical measure. Embretson (1983) coined the term *construct underrepresentation* to describe the situation where an instrument insufficiently represents all the domains and dimensions relevant for a given construct in a given culture. There is an important difference between our term *construct bias* and Embretson's term. Whereas construct underrepresentation is a problem of instruments measuring broad concepts with too few indicators which can usually be overcome by adding items relating to these domains/dimensions, construct bias can only be overcome by adding items relating to new domains/dimensions. Identification of construct bias calls for detailed culture-specific knowledge.

Cross-cultural differences in the concept of depression are one example. Another empirical example can be found in Ho's (1996) work on filial piety (defined as a psychological characteristic associated with being "a good son or daughter"). The Chinese conception, according to which children are expected to assume the role of caretaker of elderly parents, is broader than the Western. An inventory of filial piety based on the Chinese conceptualization covers aspects unrelated to the concept among Western subjects, whereas a Western-based inventory will leave important Chinese aspects uncovered.

TABLE 10.1. Sources of Bias in Cross-Cultural Assessment

Type of Bias	Source of Bias
Construct bias	Only partial overlap in the definitions of the construct across cultures Differential appropriateness of the behaviors associated with the construct (e.g., skills do not belong to the repertoire of one of the cultural groups) Poor sampling of all relevant behaviors (e.g., short instruments) Incomplete coverage of all relevant aspects/facets of the construct (e.g., not all relevant domains are sampled)
Method bias	(a) Incomparability of samples (e.g., caused by differences in education, motivation) (b) Differences in environmental administration conditions, physical (e.g., recording devices) or social (e.g., class size) (b) Ambiguous instructions for respondents and/or guidelines for administrators (b) Differential expertise of administrators (b) Tester/interviewer/observer effects (e.g., halo effects) (b) Communication problems between respondent and interviewer (in the widest sense) (c) Differential familiarity with stimulus material (c) Differential familiarity with response procedures (c) Differential response styles (e.g., social desirability, extremity scoring, acquiescence)
Item bias	Poor translation and/or ambiguous items Nuisance factors (e.g., item may invoke additional traits or abilities) Cultural specifics (e.g., incidental differences in connotative meaning and/or appropriateness of the item content)

NOTE: (a) sample bias (b) administration bias (c) instrument bias.

Yang and Bond (1990) presented indigenous Chinese descriptors and a set of American descriptors to a group of Taiwanese subjects. Factor analyses showed differences in the Chinese and American factor structures. Similarly, Cheung et al. (1996) found that the Western-based five-factor model of personality (McCrae and Costa 1997) does not cover all the aspects deemed relevant by the Chinese to describe personality. In addition to the Western factors of *extraversion*, *agreeableness*, *conscientiousness*, *neuroticism* (emotional stability), and *openness*, two further factors were found relevant for the Chinese context: *face* and *harmony*.

Construct bias can also be caused by differential appropriateness of the behaviors associated with the construct in the different cultures. An example of this comes from research on intelligence. Western intelligence tests tend to focus on reasoning and logical thinking (e.g., Raven's Progressive Matrices), while omnibus tests also contain subtests that tap into acquired knowledge (e.g., vocabulary scales for the Wechsler scales). When Western respondents are asked which characteristics they associate with an intelligent person, skilled reasoning and extensive knowledge are frequently mentioned as well as social aspects of intelligence. These social aspects are even more prominent in everyday conceptions of intelligence in non-Western groups. Kokwet mothers (Kenya) expect that intelligent children know their place in the family and the fitting behaviors for children, such as proper forms of address. An intelligent child is obedient and does not create problems (Segall et al. 1990).

Method bias – An important type of bias, called *method bias*, can result from such factors as sample incomparability, instrument differences, tester and interviewer effects, and the mode of administration. Method bias is used here as a label for all sources of bias emanating from factors often described in the methods section of empirical papers or study documentations. They range from differential stimulus familiarity in mental testing to differential social desirability in personality and survey research. Identification of methods bias requires detailed and explicit documentation of all the procedural steps in a study.

Examples – Among the various types of method bias, sample bias is more likely to jeopardize cross-cultural comparisons when the cultures examined differ in more respects. Such a larger cultural distance will often increase the number of alternative explanations for cross-cultural differences to be considered. Recurrent rival explanations are cross-cultural differences in social desirability and stimulus familiarity (testwiseness). The main problem with both social desirability and testwiseness is their relationship with country affluence; more affluent countries tend to show lower scores on social desirability (see Chapter 13). Subject recruitment procedures are another source of sample bias in cognitive tests. For instance, the motivation to display one's attitudes or abilities may depend on the amount of previous exposure to psychological tests, the freedom to participate or not, and other sources that may show cross-cultural variation.

Administration method bias can be caused by differences in the procedures or mode used to administer an instrument. For example, when interviews are held in respondents homes, physical conditions (e.g., ambient noise, presence of others) are difficult to control. Respondents are more prepared to answer sensitive questions in self-completion contexts than in the shared discourse of an interview. Examples of social environmental conditions are individual (versus group) administration, the physical space between respondents (in group testing), or class size (in educational settings). Other sources of administration that can lead to method bias are ambiguity in the questionnaire instructions and/or guidelines or a differential application of these instructions (e.g., which answers to open questions are considered to be ambiguous and require follow-up questions). The effect of test administrator or interviewer presence on measurement outcomes has been empirically studied; regrettably, various studies apply inadequate designs and do not cross the cultures of testers and testees. In cognitive testing, the presence of the tester is usually not very obtrusive (Jensen 1980). In survey research there is more evidence for interviewer effects (Groves 1989; Singer and Presser 1989; Fowler 1991; Hox, de Leeuw, and Kreft 1991; de Leeuw and van der Zouwen 1988; Lyberg et al. 1997). Deference to the interviewer has been reported; subjects were more likely to display positive attitudes to a particular cultural group when they are interviewed by someone from that group (e.g., Cotter, Cohen, and Coulter 1982; Reese et al. 1986; Aquilino 1994). A final source of administration bias is constituted by communication problems between the respondent and the tester/interviewer. For example, interventions by interpreters may influence the measurement outcome. Communication problems are not restricted to working with translators. Language problems may be a potent source of bias when, as is not uncommon in cross-cultural studies, an interview or test is administered in the second or third language of interviewers or respondents. Illustrations for such miscommunications between native and nonnative speakers can be found in Gass and Varonis (1991).

Instrument bias is a common source of bias in cognitive tests. An interesting example comes from Piswanger's (1975) application of the Viennese Matrices Test (Formann and Piswanger 1979). A Raven-like figural inductive reasoning test was administered to high-school students in Austria, Nigeria and Togo (educated in Arabic). The most striking findings were cross-cultural differences in item difficulties related to identifying and applying rules in a horizontal direction (i.e., left to right). This was interpreted as bias in terms of the different directions in writing Latin as opposed to Arabic.

Item bias – The third type of bias distinguished here refers to anomalies at item level and is called *item bias* or *differential item functioning*. According to a definition that is widely used in education and psychology, an item is biased if respondents with the same standing on the underlying construct (e.g., they are equally *intelligent*), but who come from different cultures, do not have the same mean score on the item. The score on the construct is usually derived from the total test score. Of all bias types, item bias has been the most extensively studied; various psychometric techniques are available to identify item bias (e.g., Holland and Wainer 1993; Camilli and Shepard 1994; Van de Vijver and Leung 1997b).

Examples – Although item bias can arise in various ways, poor item translation, ambiguities in the original item, low familiarity/appropriateness of the item content in certain cultures, and the influence of cultural specifics such as nuisance factors or connotations associated with the item wording are the most common sources. For instance, if a geography test administered to pupils in Poland and Japan contains the item "What is the capital of Poland?," Polish pupils can be expected to show higher scores on the item than Japanese students, even if pupils with the same total test score were compared. The item is biased because it favors one cultural group across all test score levels. Even translations which are 'correct' can produce problems. A good example is the test item "Where is a bird with swimming feet most likely to live?" which was part of a large international study of educational achievement (cf. Hambleton 1994). Compared to the overall pattern, the item turned out to be unexpectedly easy in Sweden. An inspection of the translation revealed why: the Swedish translation of the English was "bird with swimming feet" which gives a strong clue to the solution not present in the English original.

10.2.2 How to Deal with Bias

The previous section contains real and fictitious examples of bias. It is important to note that bias can affect all stages of a project. Minimizing bias is thus not an exclusive concern of question developers, test or questionnaire administrators, or data analysts. Since bias can challenge all stages of a project, ensuring quality is a matter of combining good theory, questionnaire design, administration, and analysis. The present section presents various ways in which the types of bias discussed above can be dealt with. Table 10.2 summarizes strategies available to identify and deal with bias.

TABLE 10.2. Strategies for Identifying and Dealing with Bias

Type of Bias	Strategies
Construct bias	Decentering (i.e., simultaneously developing the same instrument in several cultures)
	Convergence approach (i.e., independent within-culture development of instruments and subsequent cross-cultural administration of all instruments)
Construct bias and/or method bias	Use of informants with expertise in local culture and language
	Use samples of bilingual subjects
	Use of local pilots (e.g., content analyses of free-response questions)
	Nonstandard instrument administration (e.g., "thinking aloud")
	Cross-cultural comparison of nomological networks (e.g., convergent/discriminant validity studies, monotrait-multimethod studies
	Connotation of key phrases (e.g., examination of similarity of meaning of frequently employed terms such as "somewhat agree")
Method bias	Extensive training of interviewers
	Detailed manual/protocol for administration, scoring, and interpretation
	Detailed instructions (e.g., with sufficient number of examples and/or exercises)
	Use of subject and context variables (e.g., educational background)
	Use of collateral information (e.g., test-taking behavior or test attitudes)
	Assessment of response styles
	Use of test-retest, training and/or intervention studies
Item bias	Judgmental methods of item bias detection (e.g., linguistic and psychological analysis)
	Psychometric methods of item bias detection (e.g., Differential Item Functioning analysis)

A taxonomy of the main approaches to deal with bias is presented in Table 10.2 (cf. Van de Vijver and Tanzer 1997). It would be pointless to attempt to be exhaustive in such a taxonomy; therefore, the goal is more modest. An attempt is made to provide an overview of solutions that have been presented in the past and to suggest directions in which a possible solution may be found in the event that the table does not provide a ready-made answer.

It should be emphasized that the focus of this chapter is on comparative studies. Within this context, culture-specifics constitute a potential challenge to be overcome. This focus on similarity is sometimes seen as a focus on universal aspects and the denial of culture-specifics. We do not concur with this view as some of the most interesting cross-cultural differences may reside in the cultural specifics. However, from a methodological vantage point, cultural specifics need to be handled with care as, by definition, they are difficult or even impossible to compare across cultures. So, the focus on bias in comparative research is not meant to eliminate culture-specifics but to tell these apart from more universal aspects and to ascertain which aspects are universal and which are culture specific.

The first example of dealing with *construct bias* is cultural decentering (Werner and Campbell 1970). A modified example can be found in the study of Tanzer, Gittler, and Ellis (1995). Starting with a set of German intelligence/aptitude tests, they developed an English version of the test battery. Based on the results of pilot tests in Austria and the United States, both the German and English instructions and stimuli were modified before the main study was carried out. In the so-called convergence approach estimates are independently developed in different cultures and all instruments are then administered to subjects in all these cultures (Campbell 1986).

A second set of remedies aims at a combination of construct and method bias. Another example is a large acculturation project, called ICSEY (International Comparative Study of Ethnic Youth). The project studies both migrant and host adolescents and their parents in 13 countries, including migrants from about 50 different ethnic groups. Prior to the data collection, researchers met to decide on which instruments would be used. Issues like adequacy of the instrument vis-à-vis construct coverage and translatability (e.g., absence of colloquialisms and metaphorical expressions) were already factored into the instrument design, thereby presumably avoiding various possible problems in later stages. Other measures taken include using informants with expertise in local culture and language, samples of bilingual individuals, local pilots (e.g., content analyses of free-response questions), nonstandard instrument administration (e.g., 'thinking aloud'), and a pretest study of the connotation of key phrases.

The cross-cultural comparison of nomological networks constitutes an interesting possibility to examine construct and/or method bias. An advantage of this infrequently employed method is its broad applicability. The method is based on a comparison of the correlations of an instrument that may have indicators that vary considerably across countries with various other instruments. The adequacy of the instrument in each country is supported if it shows a pattern of positive, zero, and negative correlations that is expected on theoretical grounds. For example, views towards waste management, when measured with different items across countries, may have positive correlations with concern for the environment and air pollution and a zero correlation with religiosity. Nomological networks may also be different across cultures; Tanzer and Sim (1991) found, for example, that good students in Singapore worry more about their performance during tests than do weak students, whereas the contrary was found in most other test anxiety research. For the other components of test anxiety (i.e., tension, low confidence, and cognitive interference), no cross-cultural differences were found. The authors attributed the inverted worry-achievement relationship to characteristics of the educational system, especially the "kiasu" (fear of losing out) syndrome, which is deeply entrenched in the Singaporean society, rather then to construct bias in the internal structure of test anxiety.

Various procedures have been developed that mainly address method bias. A first proposal involves the extensive training of administrators/interviewers. Such training and instructions are required in order to ensure that interviews are administered in the same way across cultural groups. If the cultures of the interviewer and the interviewee differ, as is common in studies involving multicultural groups, it is important to make the interviewers aware of the relevant cultural specifics such as taboo topics (e.g., Goodwin and Lee 1994).

A related approach amounts to the development of a detailed manual and administration protocol. The manual should ideally specify the test or interview administration and describe contingency plans on how to intervene in common interview problems (e.g., specifying when and how follow-up questions should be asked in open questions).

The measures discussed attempt to reduce or eliminate unwanted cross-cultural differences in administration conditions so as to maximize the comparability of scores obtained. Additional measures are needed to deal with cross-cultural differences that cannot be controlled by careful selection and wording of questions or response alternatives. Education is a good example. Studies involving widely different groups cannot avoid that the samples studied differ substantially in educational background, which in turn may give rise to cross-cultural differences in scores obtained. In some studies it may be possible to match groups from different groups on education by sampling subjects from specified educational backgrounds. However, this approach can have serious limitations; the samples obtained may not be representative for their countries. This problem is particularly salient when comparing countries with a population with large differences in average educational level. For example, if samples of Canadian and South African adults are chosen that are matched on education, it is likely that at least one of the samples is not representative for its population. Clearly, if one is interested in a country comparison after controlling for education, this poor representativeness does not create a problem. If the two samples are obtained using some random sampling scheme, educational differences are likely to emerge. The question may then arise to what extent the educational differences can be held responsible for observed differences. For example, to what extent could differences in attitudes toward euthanasia be explained by educational differences? If individual-level data on education is available, various statistical techniques, such as covariance and regression analysis (cf. Part IV), can be used as to determine to what extent the observed country differences can be explained by educational differences (Poortinga and Van de Vijver 1987). The use of such explanatory variables provides a valuable tool to examine the nature of cross-cultural score differences.

A perennial issue in survey research is the prevalence of response effects and styles, especially social desirability and acquiescence. Their role in cross-cultural research as a source of unwanted cross-cultural score differences should not be underestimated. For some of the response styles, questionnaires are available; for example, the Eysenck Personality Questionnaire (Eysenck and Eysenck 1975) has a social desirability subscale that has been applied in many countries. When response styles are suspected of differentially influencing responses as obtained in different cultural groups, the administration of a questionnaire to assess the response style can provide a valuable tool to interpret cross-cultural score differences.

There is empirical evidence indicating that countries differ in their usage of response scales. Hui and Triandis (1989) found that Hispanics tended to choose extremes on a five-point rating scale more often than white Americans, but that this difference disappeared when a ten-point scale was used. Similarly, Oakland, Gulek, and Glutting (1996) assessed test-taking behaviors among Turkish children, and their results, similar to those obtained with American children, showed that these behaviors are significantly correlated with the WISC-R IQ. Arvey et al. (1990), working with adults, found significant black-white differences on test taking attitudes; whites reported to be more motivated to exert effort and work hard while blacks scored higher on preparation.

Evidence on the presence of method bias can also be collected by applying test–retest, training, or intervention studies. Patterns of pretest–posttest change that are different across cultures point to the presence of method bias. Van de Vijver, Daal, and van Zonneveld (1986) administered short term training of inductive reasoning to primary school pupils from the Netherlands, Surinam, and Zambia. The Zambian subjects showed larger score increments, both in the experimental and in an untrained control group, than the other groups. The large increments in the Zambian group may be due to a lack of experience with the test procedure. As the increase was also observed in the control condition, the score increase is more likely to have been due to knowledge of the test and the testing situation than to increments in inductive reasoning

There are two kinds of procedures to assess item bias: judgmental procedures, either linguistic or psychological, and psychometric procedures. An example of a linguistic procedure can be found in Grill and Bartel (1977). They examined the Grammatic Closure subtest of the Illinois Test of Psycholinguistic Abilities for bias against speakers of nonstandard forms of English. In the first stage, potentially biased items were identified. Error responses of American black and white children indicated that more than half the errors on these items were accounted for by responses that are appropriate in nonstandard forms of English.

10.3 CONSTRUCT EQUIVALENCE, STRUCTURAL EQUIVALENCE, MEASUREMENT UNIT EQUIVALENCE, AND FULL SCORE EQUIVALENCE

Four different types of equivalence are proposed here (cf. Van de Vijver and Leung 1997a, b; for a discussion of many concepts of equivalence, see Johnson 1998a).

Construct inequivalence – This amounts to comparing *apples and oranges* without raising the level of comparison to that of *fruit* (e.g., the comparison of Chinese and Western *filial piety*, discussed above). If constructs are inequivalent, comparisons lack a shared attribute, which precludes any comparison.

Structural or functional equivalence – An instrument administered in different cultural groups shows structural equivalence if it measures the same construct in all these groups. Structural equivalence has been addressed for various cognitive tests (Jensen 1980), Eysenck's Personality Questionnaire (Barrett et al. 1998), and the so-called five-factor model of personality (McCrae and Costa 1997). Structural equivalence does not presuppose the use of identical instruments across cultures. A depression measure may be based on different indicators in different cultural groups and still show structural equivalence.

Measurement unit equivalence – The third type of equivalence is called measurement unit equivalence. Instruments show this if their measurement scales have the same units of measurement, but a different origin (such as the Celsius and Kelvin scales in temperature measurement). This type of equivalence assumes interval- or ratio-level scores (with the same measurement units in each culture). Measurement unit equivalence applies when the same instrument has been administered in different cultures and a source of bias with a fairly uniform influence on the items of an instrument affects test scores in the different cultural groups in a differential way; for example, social desirability and stimulus familiarity influence scores more in some cultures than in others. When the relative contribution of both bias sources cannot be estimated, the interpretation of group comparisons of mean scores remains ambiguous.

At first sight, it may seem unnecessary or even counterproductive to define a level of equivalence with the same measurement units, but different origins. After all, if we apply the same interval-level scale in different groups, scores may be either fully comparable or, as in the case of nonequivalence, fully incomparable. The need for the concept of measurement unit equivalence may become clear by looking at the impact of differential social desirability or stimulus familiarity on cross-cultural score differences in more detail. Differential social desirability will create an offset in the scale in one of the cross-cultural groups: a score of, say, 5 in group A may be comparable to a score of 9 in group B because of a higher social desirability in group B. Observed group differences in mean scores are then a mixture of valid cross-cultural differences and measurement artifacts. A correction would be required to make the scores comparable. It may be noted that the basic idea of score corrections needed to make scores fully comparable is also applied in covariance analysis, in which score comparisons are made after the disturbing role of concomitant factors (bias in the context of the present chapter) has been statistically controlled for.

Scalar or full score equivalence – Only in the case of scalar (or full score) equivalence can direct comparisons be made; this is the only type of equivalence that allows for the conclusion that average scores obtained in two cultures are different or equal. Scalar equivalence assumes the identical interval or ratio scales across cultural groups. It is often difficult to decide whether equivalence in a given case is scalar equivalence or measurement equivalence. For example, ethnic differences in intelligence test scores have been interpreted as due to valid differences (scalar equivalence) as well as reflecting measurement artifacts (measurement unit equivalence). Scalar equivalence assumes that the role of bias can be safely neglected. However, verification of scalar equivalence relies on inductive evidence. Thus it is easier to disprove scalar equivalence than to prove it (cf. Popper's falsification principle 1976). Measuring presumably relevant sources of bias (such as stimulus familiarity or social desirability) and showing that they cannot statistically explain observed cross-cultural differences in a multiple regression or covariance analysis is an example for falsification.

10.3.1 Equivalence Levels

Structural, measurement unit, and scalar equivalence are hierarchically ordered. The third presupposes the second, which presupposes the first. As a consequence, higher levels of equivalence are more difficult to establish. It is easier to verify that an instrument measures the same construct in different cultural groups (structural equivalence) than to identify numerical comparability across cultures (scalar equivalence). But one should bear in mind that higher levels of equivalence allow for more detailed comparisons of scores across cultures. Whereas only factor structures and nomological networks (Cronbach and Meehl 1955) can be compared in the case of structural equivalence, measurement unit and full score or scalar equivalence allow for more fine grained analyses of cross-cultural similarities and differences, such as comparisons of mean scores across cultures in *t*-tests and analyses of (co)variance.

Identification of bias and verification of equivalence are core methodological problems of cross-cultural survey research. Any comparison stands or falls on the solution of these two issues.

10.4   CONCLUSION

The typology of bias and equivalence presented in this chapter pays more attention to the process of designing and administering instruments than the taxonomy by measurement (error taxonomy) by Groves as discussed in the previous chapter. Yet the differences between the taxonomies should not be overrated. Both taxonomies attempt to model error in cross-national surveys and emphasize the systematic nature of the error. Both treat bias and error as a feature of a specific study. Bias is not an intrinsic property of an instrument but a property that emerges in the comparison of scores on a specific instrument across the nations of a study. Finally, both classification schemes emphasize that bias can arise in all stages of a project. The theoretical model that is used, the sampling procedure, the wording of items, the administration procedure may all lead to unwanted sources of score differences between nations. In the treatment of bias two kinds of approaches have been proposed. The first focuses on instrument and sample design; they can be called *a priori* techniques because they are implemented prior to the data collection. Useful tools have been described in this chapter (such as the standardized questionnaire administration). The second approach amounts to the application of statistical techniques for the identification, and in some cases, correction of bias. These procedures can be called *post hoc*, since they are applied after the data collection. The measurement error by Groves and the bias taxonomy described here emphasize the need to employ both *a priori* and *a posteriori* procedures to minimize bias. The quality of a cross-national survey is the net result of the success in dealing with bias at all stages of a project.

Chapter 11

NONRESPONSE IN CROSS-CULTURAL AND CROSS-NATIONAL SURVEYS

MICK P. COUPER
EDITH D. DE LEEUW

11.1 INTRODUCTION

Nonparticipation in surveys is a concern of survey researchers all over the world, and empirical data from an international study support this concern (de Leeuw and de Heer 2001). However, in cross-national or cross-cultural survey research, nonresponse issues have been largely ignored. As international and cross-cultural outcomes and comparisons now are becoming more important to underpin policies of international organizations (de Heer 1999b), it is of the utmost importance to pay attention to the influence of nonresponse bias in comparative research. Only if we know how data quality is affected by nonresponse in each country or culture can we assess and improve the comparability of international and cross-cultural data.

In this chapter we review the literature pertaining to nonresponse in the context of cross-cultural and cross-national studies and discuss the implications for comparative research and analyses.

In the next section, we review response rates from different surveys and countries. We focus on figures from comparative studies across countries and discuss variation between countries. This is followed by a methodological discussion of sources of nonresponse and the implications for nonresponse error. We discuss functional ways of reporting nonresponse, ending with some practical recommendations for handling nonresponse in cross-cultural and cross-national research.

There are two main types of nonresponse in surveys: unit nonresponse and item nonresponse. Unit nonresponse is the failure to obtain any information from an eligible sample unit (e.g., person, household, business). Major sources of unit nonresponse include the inability to make contact with the selected sample unit, refusal by the sample unit to be interviewed or to complete the questionnaire, and incapacity of the sample unit to complete the survey (e.g., because of physical or mental limitations). Item nonresponse or item missing data refers to the failure to obtain information for one or more questions in a survey, given that the other questions are completed. Questions may be inadvertently skipped, a respondent may not know the answer to a particular question, or a respondent may choose not to answer the item. Sometimes the information that a respondent has provided is not usable, or usable information is lost (de Leeuw 2001). At times, survey instruments with very high levels of item missing data may be reclassified as a unit nonresponse. This is often the case with a break-off during an interview or a partially completed self-administered questionnaire. This chapter is focused on unit nonresponse. For a discussion of item nonresponse, its causes, and its treatment, see Hox and de Leeuw (1999). For an introduction on how to treat item missing data, see Martin et al. (1996) or Brick and Kalton (1996).

Unit nonresponse is a feature of virtually all surveys. Nonresponse is influenced by, among other things, saliency of topic, length of interview, the survey organization, survey method, and survey design (see Groves and Couper 1998). In cross-cultural and cross-national surveys, all these factors may differ. For instance, in terms of topic saliency, in some countries a high performance on cross-cultural educational tests is seen as very important, reflecting the honor of the country. In others, these same tests are seen as a boring obligation. A less extreme example is differences in mode of data collection; in some countries the study may be done by means of a face-to-face interview, while telephone or mail surveys are used in other countries, producing different levels of nonresponse. Also, the climate for survey taking and the survey burden may vary from country to country (see Lyberg and Dean 1992).

Standardization of survey practices with regard to nonresponse has not been given the attention it deserves. Despite the pervasive nature of survey nonresponse and its potential threat to both the validity and reliability of survey results, in practice nonresponse is often dealt with as an afterthought or considered less of a problem than, say, question wording or sample design issues. The lack of attention paid to nonresponse error in cross-cultural research appears particularly striking. Most of the focus in this field is on measurement equivalence (e.g., Saris and Meurs 1990; de Leeuw and Collins 1997; Heath and Martin 1997; Johnson et al. 1997), whereas operational equivalence (sampling, coverage, nonresponse, data collection procedures, etc.) is rarely mentioned. For example, in the recent edited volume on cross-cultural survey equivalence (Harkness 1998), operational issues get only a brief mention, and the potential impact of nonresponse on cross-cultural measurement essentially is ignored. Similarly, Kamano's (1999) paper on attitudes in seven countries contains an appendix describing the sample design and data collection procedures in each country, but no mention is made of the achieved response rate. Several papers on the International Quality of Life Assessment (IQOLA) project focus on measurement equivalence (e.g., Ware and Gandek 1998), with only the occasional reference to data collection issues such as response rates. For example, Gandek et al. (1998) mention return rates for 11 countries ranging from 61% to 83%. More broadly, Jowell (1998, 168) comments, "The strict standards we apply to the evaluation of national surveys are too often suspended when it comes to cross-national studies."

Part of the reason for the lack of attention paid to nonresponse in cross-cultural research may relate to the difficulty of measuring nonresponse and nonresponse error. One of the most widely used single indicators of survey quality is the response rate, which is often given a weight far beyond its actual value as an indicator of quality. There are many different ways to calculate a response rate. Reporting a single percentage without a clear description of definitions and formulae used makes it extremely difficult to compare data from different countries. In cross-cultural and cross-national research, it is crucial to state clearly how the response rate was calculated and which components of nonresponse were used in this calculation (see Section 11.6). However, this information is often hard to find in cross-national survey enterprises (see Section 11.2).

Even when the definition of response rate is known and well described, the response rate only serves as a potential warning of deeper underlying problems in the data. Typically, efforts are expended during data collection to maximize a study's response rate. However, this often is done without consideration for the impact on nonresponse error. Simply increasing the response rate for a study does not necessarily reduce the possibility of nonresponse error. A prime example is substitution. In some countries, substitution of non-contacted sample units or refusals by available and cooperative sample units is standard fieldwork practice. In other countries, substitution is not permitted because of the potential risk of introducing bias (de Heer 1999b). For example, persons or households that cannot be contacted may differ on important characteristics from those easily contacted (e.g., more mobile persons are more difficult to contact but are of extreme importance in mobility surveys), and persons who refuse to participate may do so because they have strong feelings on the survey's topic.

11.2 RESPONSE AND NONRESPONSE ACROSS SURVEYS AND COUNTRIES

Thus far, we have discussed nonresponse in the abstract. What is the situation for major comparative research projects? Is there reason for concern about low overall response rates or about differential nonresponse across countries? For official statistical studies, de Leeuw and de Heer (2001) were able to show that response rates have been declining internationally for a large variety of surveys. Furthermore, they showed that there are differences between countries in the severity of the problem. Not only do countries differ in their response rates; they also differ in the rate in which their response rates are declining.

De Heer (1999a) compared response rates of labor force surveys, typically conducted by government statistical agencies. These show wide variation across countries. Furthermore, he reports that important components of nonresponse differ cross-nationally as well. In some countries, noncontact rates are the most important contributor to overall nonresponse (e.g., Belgium, Denmark), whereas in others, refusal rates have the largest impact (e.g., the Netherlands, the United Kingdom). This is clearly illustrated in Table 11.1. These data are all concerned with official, governmental studies. How about social and educational studies?

TABLE 11.1. Response Rates, Refusal Rates, and Noncontact Rates for 1995 Labor Force Statistics (percentages)

Mode and Country	Response Rate	Refusal Rate	Noncontact Rate
Face-to-Face:			
Australia	96	1.0	3.0
Belgium	84	1.7	14.5
Canada	94	2.0	4.0
France	91	3.1	5.7
Germany	97	—	—
Hungary	83	5.2	3.8
Ireland	91	—	—
The Netherlands	60	26.0	9.0
Poland	90	3.1	6.2
Slovenia	87	7.5	1.1
Spain	90	3.1	7.5
United Kingdom	83	13.0	5.0
United States	93	3.9	3.0
Telephone:			
Denmark	74	8.5	15.2
Finland	92	3.1	4.5
Sweden	87	5.0	7.7

SOURCE: Based on de Heer 1999a.

Table 11.2 contains response figures from the 1995 round of the International Social Survey Programme (ISSP). Twenty-three countries participated in the 1995 survey, five of which supplied no response rate information. Three additional countries used quota sampling at the last stage of selection and are excluded from the table.

Three different modes of data collection were used for the remaining countries, and these are likely to be responsible in part for differences in response rates between participating countries. For instance, in a meta-analysis Hox and de Leeuw (1994) found an average difference in response between face-to-face and mail surveys of 10% (see also Goyder 1987). For those countries that conducted the ISSP as a personal visit (face-to-face) survey, response rates (ineligibles excluded from the denominator) ranged from a low of 67% for the Netherlands to a high of 87% for Bulgaria. Even for countries with roughly comparable response rates (Japan and Russia), the source of nonresponse (refusals versus noncontacts) differed greatly. For mail surveys, only completion rates are reported; response rates, refusal rates, and noncontact rates are not reported, as it often is hard to distinguish different reasons for nonresponse and to distinguish between nonrespondents and ineligible sample units.

TABLE 11.2. Completion Rate and Response Rate Information for 1995 ISSP Countries (percentages)

Mode and Country	Completion Rate[a]	Response Rate[b]	Refusal Rate	Noncontact Rate
Face-to-Face:				
Austria	65.0	68.7	13.1	17.0
Bulgaria	83.6	86.6	4.1	0.6
Japan	69.8	74.7	7.6	15.4
Latvia	54.9	79.3	19.3	0.0
The Netherlands	50.7	66.9	26.0	0.0
Russia	56.0	67.1	30.2	4.0
Slovenia	75.5	80.6	9.9	6.6
Drop-off:				
Canada	51.9	74.3		
Great Britain	52.9	61.2		
Poland	79.9	81.2		
United States	63.7	76.1		
Mail:				
Germany	52.9			
New Zealand	62.0			
Norway	66.4			
Sweden	64.8			

SOURCE: Zentralarchiv für Empirische Sozialforschung 1995:
http://www.gesis.org/en/social monitoring/issp/study_monitor.htm.
[a] Completions/total sample issued (ISSP codebook).
[b] Completions/eligible sample. Estimated from information provided in ISSP codebook.

In mail surveys, sometimes information is received on moved respondents, but this depends on the specific mail system in the country; also, sometimes an explicit refusal is sent back. However, most often no questionnaire is returned, and the reason for nonresponse remains unknown.

While the ISSP is a relatively short survey module (60 substantive and approximately 20 demographic items) measuring a range of attitudes (the 1995 survey was on "National Identity"), the International Adult Literacy Survey (IALS) revolves around tests of reading ability and comprehension (NCES 1998). Response rates from several countries participating in the IALS are presented in Table 11.3. In all instances, face-to-face interviews were conducted. While the range of countries participating in the IALS is much narrower than the ISSP (Western Europe and North America), the range of response rates is quite varied. For example, a response rate of 45% was achieved for the Netherlands, raising concerns not only about

TABLE 11.3. Response Rates for IALS

Country	Age Range	Number of Respondents	Response Rate (%)
Canada	16+	5,660	69
France	16–65	2,996	55
Germany	16–65	2,062	69
The Netherlands	16–74	3,090	45
Poland	16–65	3,000	75
Sweden	16+	3,038	60
Switzerland	16+	3,000	55
United States	16–65	3,053	60

SOURCE: National Center for Education Statistics 1998, Table 3.2.

inference to the Dutch population but also about comparisons of literacy levels between the Netherlands and other countries. The IALS report is devoid of additional details on nonresponse, raising concerns about the comparability of the data.

In the European context, the survey quality of the IALS was studied by the Eurolit Expertgroup. Part of this study was an inventory of survey practices in eight European countries by de Heer (1999b). Representatives of survey organizations (both in the government or public sector and the private sector) in several countries were interviewed on the survey procedures used (e.g., sampling procedures, contact strategies, fieldwork practices, data processing procedures, weighting procedures). Again, we see variation in the average reported response rates between the countries; this variation is especially large in the public (i.e., government) survey organizations. Table 11.4 presents average response rates for both the public and the private sector in several European countries. It should be noted that these figures are not calculated averages but informed estimates provided by key contacts in each research organization.

Response and nonresponse vary across countries and surveys in Europe. In most cases, public organizations, the majority being official statistical offices, have higher response rates than private organizations. This can be explained partly by the mandatory character of some official surveys and partly by the fact that people are more willing to participate in official government surveys than in commercial surveys. In most countries, it is reported that the survey-taking climate is worsening. Response rates are either decreasing or more effort is needed to keep the response at a stable level. The worsening climate for surveys is blamed to some extent on the increasing survey burden (i.e., more and longer surveys) and the increased activities of telemarketing organizations, as well as changing labor market conditions, which affect the quality of the interviewer labor pool, and other macro-level changes (e.g., increasing levels of distrust in public institutions).

TABLE 11.4. Response Indications[a] for Public and Private Survey Organizations in Eight European Countries (M = mandatory participation, V = voluntary participation)

Country	Average Response Public Sector (%)			Average Response Private Sector (%)		Reported Survey Taking Climate
France	90	M		60	V	Moderate
Germany	97	M	60 V	55–60	V	Worsening
Greece	90	M		—		Moderate
Italy	90	M		60	V	Worsening
The Netherlands	60	V		50–60	V	Not positive
Portugal	80–90 M			—		—
Sweden	80–90 V			60–70	V	Fairly good, but...
United Kingdom	70–85 V			60–70	V	Worsening

SOURCE: From de Heer 1999b, Table 3.2.
[a] These figures are informed estimates provided by key contacts in research organizations.

De Heer (1999b) concludes that a very high response rate of at least 80% does not seem to be realistic for the European component of an International Literary Survey. He notes, however, that "decent" response rates in Europe are still possible, provided that special attention is paid in each country to the response problem and that current best methods for nonresponse reduction are implemented and monitored.

Table 11.5 presents data for a survey of a very different type. The Third International Mathematics and Science Study (TIMMS) is a school-based study where pupils in comparable grades are tested using standardized measures of science and mathematics knowledge (Foy, Martin, and Kelly 1996). In such a design, nonresponse can arise at the school level and at the student level within schools. In addition, school districts may choose not to participate, and lack of cooperation by teachers may exclude entire classes of students. In the TIMMS design, nonparticipating schools were replaced by schools of similar character, a form of substitution. Table 11.5 contains the participation rates before substitution (replacement). The overall participation rate is the product of the school rate and the student rate.

Some of these rates are improbably high—for example, Thailand at 99% and Iran at 98%—suggesting coercion (mandatory surveys), substitution or other non-probability based designs, or unusually cooperative societies. On the other hand, the Netherlands again has the lowest overall response rate (23%), with the next lowest being Austria at 39%. This wide variation in response rates across countries suggests major cross-national differences in survey data collection procedures and/or people's and institutions' reactions to survey requests. Unfortunately, details on fieldwork procedures in the countries and on the calculation of the final response rates are not given. Either way, the validity of comparative analyses based on these data may be brought into serious question because of the potential for differential nonresponse bias (see Section 11.3.1).

TABLE 11.5. Participation Rates for TIMMS (weighted percentages)

Country	School Participation Before Replacement	Within-School Student Participation	Overall Participation Before Replacement
Australia	75	92	69
Austria	41	95	39
Belgium (Fl)	61	97	59
Belgium (Fr)	57	91	52
Bulgaria	72	86	62
Canada	90	93	84
Columbia	91	94	85
Cyprus	100	97	97
Czech Republic	96	92	89
Denmark	93	93	86
England	56	91	51
France	86	95	82
Germany	72	87	63
Greece	87	97	84
Hong Kong	82	98	81
Hungary	100	87	87
Iceland	98	90	88
Iran	100	98	98
Ireland	84	91	76
Israel	45	98	44
Japan	92	95	87
Korea	100	95	95
Kuwait	100	83	83
Latvia	83	90	75
Lithuania	96	87	83
The Netherlands	24	95	23
New Zealand	91	94	86
Norway	91	96	87
Philippines	96	91	87
Portugal	95	97	92
Romania	94	96	89
Russian Federation	97	95	93
Scotland	79	88	69
Singapore	100	95	95
Slovak Republic	91	95	86
Slovenia	81	95	77
South Africa	60	97	58
Spain	96	95	91
Sweden	97	93	90
Switzerland	93	98	92
Thailand	99	100	99
United States	77	92	71

SOURCE: From Foy, Martin, and Kelly 1996, Tables 2.7, 2.8, and 2.11.
NOTE: Percentages based on upper grades only.

In summary, we have shown that there is wide variation in response rates across countries and surveys. There is a declining trend in response rates, but this too varies across countries and surveys, indicating differences in survey design and effort as well as societal differences. It should be noted that the quality of the response information provided varied; often few details beyond basic response rate information were provided in the reports. This lack of documentation makes it difficult for users of the data to make judgments about the quality and comparability of these cross-national surveys.

Understanding the cause of the differences in response rates across countries and the changes over time requires an understanding of societal shifts in the value of surveys and is itself an interesting cross-national research question. It also requires a profound understanding of the influence of design and fieldwork procedures on nonresponse reduction. More importantly, although these wide differences in response rates are a cause for concern in itself, it is plausible that substantive measures taken from cross-national surveys may be susceptible to nonresponse biasing effects.

11.3 EFFECTS OF NONRESPONSE

11.3.1 Nonresponse and Bias

How may the differences in response between countries discussed above threaten the validity of comparative research? How can differential nonresponse bias substantive measures? To answer this question, we first have to further define nonresponse error. Nonresponse error is a function of the nonresponse rate and the differences between respondents and nonrespondents. If nonresponse is the result of pure chance—in other words, if nonresponse is completely random—there is no real problem. Of course, the realized sample is somewhat smaller, resulting in slightly larger confidence intervals around estimators. However, the conclusions will not be biased due to nonresponse. Only when respondents and nonrespondents differ from each other on the variable(s) of interest is there a serious problem. In this case, nonresponse is selective, and certain groups may be underrepresented. In the worst-case scenario, there is a substantial association between nonresponse and an important variable of the study, causing biased results.

For example, people who travel a lot are difficult to contact. For a survey on mobility, this may lead to an underestimate of mobility. People living in large cities refuse more often than people living in rural areas, and in light of the higher crime rate in large cities compared to that of rural areas, these refusals may threaten the validity of victimization studies. Ethnic and cultural "minority" groups are often underrepresented in surveys, whether because of distrust of public institutions and information-gathering activities, different norms and experiences regarding survey participation, language difficulties, or some other reason. In the Netherlands, for example, one of the larger ethnic groups consists of descendants from Islamic North

African immigrants. Underrepresentation of this group may threaten the validity of studies of values and norms—for instance, conclusions about Sunday observances or commercial activities on religious holidays.

In its simplest form, nonresponse error has two components; the nonresponse error for a particular statistic (say, $\bar{y}$) is a product of the nonresponse rate and the difference between respondents and nonrespondents. We can represent this as follows:

$$\bar{y}_r - \bar{y}_n = \left(\frac{m}{n} \right) [\bar{y}_r - \bar{y}_m], \qquad (11.1)$$

where $\bar{y}_n$ is the full sample mean, $\bar{y}_r$ is the respondent mean, $\bar{y}_m$ is the mean for the nonrespondent (missing) cases, and m/n is the nonresponse rate.

Theoretically, then, the biasing influence of nonresponse is eliminated under two conditions: either (a) the nonresponse rate is zero (there are no nonrespondents) or (b) there are no differences between respondents and nonrespondents on the statistic of interest. Given that the values of the statistic of interest of the nonrespondents are not measurable (this is, after all, why a survey is being conducted), most survey researchers rely on minimizing m/n (the nonresponse rate) to reduce the possibility of nonresponse error. While high response rates constrain the likely size of the nonresponse error, a high response rate does not guarantee there will be no error. Even a response rate of 80% may threaten the conclusions if the nonrespondents are very different from the respondents. For instance, if the response rate was 80% and the survey findings showed 50% of the respondents were in favor of becoming a member of the European community, the true value could range from 40% in favor of membership of the EU (if all nonrespondents were opposed) to 60% in favor (if all nonrespondents were in favor of membership).

Nonresponse error is relevant not only for simple descriptive statistics such as means and proportions. Nonresponse can bias a variety of other estimates, including relationships among variables (as in regression or structural equation models) and comparative statistics (e.g., subgroup means). A staple of much cross-cultural research is analysis of differences between two or more subclass means (or proportions). For example, studies of changes in values between generations may be susceptible to nonresponse bias. Young singles typically are more difficult to reach than young couples, which may result in an underrepresentation of the singles in the youngest generation, biasing the comparison across generations. In studies of the elderly, the "oldest old" (70+) refuse more often than do the younger elderly. Since this is partly due to health reasons, comparisons on health-related questions may give too positive a picture of the health and needs of the old. Similar examples could be constructed for many other survey topics.

In comparing two subclasses, each subclass mean based on respondent cases only is subject to an error that is a function of the nonresponse rate for that subclass and the difference between respondents and nonrespondents within the subclass. An implicit assumption of much cross-cultural research is that the nonresponse error in the two subclass means is equal. While it is acknowledged that nonresponse may limit generalizability to each population, the assumption is that both the nonresponse rate *and* the difference between respondents and nonrespondents in each subclass are the same, thereby permitting valid comparisons across groups. The examples above make clear that this assumption is not always realistic. To the extent that response rates differ across countries or across subgroups within a country and/or the nonresponse mechanism differs (the difference between respondents and nonrespondents is different in the two subgroups), comparative analyses are not immune to the threat of bias due to nonresponse.

11.3.2 Other Effects of Nonresponse in Cross-Cultural and Cross-National Comparisons

In addition to affecting bias, nonresponse potentially can affect the variance of survey estimates. Even if the differences between respondents and nonrespondents can be measured and are small, nonresponse may threaten inference by effectively reducing the number of completed cases available for analysis, thereby increasing confidence intervals and standard errors relative to a survey of the same initial sample size with no nonresponse. Increasing the initial sample size to compensate for this will increase survey costs accordingly. Nonresponse also may affect variance in another way: in compensating for nonresponse through post-survey adjustment, the application of differential weighting factors tends to inflate the variances of the estimates (e.g., Lee, Forthofer, and Lorimer 1986).

More serious for potential bias is when the assumptions of probability sampling are no longer met. All of the above discussion assumes probability sampling. In cross-national surveys where one or more countries employ quota sampling, the problem of inference is much more complex. Also, substitution practices for nonresponse vary widely across countries. In academic and government surveys in North America and much of Europe, substitution is viewed as an undesirable practice. Take again the example of mobility surveys: difficult-to-reach persons often are more mobile than persons who are easy to reach (e.g., at home more). Simply substituting a "noncontact" may very well mean substituting a less mobile person for one who is highly mobile. Nevertheless, in several European countries and elsewhere in the world, substitution remains a common practice for dealing with nonresponse in the field (Vehovar 1995, 1999). It appears that in at least two cross-national European surveys (the Labor Force Survey and Family Budget Survey), substitution may even be encouraged (Verma 1992, 1993; de Heer 1999b). In terms of demographic characteristics, substitution for those not at home is likely to increase the proportion of households with nonworking women in the sample, thereby threatening the validity of labor force and family budget surveys.

As Lessler and Kalsbeek (1992, 177) note, "Substitution reduces nonresponse bias to the extent that the substitutes are collectively identical in every respect to the sample members for whom the substitutions are made." Substitutes for non-contacts should be representative of those not at home. Substitutes for refusers should be representative of these refusers on important variables under study. However, in practice, this may not be the case. For example, substituting for refusal to participate in a political survey may overrepresent those who are actively engaged and interested in politics and hence more willing to participate in a survey about these topics. Thus, substitution may increase the number of completed interviews at the expense of increased nonresponse bias.

As with quota sampling, the fact that the substitute is selected from the same geographical cluster or has similar demographic characteristics to the person being substituted does *not* mean that they share the same attitudes, values, opinions, or whatever is the substantive focus of the research. The underlying assumption is one of homogeneity of neighborhoods. In Germany, there is still some clustering of working-class people in specific areas, as there is of ethnic minorities, but this is not the case with other groups. Thus, for instance, civil service employees (i.e., Beamte) who are assumed to have a different social status from workers (although there are many different classes of Beamte) do not cluster within communities.

This also applies to the use of postsurvey adjustment to compensate for nonresponse. Usually only a few key demographic variables are available for weighting, and producing a sample resembling the population on these characteristics does not guarantee that it will represent the full population on the substantive variables of interest. Again, this is only the case when the nonrespondents are "missing at random."

Jowell (1998, citing Park and Jowell 1997) notes that although all ISSP members respected the basic rule outlawing quota sampling, ". . . by no means all nations ended up with a recognizable probability sample at the end." In several countries, substitution was used for noncontacts and refusals, and in many cases, the process used to substitute for nonrespondents and the number of substituted cases is not reported. Jowell concludes, "The result is not only that the recorded response rates in these countries are improbably high, but that the availability biases in their sample compositions are also likely to be larger than those in other countries" (1998, 171).

In summary, the seriousness of biasing effects of nonresponse depends on the nonresponse rate and the difference between respondents and nonrespondents on key variables. If the difference between respondents and nonrespondents is large, the biasing effects may be large, even with a relatively high response rate. Differential nonresponse may affect the validity of subgroup comparisons, because the answers of differently composed groups are being compared. Finally, the same response rate in countries or in subgroups does not guarantee comparability. Again, the composition of the groups to be compared may differ because of differences in fieldwork procedure, such as allowing for substitution of noncontacts or refusers.

11.4   REASONS FOR NONRESPONSE AND ITS EFFECTS ON CROSS-CULTURAL AND CROSS-NATIONAL RESEARCH

In the examples given, we have differentiated between different sources of nonresponse, implying that nonresponse cannot be treated as a unitary phenomenon. In fact, evidence is mounting that these different sources of nonresponse may lead to different biases (see Groves and Couper 1998). In household surveys, the two main types of nonresponse are failure to contact the selected unit and refusal to participate. Other sources of nonresponse error may become more or less salient depending on the nature of the survey. For example, inability to participate because of physical or mental incapacity may be a particular problem in surveys of the elderly; failure to participate because of language difficulties may be a particular concern for analysis focusing on immigrant communities. In mail or other self-administered surveys, literacy may be an inhibiting factor in participation, even if sample persons are willing to participate (Couper, Singer, and Kulka 1998). Even the two main types of nonresponse may have different implications for nonresponse bias, depending on the nature of the survey. In travel or mobility surveys, the ability to find people at home may be directly related to their travel behavior, especially in surveys with relatively short field periods (e.g., one week). Similarly, refusal to participate in a survey on political issues may be directly related to sample persons' interest in, knowledge of, and participation in politics.

The differential nature of nonresponse (e.g., noncontact, refusal, inability to participate) has several implications for cross-cultural research. First, it is important to provide detailed information on the composition of the nonrespondent group—that is, to distinguish between noncontacts, refusals, and ineligibles. We return to this point in the next section. Second, response rates that appear similar across countries or cultures may in fact mask differences in the composition of the nonresponse. For example, if the nonresponse group in one society is made up almost entirely of noncontacts, while in another it is mostly refusals, the likelihood of nonresponse bias in estimates of differences between the two societies may be greater than if the source of nonresponse (or the nonresponse mechanism) is similar in both societies. Therefore, attempts to produce equivalent response rates by attacking noncontacts in one society while attempting to convert refusals in another may increase rather than reduce nonresponse error. Evidence of large differences in noncontacts across countries due mainly to differences in fieldwork procedures is well documented by de Heer (1999a) and de Leeuw and de Heer (2001). For some examples and illustrations, see Tables 11.1 and 11.2. This again argues for the need to understand the causes of nonresponse and how they may differ across survey designs, topics, and populations of interest.

Space does not permit a thorough discussion of the underlying causes and correlates of nonresponse. These are dealt with at length by Groves and Couper (1998) and others (e.g., Goyder 1987; Brehm 1993; Morton-Williams 1993; Hox, de Leeuw, and Vorst 1995). Comparative research further complicates the matter because the reasons for nonresponse may vary across countries, regions, and cultures. To our knowledge, there is a paucity of research comparing sources of nonresponse cross-nationally.

However, there is evidence that survey nonresponse is not uniform across countries. Lyberg and Dean (1992) first introduced the notion of a "survey-taking climate" that may differ between and within societies and may differentially affect response rates and the mix of respondents and nonrespondents. While such macro-level influences are hard to detect, it is important to understand how the climate for survey taking may differ and the effect this may have on response rates. Several case studies point to the effect of changing climate on response rates. For example, the rapid spread of telemarketing and the increasing warnings to consumers in the United States are likely to be partly responsible for the recent decline in telephone survey response rates. Two examples (in Sweden and Germany) of the effect of broad public debate about research intrusion and privacy issues on response rates of surveys unrelated to the debate have shown that the climate may change fairly rapidly within a society (see Groves and Couper 1998).

In the Netherlands, a different mechanism is probably at work. Privacy and voluntariness are no longer stressed and are taken for granted. The well-documented problem of high nonresponse in the Netherlands (de Heer 1999a; de Leeuw and de Heer, 2001) relative to other countries may well be an indicator of a different climate for survey-taking in that country, of a different attitude toward surveys. This can be illustrated by the following example: in the Netherlands, Hox, de Leeuw, and Vorst (1995) tried to replicate an American study on determinants of survey participation (Groves, Cialdini, and Couper 1992). Some questions from the "responsibility for surveys scale" (e.g., "I think it is my civic duty as a citizen to agree to participate when asked in most surveys that come from a government organization") provoked an amused reaction from respondents. In the mid-nineties, participation in official surveys definitely was not seen as a civic duty in the Netherlands, while in the United States, an appeal to one's duty as a citizen still appeared to have some power.

Kuechler's (1998) discussion of survey research in Germany before and after reunification serves as another example of changing survey research climates. Although the response rates in former East Germany are still higher than in former West Germany, there appears to be a decline. Kuechler attributes this to a change of survey climate in the East after reunification, from an initial eagerness to participate in surveys to more reluctance and a changing attitude toward the value of these surveys. Bronner (1988), who investigated response trends of immigrant groups, drew a similar conclusion. He found that recent immigrants to the Netherlands (mainly from North Africa) had a much higher response rate than immigrants who had been there for years (mainly from Southern Europe). With an increase of "minority" research and a longer stay in the Netherlands, the novel and flattering experience of being asked for one's opinion turns into a nuisance, as the recent decreasing response rate among North African immigrants appears to indicate (Bronner, personal communication). Thus, possible changes in response rates and reasons for participation for some groups who have experienced or are experiencing this sort of transition raises concerns about comparisons between these groups and others that have not experienced such a transition over time. The same concern exists for countries and regions in transition.

However, even within relatively stable countries, differences in response rates and response reasons can be seen over time. In the United States, Harris-Kojetin and Tucker (1999) find evidence that response rates to the Current Population Survey over a 28-year period varied systematically with changing economic and political conditions within the country. Similarly, the urban-rural difference in response rates is a well-documented phenomenon in several countries (e.g., Goyder 1987; Foster and Bushnell 1994; Bros et al. 1995; Groves and Couper 1998).

In summary, there are different sources or types of nonresponse. The contribution of each source to the total nonresponse varies across countries and over time. Within countries, there are variations between cultural/ethnic groups and between sociodemographic regions. Studies for which comparisons across countries and divisions (regions, cultural/ethnic groups, or time) are central may be especially vulnerable to differential effects of nonresponse. The danger is especially great for deeply divided societies where attitudes towards the central institutions of the society, which are often those conducting national surveys, may be highly divergent. Similarly, societies undergoing rapid political and/or economic transition may experience changes in the survey-taking climate that may threaten both trend analyses and comparative studies.

11.5   REDUCING NONRESPONSE

Not all of the factors that influence response are under the control of the survey designers or researchers. For instance, we know that saliency of topic influences response. Obviously, researchers cannot simply change the topic; they only can try to make the questionnaire or the interview as interesting and salient as possible within the limitations of the topic. We also know that mandatory surveys have a higher response than voluntary surveys, but again researchers cannot simply declare a survey mandatory. However, there are factors that affect the likelihood of response and that can be influenced by researchers through the survey design and fieldwork procedures. In this section, we concentrate on these controllable factors—that is, on fieldwork procedures and design.

A relatively simple and straightforward way to reduce nonresponse is reducing the number of *noncontacts* by increasing fieldwork effort. Increasing the number of contact attempts, varying the times at which contacts are attempted, and following a prescribed schedule will reduce the number of noncontacts (Purdon, Campanelli, and Sturgis 1999). The optimal time of day to reach a potential respondent at home may depend on the culture or country. For example, the early evening (around six o'clock) is a good time to contact the Dutch, who prepare and eat dinner early. In other countries (e.g., Spain, Italy, France) other times may be optimal. What is important is that the contact attempts are many and varied in time of day and day of week, depending on the local culture; a good survey design will take this into account. Another important point is that the timing of contact attempts is not left to the interviewer and that the timing and number of attempts are monitored or checked by supervisors. Regardless of timing, there are many barriers to contact—gatekeepers, security systems, telephone screening devices, and so on—that interfere with the process of making contact with a sampled household. Especially in comparative research across countries and cultures, it is important to pay attention to the fieldwork procedures and interviewer instructions for contacting and selecting respondents. For example, in some cultures the head of the household acts as gatekeeper and has to give permission before other household members may be interviewed. Researchers have to train and instruct interviewers how to cope effectively with gatekeepers and how to get permission to conduct an interview.

It is much harder to reduce nonresponse by reducing the number of refusals. Still, there are various strategies that have been found to be effective in doing this, and we briefly review some of the key findings.

Although researchers cannot change the topic and thus the saliency, they can try to make responding as attractive as possible. To do so, one has to reduce the response burden by making the questionnaire as short and as easy to answer as possible, thereby bringing the "costs" and "benefits" more in balance (Dillman 1978a, 2000; Groves, Cialdini, and Couper 1992). To reduce cognitive and emotional burden, it is necessary to pretest questionnaires and interview schedules using cognitive laboratory procedures (Forsyth and Lessler 1991) and to check whether terms and phrases used are easily understood and do not cause embarrassment. In addition, any fears or apprehensions of potential respondents have to be laid to rest. Reassurance that the organization is trustworthy and that the survey is real and not "selling in disguise" will reduce refusals (de Leeuw and van Leeuwen 1999). Also, one should avoid any real costs for the respondent, be it travel costs in special-site surveys (e.g., health tests), postage in mail surveys, or connect time in Internet surveys.

On the other side of the balance, one can increase the benefits by rewarding respondents. Rewards can be intangible, like the attention of an interviewer, the appreciation of a research organization, or the assurance that respondents have contributed to the general good of society; or they can be more tangible, like a summary of results (Dillman 1978a, 2000). Monetary incentives and nonmonetary gifts have proved to increase response in mail, telephone, and face-to-face interviews, with prepaid incentives being more effective than promised incentives (Singer et al. 1999; Singer 2001). What is important is that an "incentive" really acts as an incentive for the respondent. In other words, an incentive must be regarded as a benefit or reward. What is rewarding differs from culture to culture and from country to country. To decide which incentive to use, one should closely follow the cultural customs in a country. For instance, in the Netherlands, one usually gives a bouquet of flowers to the hostess when visiting a friend for dinner. Hence, flowers or a "flower token" work well as an incentive in the Netherlands. Promised summaries of survey results, on the other hand, have only an incentive value in literate, information-oriented countries.

Another design feature that may influence a potential respondent to reciprocate by responding is an advance letter. A well-written advance letter reduces refusals by legitimizing the survey and reassuring and motivating potential respondents. However, much depends on the content, style, and length of the letter, and advance letters can have a negative effect on response (for an overview, see Groves and Couper 1998). Most effective is a carefully drafted, simple, short letter (Lynn, Turner, and Smith 1997; Dillman 2000).

Less under the control of the survey organization is the interviewer-respondent interaction. A positive and friendly approach by the interviewer (Snijkers, Hox, and de Leeuw 1999) and a social, flexible, and tailored interviewer-respondent interaction (Morton-Williams 1993; Groves and Couper 1998) are useful tools in reducing refusals. Although actual interviewer behavior is not under the strict control of the researcher, interviewer training and monitoring are. Recent research (Groves and McGonagle 2000; de Leeuw and de Heer 2001) suggests that survey response can be increased by special interviewer training and monitoring. For mail surveys, the design of the questionnaire, the tone of the prenotification and cover letter, and the number of reminder contacts are crucial to obtaining a high response (for details, see Dillman 1978a, 2000). Dillman's *Total Design Method* has proved to be effective in such different cultures as North America, Canada, Japan, and in many parts of Europe (de Leeuw and Hox 1988; Hox and de Leeuw 1994).

Reducing nonresponse as such is not the most important issue in cross-cultural research. What is important is the reduction of nonresponse *bias*. If the response inducing measures raise the response rate *without* reducing the nonresponse bias, the effect is only "cosmetic." A clear example is substitution; in several countries, substitution is not permitted because of the potential risk of introducing bias. For example, as discussed earlier, persons or households that cannot be contacted may differ on important characteristics from those easily contacted. Thus, in these cases substitution of "noncontacted" by "easy to contact" respondents will not reduce the nonresponse error (see also Section 11.3.2).

To summarize, there are many factors that influence response (for a conceptual framework, see Groves and Couper 1998). Some of these factors (e.g., saliency of topic, characteristics of sample units) are not under the control of the researcher, but many others are. Some of these are centered on reducing noncontacts; others are focused on reducing refusals. It should be noted that each measure in itself only has a small effect; for a considerable increase of response, a whole battery of measures is needed. Such considerable efforts may well increase survey costs. However, in serious cross-cultural and cross-national research reduction of variation in fieldwork and survey design between countries is of the utmost importance for valid comparisons. For this reason, it is important that budget constraints do not severely limit the use of an array of response-inducing measures (de Heer 1999b).

11.6   CALCULATING AND REPORTING NONRESPONSE AND RESPONSE RATES FROM A COMPARATIVE PERSPECTIVE

Selective nonresponse may seriously threaten the validity of cross-cultural studies. Furthermore, as indicated, nonresponse and sources of nonresponse vary across countries, subgroups, and surveys. For valid cross-cultural and international comparisons, we must aim to avoid unnecessary variation. Above we discussed how optimization of fieldwork procedures and total quality management (TQM) would help. Still, there always will be some degree of nonresponse. For valid cross-cultural and cross-national comparisons, it is of the utmost importance that the various sources of nonresponse are reported. This serves two goals: first, it will guide a more thorough understanding of response differences between countries; second, it will provide an empirical basis for the necessary statistical adjustment.

Nonrespondent cases can be divided into noncontacted cases and refusals. There are likely to be different causes underlying these two major sources of nonresponse. Those who are never at home during the times the interviewer calls over the survey period (and who may well be willing to participate if they were given the opportunity) are likely to have different values on the variables of interest than those who were contacted but refused to participate. Those who were contacted but refused may have done this for a variety of reasons—for instance, in response to some factor external to the survey topic or directly because of the stated purpose of the survey, the sponsoring organization, the approach and demeanor of the interviewer, and so on. Groves and Couper (1998) and others have shown that the "noncontacted" and the "refusers" differ in terms of several demographic measures and hence should not be considered as a single group of nonrespondents but as two distinct groups.

Other types of nonresponse may be identified depending on the specific nature of the survey. For example, in surveys of the elderly, nonresponse due to incapacity or illness may be particularly important (Groves, Raghunathan, and Couper 1995). In surveys of immigrants or ethnic subgroups, the inability to participate for reasons of language may produce bias. In many surveys, however, a catchall "other nonresponse" category often is used instead of a more detailed system. In comparative research, we strongly advise *against* using broad categories such as "other nonresponse." For valid comparisons across countries and cultures, it is important not only to distinguish nonrespondent cases from ineligible cases but also to identify the reason for nonresponse.

However, for many surveys, the response rate is the only indicator of nonresponse error provided in reports. Moreover, the response rate often is calculated differently across surveys, or insufficient information is provided, making evaluation difficult. For comparative evaluations of the quality of cross-national surveys, it is especially important that the same response rate definitions be used in all countries and that the report clearly states the sources of nonresponse. For details of response rate calculation and sources of nonresponse, see Groves (1989). A more recent set of response rate standards was promulgated by the American Association for Public Opinion Research (AAPOR 2000). A regularly updated version can be found on the AAPOR Web site (http://www.aapor.org).

In its simplest form, a response rate is the ratio of completed interviews or questionnaires over eligible reporting units. But what are the eligible reporting units? The complexities of modern sample surveys (such as mixed-mode designs, longitudinal surveys, subsampling within units, and over- or undersampling of certain groups) allow for many variations of this basic form. In a typical cross-sectional survey, there are four main outcomes or dispositions:

(1)   Completed interviews/questionnaires;
(2)   Nonrespondents (eligible cases not interviewed due either to noncontact or overt refusal);
(3)   Cases of unknown eligibility (e.g., not attempted to interview, unsafe area, not worked because of short fieldwork period);
(4)   Ineligible cases (vacant dwelling, outside defined population).

Much of the variation in reporting of response rates arises from definitions of completed versus partial interviews and eligible versus ineligible reporting units. Explicit rules for determining interview completion status should be used. Some organizations define partial interviews in terms of proportion of the interview complete, while others use several key items in the survey as indicators of completion status.

Ineligibility may arise for any number of reasons. For example, in household surveys, vacant units and businesses or institutions (e.g., college dormitories, jails, hospitals) should be excluded. In telephone surveys using random-digit dialing (RDD), many telephone numbers are generated that are nonworking numbers (and this definition may vary across countries), dedicated fax or modem lines, business numbers, and so on. Sample cases known to be ineligible are removed legitimately from the denominator of the response rate. In mail surveys, it is very difficult to discover whether questionnaires are not returned because of refusal or ineligibility. A special follow-up (e.g., by telephone or certified mail) is often necessary to establish whether nonresponding cases were ineligible and can be removed from the denominator. As this often is not feasible, response rates in mail surveys often slightly underestimate the response compared to interview surveys; in interview surveys, the ineligibles are more easily identified and the response rate can be adjusted accordingly.

Response rates often are falsely inflated by removing all cases of unknown eligibility from the denominator or, in extreme cases, all non-contacted reporting units. For cases with unknown eligibility, there are several ways of estimating eligibility, varying by mode of data collection and sample design. The most conservative approach is to assume they are all eligible, producing a lower-bound response rate. In many political polls, it is a common (but incorrect) practice to assume all such cases are ineligible, essentially producing a cooperation rate (interview divided by contacted and eligible sample units) rather than a response rate. Whatever approach is used to estimate eligibility rates, it is extremely important to be explicit about the assumptions used and to provide the component counts so data users and readers can verify the rate or compute their own rates.

In summary, there are many sources of nonresponse that can threaten the validity of a survey. Each source contributes to the overall nonresponse and hence to the response rate. Therefore, in cross-cultural and cross-national research, it is imperative to state clearly the way the response rate was calculated. Further, researchers should report figures for each different source of nonresponse. Only then is it possible to make fair comparisons. Finally, within cross-national studies, there may be considerable variation in mode and procedures of data collection (e.g., use of substitution) that influences the comparability of overall response rates. Reporting these details is a prerequisite for quality in international studies.

11.7 CONCLUSION

The goal of this chapter is to raise awareness of a largely ignored issue in cross-national and cross-cultural survey research. Although attention has been paid to response and nonresponse in comparative survey research, there is growing evidence that countries differ not only in their response rates but also in the composition of the nonresponse due to differences in design and fieldwork procedures, among other things. Given the paucity of strict comparative cross-cultural and cross-national research, we are forced to integrate findings of a limited number of between-country studies with a large number of within-country studies and extrapolate these to the likely effects of nonresponse bias on comparative research. In our view, more attention needs to be paid to nonresponse bias in comparative research.

Thus we recommend special studies to explore the extent and implications of cross-national and cross-cultural studies for comparative analysis. If response rates are very different across different nations or cultures, it is important to understand why this may be so and what impact this may have on comparative analysis, the staple of cross-national research. It also is important to distinguish between the different components of nonresponse—noncontact, refusal, and ineligibility; the last can be subdivided in relevant subcategories for individual studies, such as language problems in comparative studies of immigrants and health-related problems in comparative studies of the elderly. However, even if the response rates are similar, it is necessary to learn how respondents may differ from nonrespondents both within and between groups. This is the only way to assert the potential for nonresponse bias in comparative studies.

Much nonresponse research has been conducted on single surveys within one country. We argue for extending this work to include major cross-national studies such as the ISSP, the IALS, the Eurobarometer surveys, and others.

At a minimum, we recommend reporting components of nonresponse and details of fieldwork and the sample realization, allowing readers to calculate their own response rates. If substitution is used, identify the substitutes in the data file and report the reasons for substitutions (no one home, refusal, etc.). Any information on differential nonresponse in key subgroups also should be reported. All of the above information can be collected readily as part of the data collection process with little additional cost in time or money. Often this information is already available in some form at the data collection department.

Furthermore, one should try to reduce variation in response rate *and* in the components of nonresponse (i.e., noncontacts, refusals). This can be done by optimizing fieldwork procedures and using a total quality management approach including a whole battery of response-inducing measures instead of just one or two. Of course, this may affect the budget needed for international research.

The last step is designing for nonresponse, a perspective espoused by Groves and Couper (1998). Nonresponse reduction requires careful planning, the collection of auxiliary data to inform decisions about nonresponse follow-up, and the use of such information to enhance adequate postsurvey adjustment for nonresponse. In cross-cultural research, this implies much more careful consideration of the likely impact of nonresponse on key statistical estimates than simply maximizing or reporting response rates in each country or subgroup.

Finally, important for international research and the handling of nonresponse is a thorough conceptual framework or theoretical basis (see Hox, de Leeuw, and Vorst 1995; Groves and Couper 1998). Only if a theoretical rationale for response-inducing strategies is available is it possible to implement successfully this strategy in diverse cultures by tailoring theoretical insights to the local sociocultural situation. An optimal strategy for reducing nonresponse in cross-cultural and cross-national studies requires a thorough knowledge both of nonresponse reduction strategies and of the local culture and social rules. This integration of knowledge will result in improved fieldwork methods for reducing nonresponse and will, furthermore, stimulate the development of more sophisticated data-driven methods for postsurvey adjustment.

Chapter 12

DATA COLLECTION METHODS

KNUT KALGRAFF SKJÅK
JANET HARKNESS

12.1 INTRODUCTION

Data collection methods, also called 'modes,' are important for both monocultural and cross-cultural research because they are a source of survey error (e.g., Kish 1965; Groves 1989; Lyberg and Kasprzyck 1991; de Leeuw 1992; Saris and Kaase 1997) and because the proper tailoring of modes can help improve survey quality (de Leeuw and Nicholls II 1996; Dillman 2000; Snijkers and Luppes 2000). This chapter discusses modes, response effects, and mode tailoring for the cross-national context, focusing in particular on two issues: response effects and questions of comparability in 'use one mode' models. It begins with a brief introduction to modes and to mode-related effects and error, then presents findings from a seven-country modes experiment using regular survey instruments and finishes by considering issues related to standardizing modes across countries versus mode tailoring for local contexts.

12.2 MAJOR DATA COLLECTION MODES

De Leeuw (1992) distinguishes three major methods of survey data collection: face-to-face interviews, telephone interviews, and mail surveys (cf. Groves and Kahn 1979; Lyberg and Kasprzyk 1991). Procedures within each of these three survey types have changed and diversified considerably in recent years (Lyberg and Kasprzyk 1991; de Leeuw 1992; de Leeuw and Nicholls II 1996). Computer-assisted procedures have been developed within each mode, establishing CAPI (computer-assisted personal interviews) and CATI (computer-assisted telephone interviews) alongside more traditional forms of face-to-face and telephone interviews. The computer-based self-completion format with greatest currency to date is the Web survey. Although a comparative newcomer, Web surveys of various kinds seem set to become a fourth major method of survey data collection for some types of survey.

Although degrees of use vary worldwide, computer-assisted techniques are generally seen as advantageous. CAPI and CATI modes, for example, can eliminate filter errors, provide automatic consistency checks, and enable a range of tailored optimizing features to be produced automatically for an application. This technology is increasingly replacing traditional methods (de Leeuw and Nicholls II 1996).

12.2.1 Modes as Components of Study Design

A survey's collection methods, sample design, and organizational procedures are interdependent (cf. de Leeuw 1992; Grube 1997). The choice of data collection method(s) depends on study design factors, such as the budget and the fielding time available, the population to be sampled and the sample design, as well as the saliency and sensitivity of the survey topic and the complexity of the questionnaire. The response rate that can be expected for a given mode and context also is highly pertinent, especially as response rates decline in many countries (de Leeuw and de Heer 2001).

Mode options can differ greatly depending on context. In some developed countries, personal interviews, for example, may be the most expensive mode and mail surveys may offer an inexpensive alternative with healthy response rates. In others, response to mail surveys may regularly be significantly lower than for other modes.

12.3   RESPONSE DIFFERENCES

12.3.1 Sources of Response Effects

Each method of data collection has specific and different effects on the data collected. Some are positive. Privacy of disclosure, as provided by self-completion formats, for example, can reduce social desirability effects. Other effects are negative. For example, telephone interviews may encourage satisficing (Krosnick 1991; Green and Krosnick 2001). The term 'mode effects' generally is reserved for negative effects of various kinds, from response effects to effects on contact potential. The term 'response effects' will be used here to refer to mode effects that negatively affect response behavior, such as satisficing. At the same time, it often is difficult to distinguish clearly between effects directly related to the mode(s) used and effects related to other aspects of a study design or to the fielding context that may or may not be connected with mode.

Monocultural research on response behavior associated with the three traditional modes has shown that different modes of administration can affect responses (Sudman and Bradburn 1974; Groves and Kahn 1979; Sykes and Collins 1988; Körmendi and Noordhoek 1989; Ayidiya and McClendon 1990; de Leeuw 1992; Krysan et al. 1994; Lass, Saris, and Kaase 1997; Saris and Hagenaars 1997).

Dillman et al. (1996) and Rockwood, Sangster, and Dillman (1997) identify four potential sources of response differences: (1) social desirability, (2) response style bias (acquiescence), (3) question order effects, and (4) other response effects, such as recency and primacy (Krosnick and Alwin 1987). The first pair belong to sociocultural considerations in responding, the second more to cognitive processing. Dillman (2000) considers these sociocultural and cognitive processing factors to be chiefly responsible for response differences across modes.

The cognitive stimulus that people receive differs from mode to mode and is related to how the medium of communication affects the interaction between respondents and the interviewer or the questionnaire (Dillman 1978a; Sykes and Collins 1988; Groves 1989; Dillman 1991; Lyberg and Kasprzyk 1991; Schwarz et al. 1991b; de Leeuw 1992; Couper and Groves 1992). Leaving cultural priming considerations aside, the chief effects on cognitive processing are connected to whether respondents only hear questions and answer options, whether they can read the questions and answer options at leisure and in the order they decide, or whether, as in many face-to-face interviews, they hear questions and answer options but are also offered written or pictorial back-up materials.

Cognitive processing factors and sociocultural considerations are related to three key characteristics of each mode described in de Leeuw (1992): (1) the medium of communication (mainly oral/aural in interviews, visual/written in most self-completion); (2) the 'locus of control'—who has the most control over the questioning and answering process, respondent or interviewer; and (3) whether disclosure of answers is private and anonymous, as in mail self-completion, or shared, as in interviews. Each of these three characteristics is related to the others. Privacy of disclosure, for example, differs across modes and correlates negatively with interviewer locus of control and with socially desirable answers. In other words, the more control over the questions and answers respondents have, the greater their willingness to disclose on sensitive matters. At the same time, the greater the control of respondents, the slimmer the chance is that they can be persuaded to respond if they are reluctant to do so. Thus self-completers may be less prone to providing socially desirable answers because no one is privy to their responses (Gove and Geerken 1977; Groves and Kahn 1979; DeMaio 1984; Ross and Mirowsky 1984; Sykes and Hoinville 1985; de Leeuw and van der Zouwen 1988; Krysan et al. 1994; Moum 1998; Presser and Stinson 1998) and provide more information on sensitive issues (Aquilino 1992, 1994; Grube 1997). However, they also may choose more often not to respond.

Interview speed, as controlled by the interviewer, has been connected with 'top of the head' response strategies (Schwarz et al. 1991). Under pressure of time, respondents resort to providing the first response that comes to mind. If they feel that time is short, respondents may resort more to 'top of the head' response strategies. In telephone interviews, interviewers tend toward faster delivery than in face-to-face settings (Miller and Cannell 1982; Sykes and Collins 1998), perhaps because they only need to deal with vocal social interaction. Telephone surveys are usually conducted without written support materials (Green and Krosnick 2001). Self-completion formats are usually based entirely on visual/written materials, sometimes with an auditory component. Face-to-face interview settings with literate populations thus offer the greatest flexibility; written and visual support materials can be used to help respondents process questions and response options presented orally. With low literacy populations, certain aids cannot be used.

Much of the research on response effects (not mode effects in general) suggests that the net differences usually are not large and that only particularly sensitive questions or specific question designs tend to be involved (Groves 1989; de Leeuw, Mellenbergh and Hox 1996; but see, for example, Silberstein and Scott 1991; Green and Krosnick 2001). De Leeuw, Mellenbergh and Hox (1996) point out, however, that small differences in marginals may be important for other analyses. Few unintended side effects have been noted for computer-assisted applications (Bradburn et al. 1991; de Leeuw and Nicholls II 1996; Wright, Aquilino, and Rasinski 1998; Richman et al. 1999; Scherpenzeel 2000), provided the technological aspects are properly handled and all other things are equal. Research on electronic communication versus interpersonal exchange does, however, note differences across cultures (Tan et al. 1998).

12.3.2 Response Effects across Cultural Contexts

Cognitive processing is more a matter of innate propensities than culture-dependent factors. Nonetheless, culture and experience can be expected to have priming effects on respondents' performance (see Chapters 4 and 6), much in the way test familiarity can influence test performance.

Unfortunately, little research currently is available on ways in which mode-related response effects might differ from culture to culture. It also remains to be clarified how differences within modes that are related to culture and social norms affect response behavior. For example, compliance with social norms affects more areas of life and affects them more strongly in some cultures than in others. Thus the saliency of certain response behaviors associated with specific modes—such as social desirability effects—can be expected to differ in degree across countries. In Western contexts, private disclosure in surveys reduces socially desirable answers. We do not know whether privacy of disclosure has greater effects in cultures more likely to give more pronounced socially desirable answers than it does in countries less affected by the need for this kind of impression management. Are there, for example, patterned differences as to which kind of questions show effects or greater effects across cultures?

While self-completion formats are assumed to enhance disclosure, putting information down on record (paper or electronic) may be perceived in some cultural settings as far more permanent and threatening than providing information orally. In Western contexts, person-to-person interviews encourage rapport and increase respondent cooperation and effort. In other cultural contexts, personal interviews may have to be conducted in the presence of a chaperone—that is, an officially monitoring 'third party presence'—which may eliminate any rapport as well as the opportunity to probe. Equally, we do not know whether the effect is reduced in cultures where physical privacy is not a well-defined or desired notion. Little research is available to date on such aspects of data collection methods where mode and fielding issues merge.

12.3.3 Nonresponse Related to Mode

Estimations of *unit nonresponse* are estimations of the nonparticipation of sample units in a study (see Chapter 11). Survey response rates differ within countries; people in rural areas in Western countries are more likely to participate than are people in towns. In some developing countries, the opposite could be the case, with urban populations more open to surveys.

Response rates in general also differ across countries (de Leeuw and de Heer 2001; Harkness, Langfeldt and Scholz 2001). In some cultural and political contexts, the survey climate is healthier (or survey satiation lower) than in others—that is, people in some contexts are more willing to participate than are people in other contexts (see Chapter 11).

Further, response rates *for given modes* differ within countries (Groves and Kahn 1979; de Leeuw and van der Zouwen 1988; Hox and de Leeuw 1994; Aquilino and Wright 1996) and across countries (de Heer 1999a; Harkness, Langfeldt and Scholz 2001). In numerous countries, for instance, unit nonresponse is higher in mail and telephone surveys than in studies conducted as face-to-face interviews. Various factors contribute to these findings (see Chapter 11).

Item nonresponse is nonresponse that occurs through respondents not answering questions, either intentionally or unintentionally. The frequency of item nonresponse and the kind of items affected differ across modes (de Leeuw 1992; de Leeuw, Mellenbergh and Hox 1996; Owens, Johnson and O'Rourke 1999). Research indicates that item nonresponse tends to be higher in traditional self-completion formats.

People are more reluctant to answer sensitive questions than innocuous questions and, depending on survey exposure, populations differ in their awareness that they have the right not to answer individual questions. Fowler, Roman and Di (1998) illustrate for a monocultural context how difficult it is to identify sensitive questions before the event. However, questions sensitive in one context may not be sensitive in another. We thus can expect cultural differences both in which questions respondents prefer not to answer and in the confidence with which people know or decide they have the 'right' not to answer. Populations reluctant to answer but uncertain of the social acceptability of not answering may systematically opt for an indirect form of nondisclosure, such as selecting 'don't know' categories. Populations aware they are not compelled to answer may fulfill the social contract by satisficing. As a result, item response and item nonresponse can be culturally patterned (e.g., Owens, Johnson and O'Rourke 1999; Meloen and Veenman 1990).

Weaknesses in layout and design can prompt unintentional item nonresponse, for example, through incorrect filter processing. Respondents processing instructions in self-completion formats may make more mistakes with poor designs than practiced interviewers, thus these errors can be mode related. As a result, respondents may fail to answer questions they should answer or, in overlooking filters, answer questions they were not intended to answer. Experience with questionnaires differs across societies, and it seems likely that respondents and interviewers less familiar with questionnaire formats could make more such errors. Computer-assisted applications are expected to minimize such problems but raise their own set of problems and errors.

12.4   A SEVEN-COUNTRY INVESTIGATION ON MODES

This section presents findings from an International Social Survey Programme (ISSP) experiment conducted in seven countries on mode-related response effects. The ISSP operates on the principle of 'keep as much the same as possible' across implementations. When the Programme began, the required mode was self-completion. As new members joined, the self-completion format proved unsuitable for populations with low literacy levels and personal interviews were allowed. Members now field the same questions and answer options in a required sequence, using a format *suitable* for self-completion. By 1995, the ISSP had twenty-three members, of which four regularly fielded a mail survey, ten conducted personal interviews, five had self-completion drop-offs or with interviewer attending, while one country alternated between mail and self-completion with interviewer attending. A methods group was set up to investigate what effects these different implementations might have on the comparability of ISSP data. They designed an experiment to compare the data from the regular 1996 ISSP fielding in each country with those from a second comparable fielding using a different mode. The questionnaire was thus not one designed to maximize mode-related response effects. We could therefore expect that differences in the experiment might be smaller than in specially designed experiments, but also that our study might point to differences not yet found in other modes research.

12.4.1 Research Design

Seven ISSP member countries took part in the experiment: Canada, Germany, Hungary, Norway, the Philippines, Slovenia and the United States.[1] In addition to their regular 1996 ISSP study (minimum N 1,000), participants fielded additional cases (minimum N 200–300) using a different mode of individual choice. Table 12.1 gives an overview of modes and samples. Slovenia collected its data late in 1995, the other six countries in 1996. Germany also collected data on a mail mode from December 1996 through February 1997.

Nationwide sampling frames were used, except in the Philippines, where the experiment was conducted in the greater Manila area. Response rates differed widely in some instances. Apart from the United States and Slovenian net samples, significant differences resulted across samples with respect to age, educational level, and, in some cases, gender. Controlling for these differences left our findings unchanged (ANCOVA for unbalanced design).

TABLE 12.1. Modes of Administration and Samples by Country

Country	Modes	Cases
Germany	**Self-completion, interviewer attending**	2,340
	Face-to-face	1,128
	Self-completion, mail	384
USA	**Self-completion, interviewer attending**	894
	Face-to-face	438
Philippines	**Face-to-face**	300
	Self-completion, interviewer attending	300
Hungary	**Face-to-face**	1,500
	Self-completion, interviewer collecting	307
Slovenia	**Face-to-face**	1,004
	Self-completion, mixed	202
Canada	**Self-completion, interviewer collecting**	1,182
	Telephone	286
Norway	**Self-completion, mail**	1,344
	Telephone	600

NOTE: Standard ISSP mode mentioned first, in bold.

Sixteen of the sixty substantive items in the module are analyzed here which were contained in two Likert-statement batteries of eight items each (see Appendix for items). Cultural bias effects with Likert agree-disagree designs are not discussed here, but see, for example, Tanzer (1995), Javeline (1999) and Skinner (2001). The first battery measures attitudes to government spending (5-point scale with status quo as middle category), the second measures external and internal political efficacy and trust (5-point agree–disagree scale). In the Efficacy battery, five statements assert high efficacy, and three assert low efficacy. The statements are longer and more complicated than those in the Spending battery and may make higher cognitive demands on respondents.

12.4.2 Analysis

Modes research often is designed to maximize the chances of detecting effects; comparisons and replications of studies also are mostly monocultural, whereas the findings presented in Table 12.2 are cross-national. Even so, the literature points to an entanglement of factors and their effects, making it difficult to generalize results and determine causes. The most consistent finding is that social desirability has less of an effect on response patterns in self-completion formats than in face-to-face interviews. The research literature comparing effects across modes tends to agree that differences are small between self-administration and interviewer modes.

TABLE 12.2. Response Styles. Differences between Number of Responses

| | Government Spending | | | | | | Political Efficacy | | | | | |
| | Categories | | | | | | Categories | | | | | |
	1	2	3	4	5	dk/na	1	2	3	4	5	dk/na
Germany												
FtF - SCA												
Diff(means)	0.09	0.05	-0.09	-0.14	-0.03	0.12	0.18	-0.05	-0.21	-0.40	0.34	0.15
t Value	1.60	0.79	-1.29	-4.18	-1.14	2.90	4.85	-1.02	-3.96	-7.90	7.63	2.83
p	0.11	0.43	0.20	<.0001	0.26	0.004	<.0001	0.31	<.0001	<.0001	<.0001	0.005
FtF- SCM												
Diff(means)	0.25	0.50	-0.49	-0.23	-0.09	0.05	0.12	-0.22	0.10	-0.35	0.31	0.05
t Value	3.20	5.98	-4.53	-3.88	-2.35	0.75	2.08	-2.56	1.14	-4.32	4.71	0.57
p	0.001	<.0001	<.0001	0.0001	0.02	0.45	0.04	0.01	0.26	<.0001	<.0001	0.57
SCA - SCM												
Diff(means)	0.16	0.46	-0.39	-0.09	-0.07	-0.07	-0.06	-0.17	0.30	0.05	-0.03	-0.10
t Value	2.31	6.06	-4.00	-1.57	-1.73	-1.28	-1.01	-2.19	3.87	0.59	-0.51	-1.37
p	0.02	<.0001	<.0001	0.12	0.08	0.20	0.31	0.03	0.0001	0.56	0.61	0.17
USA												
FtF - SCA												
Diff(means)	0.04	0.03	0.02	-0.08	0.02	-0.04	0.12	0.09	-0.31	-0.07	0.13	0.04
t Value	0.48	0.29	0.16	-1.22	0.48	-0.44	1.84	1.06	-3.49	-0.79	2.18	0.49
p	0.63	0.77	0.88	0.22	0.63	0.66	0.07	0.29	0.0005	0.43	0.03	0.62
Philippines												
FtF – SCA												
Diff(means)	-2.32	1.67	1.09	0.48	-0.02	-0.29	-1.13	0.64	0.18	1.03	-0.26	-0.45
t Value	-16.9	10.7	3.80	6.29	-0.47	-4.22	-10.8	4.35	1.15	8.97	-4.78	-3.99
p	<.0001	<.0001	0.0002	<.0001	0.64	<.0001	<.0001	<.0001	0.25	<.0001	<.0001	<.0001

(table continues on next page)

TABLE 12.2. (continued)

| | Government Spending | | | | | | Political Efficacy | | | | | |
| | Categories | | | | | | Categories | | | | | |
	1	2	3	4	5	DK/NA	1	2	3	4	5	DK/NA
Hungary												
FtF – SCC												
Diff(means)	0.14	0.55	-0.35	0.02	-0.04	-0.32	0.08	0.08	-0.16	-0.14	0.08	0.06
t Value	1.39	5.06	-3.28	0.39	-1.22	-2.74	1.27	0.86	-1.73	-1.44	1.04	0.61
p	0.17	<.0001	0.001	0.69	0.22	0.007	0.21	0.39	0.08	0.15	0.30	0.54
Slovenia												
FtF – SCM												
Diff(means)	0.37	0.50	-0.14	-0.13	-0.11	-0.49	0.16	0.16	-0.16	0.30	-0.03	-0.43
t Value	3.01	3.73	-1.06	-2.04	-1.94	-3.53	1.71	1.33	-1.47	2.85	-0.29	-2.98
p	0.003	0.0002	0.29	0.04	0.05	0.0005	0.09	0.18	0.14	0.005	0.77	0.003
Canada												
Tph – SCC												
Diff(means)	0.14	0.56	-0.31	-0.10	-0.19	-0.11	0.12	0.55	-0.78	0.27	-0.11	-0.04
t Value	1.66	5.42	-2.51	-1.40	-4.14	-2.28	1.87	5.16	-8.50	2.84	-1.76	-1.02
p	0.10	<.0001	0.01	0.16	<.0001	0.02	0.06	<.0001	<.0001	0.005	0.08	0.31
Norway												
Tph – SCM												
Diff(means)	-0.03	0.47	0.01	-0.04	0.36	-0.05	-0.07	0.83	-0.76	0.20	-0.09	-0.11
t Value	-0.54	6.77	0.14	-1.02	-13.2	-1.10	-1.75	11.2	-11.4	2.99	-2.51	-2.33
p	0.59	<.0001	0.89	0.31	<.0001	0.27	0.08	<.0001	<.0001	0.003	0.01	0.02

FtF = face-to-face; SCA= self-completion interviewer attending; SCM = self-completion mail; Tph = Telephone; SCC = self-completion, collected later.

There is less consensus for comparisons of face-to-face and telephone interviews; recent monocultural research points to important (e.g., Green and Krosnick 2001) or to small (e.g., Scherpenzeel 2000) differences. There is also disagreement on why and when other mode effect differences, such as acquiescence, response style extremity, or primacy/recency effects, occur (Dillman 2000; see overview in Krosnick and Fabrigar 1997).

In cross-cultural comparisons, effects and their causes for one context are no easier to disentangle. Indeed, given that cultural considerations and cognitive processing factors may interact to produce different response effects for different questions in different contexts, disentanglement across results for countries is especially problematic. Social desirability, primacy/recency effects, and response style biases all could be involved in the response patterns to questions; some of the effects will be culture-specific, others perhaps not, or less so. The findings presented below, as said, are not culled from a special modes design; effects consequently may be smaller.

Table 12.2 presents the average number of times respondents selected the different response categories in the two instruments. The most striking feature in Table 12.2 is the similarity of the results in Canada and Norway, the two countries comparing self-completion and a telephone mode. Response distributions between the telephone interviews and the self-completion modes differ in the same way in both countries. In the Efficacy instrument, self-completion respondents select the middle response more often than they do in the Spending instrument. In the Spending instrument the middle category represents the status quo: 'spend the same amount as now.' In the Efficacy instrument the middle category is "neither/nor." There is no one obvious explanation for the difference in selection of status quo versus "neither/nor." Self-completion respondents may be trying hard to cooperate and optimize their answers (Krosnick 1991). If they find it difficult to decide, the optimizing quality answer might be the middle category; when deciding is simpler, an optimizing response would not be the middle category. The speed of the interview or reduced rapport for respondents interviewed by telephone may mean that in either case they tend toward the middle option as a form of satisficing. Table 12.3 shows that the affirmative index is higher in telephone interviews in both countries. If social desirability, acquiescence, and primacy/recency effects are indeed entangled in the response differences for the instruments, the findings indicate that social desirability and/or acquiescence effects overrule primacy/recency effects.

Few of the patterns found for Canada and Norway (self-completion/telephone) hold for the results for Germany and the United States (personal interview/self-completion with interviewer attending). For these two countries, differences across these modes are smaller than those found for the previous two countries and modes and are smaller for the United States than for Germany. However, findings from the additional mail mode for Germany, fielded late in 1996 with a comparable sample, show that in many instances differences are even greater between the mail mode and the other two German modes. This may mean that interviewer presence overrides privacy of disclosure (contradicting Fowler, Roman, and Di 1998) or that the hour-long interview preceding the self-completion with interviewer attending affected how respondents answered.

Hungary and Slovenia each fielded a self-completion mode (drop-off, collected later) and face-to-face interviews. The results for both countries resemble those for Canada for the Spending instrument, and those for Germany and the United States for the Efficacy instrument. The affirmative indexes are significantly higher for the face-to-face samples in the Spending instrument (that is, respondents say more should be spent) but not for the Efficacy instrument.

The data for the Philippines (personal interview/self-completion) are not easy to interpret. Respondents *interviewed* preferred moderate answer categories. Respondents *self-completing* provided more extreme and affirmative responses ("strongly agree") in the Efficacy instrument than interviewed respondents did. Privacy of disclosure could lead us to expect fewer socially desirable responses in the self-completion format and more reserved and moderate answers in interviews. However,

TABLE 12.3. **Differences between Affirmative and Non-affirmative Responses**

	Government Spending	Political Efficacy		Government Spending	Political Efficacy
Germany			**Hungary**		
FtF	2.18	0.38	FtF	4.10	−0.55
SCA	1.87	0.19	SCC	3.38	−0.76
SCM	1.10	0.43	FtF − SCC	0.72***	0.21
FtF − SCA	0.31***	0.19*			
FtF − SCM	1.08***	-0.05	**Slovenia**		
SCA − SCM	0.77***	-0.24	FtF	3.59	0.05
			SCM	2.48	0.01
USA			FtF − SCM	1.11***	0.04
FtF	2.25	0.46			
SCA	2.12	0.30	**Canada**		
FtF − SCA	0.13	0.16	Tph	2.05	0.62
			SCC	1.06	0.11
Philippines			Tph − SCC	0.99***	0.51**
FtF	3.80	0.84			
SCA	4.91	2.10	**Norway**		
FtF − SCA	−1.11***	−1.26***	Tph	2.69	1.24
			SCM	1.85	0.59
			Tph − SCM	0.83***	0.65***

* p < 0.05; ** p < 0.01; *** p < 0.001.
FtF = face-to-face; SCA = self-completion interviewer attending; SCM = self-completion mail; Tph = Telephone; SCC = self-completion, collected later.

since the Efficacy items are both positive and negative, we would not expect extreme affirmative answers to every item. Thus privacy of disclosure is not an entirely satisfactory explanation for some of the results. Explaining any of these responses in terms of acquiescence bias is problematic, too, since this could be expected to be higher in the interviews than in self-completion. Finally, the increase in affirmative answers noted here for self-completion stands in contrast to the findings for self-completion versus personal interview in the four other (Western) countries in the table.

It may well be that the theme of the module (Role of Government) is much more sensitive in the Philippines than in the other countries and that different social pressures are at work, such as the 'fear factor' (socially desirable answers based on fear) as well as culturally expected interviewer accommodation, politeness, and acquiescence. Response styles and response biases also vary by culture (Ross and Mirowsky 1984; Hui and Triandis 1989; Javeline 1999), as do the saliency and procedures of face management (Brown and Levinson 1999). These, combined, may be interacting with mode and changing the features assumed elsewhere to apply for one or the other mode. Various other chapters reflect that we can expect different populations to react with different sensitivities to different topics, diachronically within one culture and synchronically across cultures. The Philippines data suggest that both the reasons why questions are sensitive and the degree of sensitivity involved can differ across contexts.

The ISSP experiment answers a few questions but poses many more. It demonstrates, for example, that mode choice matters across cultures. While the reasons may not always be clear, the analysis of univariate distributions of responses in this experiment points to substantial differences across modes for the countries conducting telephone interviews and self-completion surveys, while differences across modes for the countries administering face-to-face interviews and self-completion formats are much smaller. (The one German mail study may indicate that more difference is involved in formats with no interviewer interaction.) Unfortunately, the experiment does not provide data from each country for each mode. The findings can thus only be taken as pointers for further search. In line with other findings on data collection methods, it is not easy to identify and explain the occurrence of distinct mode effects for the experiment, with the possible exception of social desirability effects. Lastly, the meaning and effects of individual modes of administration in this study seem to vary between cultures; mode differences found in one country will not necessarily occur (replicate) in another. More information is needed on precisely how studies are implemented and what the effects of mode are on response and nonresponse. One essential step toward gathering information is the inclusion of a mode variable in data sets which records the mode(s) used per case.

12.5 MODE OPTIMIZING FOR COMPARATIVE RESEARCH

The 'keep the same mode' strategy in comparative survey research is a conservative strategy to keep as much as possible as similar as possible across participating countries. In this one-size-fits-all approach, a source questionnaire is developed (presumably tailored for the mode), and the questions are simply translated for other countries. Using the same mode is assumed to provide greater comparability of the data than would using different modes in different countries. Further assumptions are that within-mode differences are less important and that, for example, personal interviews create the same response context for respondents across cultures. The ISSP findings presented above provide some evidence that this is not the case. As things currently stand, if the same mode is required for each part of a comparative study, the options available are often limited. To some extent this depends on the spread of countries or regions involved and on technological and infrastructure differences across these. For example, methods calling for electrical equipment often are not reliable for surveying rural or poor regions, not only in developing countries. Low literacy levels in participating populations can preclude written self-completion formats, while problems with telephone density and/or access to telephones may rule out telephone studies. Thus the only *common* mode option often is face-to-face interviews.

At the same time, using one mode across countries can mean that the required mode may be a poor choice for a given local context. For instance, mail or telephone surveys can be optimal choices for some studies in Scandinavian countries, given local conditions, geography, infrastructures, and financial and time budget allocations. In Sweden, Finland, or Norway, response rates are generally good and higher than further south in Europe. Access to names, addresses, and telephone numbers of individuals is inexpensive and reliable, and a high percentage of people have telephones. In other contexts, however, telephone surveys are increasingly problematic (Green and Krosnick 2001). Face-to-face surveys, meanwhile, have become the exception in some countries, such as Sweden and Switzerland, and are expensive there by national standards. In Sweden, nationwide face-to-face representative sample surveys are currently conducted by virtually only one agency. In Switzerland, only about 15% of survey work is conducted as face-to-face interviews (www.swissresearch.org/deutsch/content/s_statistik.htm). Thus if such countries are required to collect data in face-to-face interviews so as to comply with requirements for a comparative project, they do not use the optimal local choice of mode. The first round of the European Social Survey in 2002/2003 requires face-to-face interviews in twenty-plus European countries and may result in just this situation. As indicated at the outset, deciding the best local mode option depends in part on the study requirements as a whole. However, the benefits and disadvantages of a given mode also vary across time, as illustrated by the trend away from and back to personal interviews versus telephone surveys in some Western contexts.

Tested strategies for optimizing mode choices without threatening comparability are not yet available. However, concern about falling rates of survey participation across continents (de Heer 1999a) has sharpened interest in deliberate and planned *tailoring* of collection methods to maximize quality returns. Attention has focused increasingly on planned mixed-mode designs (cf. Dillman 2000; Snijkers and Luppes 2000). The central idea is that through careful design and implementation, data collected using different methods can be combined or compared. Tailoring modes to suit the study and the population allows researchers to meet respondents' mode preferences (Voogt, forthcoming; Snijkers and Luppes 2000). Research suggests tailored questionnaires and implementations can raise response rates and reduce both unit and item nonresponse. Dillman (2000) suggests that *unimode* questionnaire designs can mitigate effects associated with one or the other mode and thus enhance data quality and data comparability across modes within the same sample. Dillman also suggests that the use of multiple modes within one survey will increase even more.

We are not aware of planned mixed-mode designs for comparative projects or deliberate tailoring of designs to optimize modes for a study across different fielding cultures. At the same time, designing survey instruments in ways that reduce both nonresponse and mode response effects seems likely to be more complicated than the nine principles for a 'unimode design' (Dillman 2000) seem to suggest. The discourse and translation issues unimode questionnaires bring with them are multiple, for example. Detailed research would be needed on how response scales might be tailored to fit culturally determined response styles but remain comparable across data sets.

Mixing data collection modes on an *ad hoc* basis to increase response rates is common practice in many Western contexts. So, too, are within-mode modifications, such as using interpreters or bilingual interviewers, whether planned or *ad hoc*, in both developed and developing countries. Lack of documentation on mode use and mode changes within a survey makes it unclear how common such *ad hoc* modifications are in comparative research. Effects go almost completely without discussion.

12.6   CONCLUSIONS AND OUTLOOK

In order to achieve comparable sample designs, Kish (1995) recommends optimal local designs within the requirements of a probability sample (see Chapter 8). In order to follow suit for modes, more must be known about mode effects across modes and countries. Modes not only have response effects that we have trouble identifying, measuring, and avoiding in comparative contexts. They are also related to sample design, contact potential, and response rates. Whatever else, information on modes should be included in data sets, and careful handling is necessary to ensure that issued and realized samples are, indeed, comparable.

It is by no means clear that keeping modes the same across countries is the way to minimize differences or to enhance data quality and comparability. Mode tailoring and mode mixing may prove in the long run to be the better bases on which to decide modes for comparative research. At the same time, we do not have enough insight into the potential of mode tailoring for comparative projects across cultures, survey cultures, and languages. The issues cannot be reduced to a forced choice between single-mode surveys with low response rates, coverage errors, and biased samples and mixed-mode designs with higher response rates, better samples, and inconsistent mode effects. Finding solutions to these questions constitutes a major challenge for survey researchers in the years to come, not least for the cross-cultural context.

Note 1. The authors take the opportunity to thank the eight ISSP member institutions involved in planning the project and the seven of these that collected the data and provided documentation. **Canada**, H. Pyman, A. Frizzell, Carleton University Survey Center , Ottawa; **Hungary**, P. Robert, M. Sagi , Tárki RT Social Research Institute, Budapest; **New Zealand**, P. Gendall , Department of Marketing, Massey University; **Philippines**, M. Mangahas, L. Luz Guerrero, Social Weather Stations, Quezon City; **Slovenia**, N. Toš , B. Malnar, Public Opinion and Mass Communications Research Centre, University of Ljubljana; **USA**, T.W. Smith, NORC, Chicago.

APPENDIX: SPENDING AND EFFICACY ITEMS

Government Spending Instrument:
Listed below are various areas of government spending. Please show whether you would like to see more or less government spending in each area. Remember that if you say 'much more', it might require a tax increase to pay for it.
(a)The environment (b)Health (c)The police and law enforcement (d)Education (e)The military and defense (f)Old age pensions (g)Unemployment benefits (h)Culture and the arts
Response scale: 1. Spend much more 2. Spend more 3. Spend the same as now 4. Spend less 5. Spend much less 8. Can't choose

Political Efficacy and Trust Instrument:
Please tick one box on each line to show how much you agree or disagree with each of the following statements.
(a)People like me don't have any say about what the government does (b)The average citizen has considerable influence on politics (c) Even the best politician cannot have much impact because of the way government works (d)I feel I have a pretty good understanding of the important political issues facing our country (e)Elections are a good way of making governments pay attention to what the people think (f)I think most people are better informed about politics and government than I am (g)People we elect as MPs try to keep the promises they have made during the election (h) Most civil servants can be trusted to do what is best for the country
Response scale: 1. Strongly agree 2. Agree 3. Neither agree nor disagree 4. Disagree 5. Strongly disagree 8. Can't choose

Chapter 13

SOCIAL DESIRABILITY IN CROSS-CULTURAL RESEARCH

TIMOTHY P. JOHNSON
FONS J. R. VAN DE VIJVER

13.1 HISTORY

The concept of social desirability derived its origin, more than 50 years ago, from a common observation by interviewers that what respondents say may not be true or not entirely true. The given answer is assumed to show a consistent distortion from reality: respondents portray themselves too positively. Scales were developed to assess this tendency, with the aim of designing measures that could indicate the level of veridicality of answers on other items. The Lie Scale of the Eysenck Personality Questionnaire and the Marlowe-Crowne Scale (discussed below) are examples of this line of thinking. Later research, however, has provided important extensions: some items are more susceptible than others to trigger socially desirable answers; also, some individuals are more likely to show socially desirable behavior than others. The consistency of individual differences in social desirability has led some theoreticians to argue that social desirability is not a response style but a personality characteristic related to conformism. Both views (social desirability as a response style versus social desirability as a stable personality characteristic) are described in this chapter. It may not be superfluous to remark that these two views, at times, hardly seem compatible and refer to seemingly unrelated research traditions.

Current cognitive models have largely failed to address the potential role of culture in evaluating target opinions and behaviors as being socially desirable, undesirable or neutral, as well as in the decision as to whether or not undesirable responses will be modified to conform to perceived norms. In this chapter, we evaluate social desirability as a source of survey measurement error and explore its implications for conducting cross-cultural research. Our focus is not on exploring how social desirability can affect answers but on determining whether social desirability has a *differential* impact on respondents from different cultural backgrounds. We argue in the present chapter that in cross-cultural survey research, the view of social desirability as a screen put in front of interviewers to prevent their view of the respondent's reality is sometimes useful, while in other studies constructs are examined that are so closely related to social desirability that any correction for it would decrease the validity of cross-cultural comparisons.

13.2   SOCIAL DESIRABILITY AS RESPONSE STYLE

Cognitive models of survey information processing that have been developed over the past 20 years generally consider a four-stage process by which respondents answer survey questions (Sudman, Bradburn, and Schwarz 1996). During the first three stages, respondents interpret questions, search for relevant information in semantic memory, and form answers. It is now understood that various sources of measurement error may be introduced at each of these stages. For example, questions may be misunderstood, respondents may be unable to retrieve necessary information, and responses may be incorrectly mapped onto survey response scales. Although serious, these are generally viewed as unintentional forms of measurement error. We contrast these with the fourth stage, during which true survey answers (or what the respondent believes to be true) are evaluated and sometimes edited (whether deliberately or not) for social desirability prior to reporting. We define social desirability in the response-style view as the tendency of individuals to "manage" social interactions by projecting favorable images of themselves, thereby maximizing conformity to others and minimizing the danger of receiving negative evaluations from them.

Concerns regarding the effects of social desirability on survey data quality led to the development of several measures designed to assess it. The developers of these measures conceptualized the tendency to provide socially desirable reports as a personality trait; some survey respondents might thus be more vulnerable to social desirability influences than others. Among American researchers, one of the most commonly employed measures has been the Marlowe-Crowne Social Desirability, or Need for Approval, Scale (Crowne and Marlowe 1960, 1964). A meta-analysis reported that over 90% of the available research literature employing a social desirability measure used the Marlowe-Crowne (MC) Scale (Moorman and Podsakoff 1992). Each of the 33 true–false items describes either culturally acceptable but improbable or culturally unacceptable but probable behaviors. Consistent with theoretical expectations, the MC Scale has been found to be inversely correlated with self-reports of numerous behaviors and conditions generally believed to be socially undesirable, including symptoms of psychiatric illness or distress, suicidal thoughts, and drug and alcohol use (Klassen, Hornstra, and Anderson 1975; Carr and Krause 1978; Strosahl, Chiles, and Linehan 1984; Welte and Russell 1993; Watten 1996). It is also found to be positively correlated with socially desirable self-evaluations, such as degree of life satisfaction and happiness (Carstensen and Cone 1983; Kozma and Stones 1987). The MC measure has also been consistently shown to have moderate correlations with other social desirability scales, including the Edwards social desirability scale (range 0.32–.042; Crino et al. 1983; Strosahl, Chiles, and Linehan 1984; Kozma and Stones 1987) and the Eysenck Lie Scale (range 0.45–0.55; McCrae and Costa 1983; Khavari and Mabry 1985). This latter measure, more formally known as the Lie Scale of the Eysenck Personality Inventory (Eysenck and Eysenck 1964), is commonly employed in countries other than the United States.

13.3   SOCIAL DESIRABILITY AS SUBSTANCE

The interpretation of negative correlations between social desirability measures and various symptoms and behaviors as evidence of mere response artifacts has been challenged. Bradburn and Sudman (1979b), for example, have interpreted the negative findings as evidence that persons who score high on tests of social desirability do, in fact, behave in an altruistic manner consistent with the underlying personality trait represented by these measures. Welte and Russell (1992) have put forth a similar argument. Empirical support for this position comes from a study that employed an external criterion. McCrae and Costa (1983) demonstrated that persons with high Marlowe-Crowne and Eysenck Lie scores were, in fact, rated more positively by their spouse across a variety of psychosocial measures. In addition, adjustments for social desirability using the MC Scale in order to get a more valid picture of the relationship between two target measures have been unsuccessful (Campbell, Converse, and Rodgers 1976; Gove et al. 1976; Gove and Geerken 1977; Kozma and Stones 1987; Welte and Russell 1993). In other instances, controlling for scores on the MC Scale has actually decreased validity coefficients (McCrae 1986). Similar findings have been reported when attempting to correct for social desirability using the Eysenck Lie Scale (McCrae and Costa 1983).

13.4   CROSS-CULTURAL STUDIES OF SOCIAL DESIRABILITY

13.4.1 Social Desirability as a Person Characteristic

Early on in the conceptualization of social desirability there was recognition that culture was important in classifying opinions and behaviors as desirable or not. Crowne and Marlowe (1964) suggested that socially desirable responding was motivated by "the need of subjects to respond in culturally sanctioned ways" in order to obtain social approval. Yet, cultural variation in social desirability and the possible impact of differential social desirability on cross-cultural surveys have never been seriously examined. We know from cross-cultural work that there are both universals and cultural specifics in social behavior. Some norms that have obvious implications for survey behavior, such as the norm of reciprocity (Gouldner 1960), are important to many cultures and might at an abstract level approach classification as etic, or pan-cultural. Yet, there are broad cross-cultural differences in specific beliefs regarding for whom, where, and when reciprocity is appropriate, suggesting that the practice of reciprocity is at least in part unique, or emic, within each culture.

Survey questions that do not activate cultural perceptions of desirability or undesirability will likely be processed and answered without deliberate misrepresentation. Culturally filtered social desirability perceptions alone, however, are not sufficient to trigger response editing. Cultural norms may also dictate those situations in which it is and is not necessary or appropriate to mislead others. Within collectivist societies, for example, the restrictions against providing misleading information to members of external groups are generally weaker than within individualist societies (Triandis 1995). Consequently, a survey question may be interpreted as having similar levels of socially desirable content by respondents from varying cultural groups, yet persons from one cultural background may be more likely to edit their responses than those from another. This suggests that culture's influence on social desirability involves a two-step process. For any given survey question, respondents will interpret, based on their cultural experiences, whether or not it contains socially desirable or undesirable content. Among those who do perceive it as containing such content, some will then determine (also based on cultural experience) that the question should nonetheless be answered as accurately as possible, while others will decide that response editing to conform with social norms is necessary and/or appropriate. Recent work that highlights cultural variations in perceptions of the appropriateness of lying in various social contexts is consistent with this process (Fu et al. 2001; Lee et al. 2001).

Warnecke et al. (1997) reported findings from Chicago, Illinois, indicating that both African-American and Mexican-American (but not Puerto Rican) respondents had higher MC scores than non-Hispanic whites after controlling for gender, age, education, and income. Several other American studies and one from South Africa have also documented higher social desirability scores, using both the MC and similar measures, among black, when compared to white, respondents (Crandall and Crandall 1965; Fisher 1967; Klassen, Hornstra, and Anderson 1975; Edwards and Riordan 1994). Recent data have also documented higher MC scores among East Asians, compared to American-born subjects (Middleton and Jones 2000; Keillor, Owens, and Pettijohn 2001). In a comparison of European-Americans, Mexican-Americans, and Mexicans, Ross and Mirowsky (1984) also found substantial cultural differences in social desirability as measured by the MC Scale. The authors reported that Mexican-Americans revealed higher social desirability scores than did European-Americans and Mexicans. They attributed the intergroup differences to the relative power of these groups in society. Groups with low power, as often is the case with immigrant groups, tend to be more concerned with impression management and hence display more socially desirable behavior. Differences in social power are an equally plausible explanation for black–white variability in MC scores. Other comparisons of MC scores across racial and ethnic groups, however, have provided counter-intuitive (Abe and Zane 1999) or no-difference findings (Tsushima 1969; Gove and Geerken 1977; Welte and Russell 1993; Okazaki 2000).

In one of the few cross-cultural studies of social desirability, Johnson et al. (1997) reported findings from a series of cognitive interviews with samples of adults representing four distinct cultural groups in Chicago: African-American, Mexican-American, Puerto Rican, and non-Hispanic white. When asked if they believed that people in general would over- or underreport behaviors that appeared to vary in their degree of social desirability (e.g., the frequency of consuming fruits/vegetables and alcohol), no cross-group differences were found. A national survey in the United

States also failed to find differences in the perceived social desirability of sets of psychiatric, self-esteem, and positive affect descriptors (Gove and Geerken 1977). In one available cross-national study, American, French, and Italian judges also showed high correlations of greater than 0.80 in their evaluations of the desirability of items included in the Adjective Check List (Gough and Heilbrun 1980).

In the United States, several researchers have documented a strong relationship between individual judgments of the desirability of personality traits and the probability that they will endorse questions about each (Edwards 1953; Phillips and Clancy 1970). Dohrenwend (1966) found this pattern to be consistent across four ethnic groups residing in New York City. Comparative data from other nations also confirm that social desirability perceptions influence behavior in a similar manner across varied cultural groups. Türk Smith, Smith, and Seymour (1993) reported high correlations between these variables among both Turkish and American students (0.86 and 0.87, respectively). In another study, Gendre and Gough (1982, quoted in Türk Smith, Smith, and Seymour 1993) administered the Adjective Check List to Italian and French subjects. This list consisted of 300 person-descriptive adjectives, such as 'clever' and 'fickle.' Subjects were asked to rate both the applicability of the adjective to themselves and the social desirability of each behavior. A high correlation was found within both cultural groups. In Japan, lower but still significant correlations were also reported (Iwawaki, Fukuhara, and Hidano 1966) between these same measures.

In a more recent study, Williams, Satterwhite, and Saiz (1998) asked students in ten countries (Chile, China, Korea, Nigeria, Norway, Pakistan, Portugal, Singapore, Turkey, and the United States) to rate the favorability of each of the items from the Adjective Check List. Favorability can be taken to be closely related to social desirability. The average cross-country correlation was 0.82. The authors also computed the correlation between the average item scores for female and male students; the mean correlation across the ten countries was 0.97. The study illustrates that cross-cultural similarities in judgments of social desirability appear to outweigh cross-cultural differences. Yet, the drop from 0.97 to 0.82 cannot be accounted for by attenuation; there appears to be a large core of cross-culturally shared views on the (un)favorability of person descriptors but there are also country-specific aspects that must be considered.

In general, the findings from these studies suggest that social desirability is likely to be a universal concept, given the strong cross-cultural similarities in ratings of social desirability and associated reporting behaviors across cultures. That some differences in patterning across countries were also found, however, suggests the possible presence of culture-specific factors.

Some studies have examined constructs that are closely related to social desirability, such as individualism–collectivism, expressiveness, and self-disclosure. According to Triandis (1995), honesty in interactions with strangers is a characteristic that is more highly valued in individualist societies, whereas concern about maintaining good relationships and face-saving are more salient (and hence socially desirable) in collectivist countries. Johnson (1998b) has reported findings from a study in the United States that documented a positive correlation between the Marlowe-Crowne and a collectivist orientation scale (0.20) and a negative correlation between the Marlowe-Crowne and a measure of individualism (−0.19). In a review of studies on cross-cultural differences in communication styles, Smith and Bond (1998) discuss studies of self-disclosure. With remarkable consistency, members of individualist societies are found to be more inclined to reveal information about themselves both to members of their in-group as well as to out-group representatives, whereas members of collectivist societies often make a sharp distinction between in-group and out-group members and show less disclosure toward the latter.

Other research has compared social desirability indices across cultural groups at the macro-level. A large database on cross-national differences in social desirability can be derived from work with Eysenck's Lie Scale. This instrument, translated into various languages, has been used in well over 300 studies, covering more than 40 countries. Van Hemert et al. (in press) examined the relationship between Lie Scale scores and the Gross National Product of various countries. Based on data for 38 countries, they found a highly significant, negative correlation of −.67, with more affluent countries tending to show lower social desirability scores. We note here that economic affluence is also positively associated with individualism at the national level (Hofstede 2001).

It can be concluded that, if conceptualized as a person characteristic, there are important cross-cultural differences in social desirability. Persons coming from more influential groups in society or from more affluent countries tend to show lower scores on social desirability. In the previous section it was found that cultures do not seem to differ greatly in what is seen as desirable behavior. A combination of these results leads us to conclude that social desirability is an important source of cross-cultural score differences and that it can be fairly adequately measured in a cross-cultural framework. The psychological meaning is less clear-cut; there is some disagreement in the literature as to whether social desirability is "mere response editing" or is associated with various other psychological traits, such as agreeableness and need for affiliation. Yet, even within the latter view, it is important to take the role of social desirability into account in cross-cultural studies, as it constitutes an important source of score differences. It is probably the most common alternative explanation of country differences in survey research and deserves to be treated accordingly (e.g., by administering a measure of social desirability).

13.4.2 Social Desirability as a Question Characteristic

A contrasting perspective that dates back several decades argues that social desirability is more properly viewed as due to question characteristics or the survey interaction process between interviewer and interviewee (Phillips and Clancy 1972; DeMaio 1984). In general, this approach is concerned with identifying question types and other elements of the survey context that are most likely to elicit socially desirable, but incorrect, responses. Here, we briefly review contextual evidence of cross-cultural differences in socially desirable responding. Consistent with some findings from comparisons of social desirability scales across ethnic groups, available evidence also suggests greater misreporting of socially desirable behavior among minority groups in the United States. In particular, validation studies have consistently documented more overreporting of voter participation among nonwhite groups when compared with whites (Katosh and Traugott 1981; Abramson and Claggett 1986). Similar validation studies have also shown that American minorities may underreport socially stigmatizing behaviors such as substance use and abortions to a greater degree than the majority white population (Jones and Forrest 1992; Hser 1997).

Some of the most compelling evidence regarding the effects of culture on the survey process comes from studies that have demonstrated that cultural distance between respondents and interviewers sometimes produces varying patterns of responses. In American studies, respondents have been shown to defer to the perceived values of other-race interviewers when answering relevant survey questions (Cotter, Cohen, and Coulter 1982; Anderson, Silver, and Abramson 1988; Finkel, Guterbock, and Borg 1991; Davis 1997). However, one experimental study that manipulated the expressed nationality of Oriental interviewers as South Vietnamese, Thai, or Japanese, found that the effects on student opinions regarding the Vietnam War might not always be in the predicted direction (Hue and Sager 1975). Another American study that conducted cognitive interviews with respondents from four cultural groups probed about their comfort discussing alcohol consumption with interviewers from their own and other race/ethnic groups (Johnson et al. 1997). They found non-Hispanic whites were equally comfortable with the idea of discussing their alcohol consumption with interviewers of their own versus other cultural groups. Minority group respondents (African-American, Mexican-American, and Puerto Rican), in contrast, expressed less comfort discussing the topic with interviewers from other cultural groups. These findings were interpreted within an individualist-collectivist framework that would expect representatives of individualist cultures to be equally comfortable discussing sensitive information with interviewers from their own and other cultures. In contrast, respondents from more collectivist cultures may be expected to have greater difficulty discussing sensitive topics with interviewers from different cultural groups.

Cultural differences in the effects of interview mode on self-reports of sensitive survey questions have also been examined. Aquilino (1994) examined the influence of mode of survey administration on answers to questions dealing with the use of psychotropic substances, including alcohol, marijuana, and cocaine across cultural groups in the United States. Face-to-face, telephone interviews, and self-administered questionnaires were administered to non-Hispanic white, black, and Hispanic respondents. Mode effects were smaller for non-Hispanic whites than for minority group members, who were more likely to report lower levels of substance use when interviewed, compared to when they completed self-administered questionnaires. Similar findings have been reported by Park, Upshaw, and Koh (1988), who found lower reporting of physical and mental health symptoms in self-administered questionnaires, compared to personal interviews, among foreign-born Koreans in the United States In contrast, no differences in reporting were found by data collection mode among American-born whites and Japanese-Americans. These findings suggest that privacy is an important consideration among respondents when deciding whether or not they are willing to report socially unpopular opinions. It also indicates that the need for privacy may be variable across cultural groups when discussing sensitive topics such as illicit drug use and mental health. Why might this be so? Drug use reporting is a good example to explore in this regard.

When validated against biochemical measures of drug ingestion, systematic differences in the accuracy of the substance use behaviors reported by African-American versus white respondents in the United States have been observed. These differences have been found when using a variety of validation methodologies, including comparisons with urine, saliva, and hair specimens (Page et al. 1977; Hser 1997; Fendrich et al. 1999). Findings have generally suggested that the reports provided by African-American respondents are less accurate.

How might social desirability account for this differential in survey response quality? It is commonly recognized that African-Americans in the United States have for centuries experienced severe racial discrimination. Subsequent to slavery, African-Americans were also exploited in the name of medical research. Most well known is the Tuskegee Syphilis Study, in which a sample of African-American males with this condition were studied but remained untreated for nearly 40 years (Jones 1981). In addition, rumors of medical exploitation, including kidnapping and grave robbing to obtain medical school cadavers (Humphrey 1973), and of eugenic attempts to sterilize African-Americans (Darity and Turner 1972), and more recently, the development of the AIDS virus as a weapon to be used against African-Americans (Stevenson 1994), are not uncommon. African-Americans are also known to be treated more harshly than whites by the American criminal justice system (Sampson and Lauritsen 1997). Consequent beliefs about group exploitation and unfair treatment by law enforcement officials are likely to create strong pressures to avoid reporting socially sanctioned behaviors. In the United States, there are currently few behaviors more stigmatizing than drug use. African-Americans are thus likely to require greater privacy assurances and protections when asked to accurately report highly sensitive drug use behaviors. Greater privacy demands are also likely to be found among vulnerable minority groups in other social contexts when asked to report socially undesirable behaviors.

In general, the empirical record clearly shows that social desirability can result from question and administration characteristics, especially when questioning socially sensitive issues. The relevance and potentially disturbing role of social desirability is compounded when the different cultures studied have different views on the acceptability of particular behaviors such as abortion, and condom or drug use.

13.5 CONCLUSIONS

In much of the literature we have reviewed, social desirability is treated as a person characteristic. Particularly the early literature treats social desirability as a nuisance variable that can distort the view on the social reality. In more recent literature, there is a growing appreciation of the limitations of this view. That there is a strong correlation between the perceived desirability of a psychological trait or behavior and the likelihood that a person is willing to acknowledge that trait or behavior (Edwards 1953) cannot be dismissed as a methodological artifact; it reveals important information about the social reality of the respondent. However, social desirability appears to be a combination of style and substance; some even maintain that there is more substance than style in social desirability (McCrae and Costa 1983). Attempts to tease out the two aspects are unlikely to enhance our understanding. Similarly, it is fruitless to treat social desirability only as a person or a survey item characteristic. The two are better seen as opposite sides of the same coin: social desirability is a personality characteristic with an influence on what a respondent wants to transpire in a survey. Like other personality characteristics, it is manifested more in some situations than in others; social desirability is more likely to affect measurement when dealing with sensitive issues or with less anonymous modes of data collection.

The cross-cultural studies of personality measures of social desirability are not numerous but are remarkably convergent in findings: social desirability shows systematic cross-cultural differences. These differences are negatively related to the level of affluence of the countries and to the level of social power of the individuals involved. Individuals from more affluent countries tend on average to show lower social desirability scores. Although it is not yet clear which factors produce cultural differences in social desirability, a picture seems to be slowly emerging. One finding is the strong relationship between GNP and social desirability. From a substantive view, GNP is merely a summary label for a host of underlying variables, such as national differences in level of education and personal income. It is unlikely that cross-cultural differences in social desirability can be accounted for by schooling. At the individual level, Warnecke et al. (1997) found that even after controlling for education and income, African-Americans and Mexican-Americans revealed higher levels of social desirability than did non-Hispanic whites.

Cross-national differences in social desirability may, however, be related to cultural value systems such as the individualism and collectivism dimensions. As discussed earlier, some evidence from individual-level research suggests that social desirability scores may be higher in collectivistic societies (Johnson 1998b). In line with this finding, van Hemert et al. (in press) reported in a sample of 23 countries a significant correlation of −0.68 between a country's individualism score and its score on Eysenck's Lie Scale. These observations are consistent with other evidence from case studies of collectivistic societies that suggest that cultural emphasis on certain modes of social interaction, such as *simpatia* in Latino cultures (Triandis et al. 1984) and the "courtesy bias" in traditional Asian cultures (Jones 1983), may encourage the production of socially desirable information during survey interviews as a byproduct of respondent need to maintain positive and harmonious relations with their interviewer. Consistent with these findings, a meta-analysis conducted by Bond and Smith (1996) found a significant relationship between a measure of conformity (in the so-called Asch paradigm) of a country and its level of collectivism. Much of the empirical evidence on intra- and cross-national differences in social desirability discussed here suggests that the need for affiliation, conformity, approval, and (lack of) self-disclosure are psychological constructs closely related to social desirability.

Despite the theoretical ambiguity of the construct of social desirability (as either related to lying or the need for conformity), the practical ramifications are clear; depending on the topic of study, social desirability can lead to various distortions of cross-cultural comparisons. A researcher comparing abortion rates across countries may face social desirability as a major obstacle: women may deliberately lie and the degree of lying may differ, depending on, among other things, the legality of abortion in the country. In this instance, the classical view of social desirability as a response style will be more helpful than its alternative, and measures such as randomized response techniques may be needed to correct for the social desirability. In other cases, however, measures to correct for social desirability may reduce the validity of the cross-cultural comparison; this holds for constructs that are related to conformity, face management, and deference.

Social desirability is also related to the cultural distance of the groups studied. If social desirability affects all items of a scale and the countries compared are widely different (e.g., in terms of affluence), there is a fair chance that observed score differences are due, at least in part, to social desirability rather than to the construct of the scale. The need to correct for social desirability may also depend on its expected size. If social desirability affects only a few items, item bias techniques may identify the anomalous items. In sum, there is no simple safeguard against social desirability. However, depending on factors such as cultural distance and topic of study, it is often possible to evaluate whether there is reason to be cautious in interpreting the cross-cultural differences or even a score correction or whether such corrections would be countereffective.

Much cross-cultural research on attitudes and beliefs seems to be implicitly geared toward the observation of significant differences. It is remarkable that cross-cultural differences in the social-psychological domain are often taken for granted, whereas cross-cultural score differences in the cognitive domain are often "explained away" by referring to measurement artifacts (cf. Van de Vijver and Leung 1997b). The recognition of the impact of social desirability would be an important first step toward establishing an impartial treatment of cross-cultural differences. The development of theories that explain cultural differences in social desirability tendencies and of methods that can measure and control for them during the conduct of cross-cultural survey research are important challenges for future research.

PART FOUR

ANALYSIS OF COMPARATIVE DATA

Chapter 14

BIAS AND SUBSTANTIVE ANALYSES

FONS J. R. VAN DE VIJVER

The analysis of cross-national survey data typically focuses on substantive issues and on issues of bias and equivalence. The analysis of bias and equivalence attempts to determine the level of comparability of data across cultures and the presence of any nuisance factors. In many studies the two are treated sequentially. Bias is often the first concern, since its presence may have implications for substantive analysis. Items which prove to be inequivalent across the populations studied may, for example, need to be excluded from analysis. At the same time, these two steps cannot always be neatly separated. Bias analyses are sometimes fully integrated in the main substantive analyses. In some studies, bias itself may be the major concern; in numerous studies (such as studies of item bias) substantive issues are not, or are only marginally, addressed (e.g., Holland and Wainer 1993). In order to simplify the presentation, however, bias analysis and substantive analyses are discussed here separately. The first part of the chapter addresses the study of bias and equivalence, as it has been developed in cross-cultural psychology, while the second part presents frequently employed substantive analyses.

14.1   ANALYSIS OF BIAS

Statistical techniques are considered here which can be used to identify bias. The concepts of bias and equivalence have been described in Chapter 10. Bias refers to the presence of nuisance factors on cross-cultural comparisons, while equivalence involves the consequences of the presence of bias on cross-cultural comparisons. All surveys, not just cross-national research, can be affected by bias of certain kinds, such as social desirability and acquiescence. In comparative research, an examination of bias focuses on the *differential* impact of bias in the cultures examined.

In line with this tradition, this chapter does not describe ways to analyze the influence of factors such as social desirability on questionnaire response *per se*, but, instead, considers ways to investigate the differential impact of these factors in different cultural groups. Various techniques have been developed to examine bias, the most frequently employed of these are presented in the first part of this chapter.

Various statistical techniques can be used to discover whether the same underlying construct is measured across cultural groups. The most commonly applied statistical procedures are exploratory and confirmatory factor analysis, item response theory, and various item bias detection procedures (also known as differential item functioning). The statistical procedures for analyzing the equivalence of data follows the distinction between functional or structural equivalence, measurement unit equivalence and full score equivalence.

Before turning to the statistical procedures, three caveats are warranted. First, the focus on the statistical analysis of bias should not be taken as implying that statistical bias analysis can identify and remedy all forms of bias. Only the combination of appropriate design and proper statistical analysis can help to maximize the validity of cross-cultural comparisons. Second, most bias detecting statistics require the presence of multiple indicators of a construct. This is, however, a major obstacle for survey research, where the number of indicators is often very small. Third, two extreme and equally counterproductive views on bias are widely held. On the one hand, researchers commonly assume that questionnaires 'travel well' and proceed to compare means across cultural groups without giving any consideration to equivalence. This assumption can easily lead to incorrect interpretations of cross-cultural differences if, for example, one country is held to score higher than another, ignoring the fact that the questionnaire did not measure the target construct in one country, thereby rendering the comparison of scores meaningless. On the other hand, bias is sometimes discussed as a source of error to be eliminated at all costs. The standpoint taken here is that instruments that are not equivalent across cultures cannot be used for cross-cultural comparisons. At the same time, a lack of equivalence can point to the presence of interesting cross-cultural differences.

There are various reasons why an instrument might yield inequivalent scores. The first set of reasons is technical and relates to item or questionnaire characteristics such as poor translations. Cross-cultural differences obtained are artificially enlarged, or, less commonly, reduced and a better designed or translated instrument would reveal different cross-cultural findings. The second set of reasons is substantive and involves real cross-cultural differences which an instrument fails to capture. For example, suppose that a standard questionnaire to assess depression is administered to various cultural groups. These inventories usually contain a mixture of psychological and somatic symptoms. There is evidence that the differentiation between these two sets of symptoms is not equally strong across cultures, as illustrated by work on depression among American Indians and Alaskan Inuits using the Center for Epidemiological Studies Depression Scale (Radloff 1977). Factor analyses of Radloff's scale were carried out on data obtained from these groups. Whereas Radloff reported four independent factors among Anglo-Americans, applications among American Indians and Alaskan Inuits found highly correlated Depressed Affect and Somatic Complaint factors (Allen 1998). Tanaka-Matsumi and Marsella (1976) also found evidence that cultures differ in their relative contribution and distinctiveness of psychological and somatic complaints in depression. For instance, when Japanese and Americans were asked to generate words associated with depression, the Americans referred more often to mood states while the Japanese group gave more somatic responses. Because of the differential contribution of mood states and somatic symptoms to depression, a statistical analysis of the bias in the inventories can be expected to show that the instrument is inequivalent across countries. This is exactly what a good analysis of equivalence should do in this case, that is, make clear that the scores cannot be compared across cultures. It is clear that the inequivalence cannot be dismissed as an anomaly of instrument design or administration, since the inequivalence points to an important cross-cultural difference in the conceptualization of depression.

14.1.1 Functional and Structural Equivalence

One way to examine the extent to which an instrument measures the same constructs across different countries or cultures is to use nomological networks, a specific form of construct validity (Cronbach and Meehl 1955). These reveal the degree to which a target measure, administered in various countries, shows the expected pattern of positive correlations with other measures measuring identical or related constructs (convergent validity) and has zero correlations with measures of presumably unrelated measures (discriminant validity). Convergence and discriminant validity are both needed to demonstrate construct validity. By examining the patterning of the correlations in each country, we can find out whether an instrument measures the same underlying construct across countries. If the expected pattern of correlations is borne out, the equivalence of the target measure is supported. Paunonen and Ashton (1998) give examples from the area of personality psychology.

Exploratory Factor Analysis
Exploratory factor analysis was probably the first technique with which equivalence was systemically investigated in studies by Burt, Ahmavaara, Cattell, and Tucker in the 1950s and 1960s (Harman 1976). This work focused on assessing the identity of factors obtained in independent studies. Later research refined and generalized earlier work, but the basic ideas have remained the same. Independent factors are rotated to each other (in order to compensate for the rotational freedom in factor analysis) and factor loadings are compared. It is common in the literature on exploratory factor analysis to allow for differences in eigenvalues (reliability) of the factors. For example, if four variables have loadings of .10, .20, .30, and .40 in one group and .20, .40, .60, and .80 in a second group, the factor is assumed to be identical. To optimize agreement, the statistics used penalize for deviations from identity after a multiplying constant has been computed. In other words, "agreement up to a multiplying constant" is examined, not absolute identity.

The most commonly applied statistic is Tucker's (1951) coefficient of congruence, also known as Tucker's phi. Historically it would be more accurate to call it Burt's coefficient, since his "unadjusted correlation," proposed in 1948, is exactly the same as Tucker's. Wrigley and Neuhaus (1955; see Harman 1976, 344) independently proposed the same coefficient.

The application of exploratory factor analysis to examine structural equivalence begins with factor analyses of the questionnaire in each of the populations studied. Common procedures for determining the number of factors, such as the scree plot, eigenvalue-larger-than-one criterion (which often overestimates the dimensionality), and interpretability of the rotated solution, can be used to determine the number of factors. Many introductory sources are available for further consultation, such as Harman (1976), Kim and Mueller (1978), Frankfort-Nachmias and Leon-Guerrero (1999), and Tabachnik and Fidell (2001).

At the next stage, equivalence is examined. The stage begins with a target rotation. One cultural group is designated as the target. In some cases there may be a clear reference group (e.g., an existing instrument, frequently employed in a country, is translated into various languages, the group in which the source instrument was developed is then the reference group), but in most cases the choice of a target group is arbitrary. If only one factor is extracted, no rotation is possible and the factor congruence can be easily computed. If more than one factor is extracted, the computation of factorial congruence should be preceded by a target rotation. Dedicated software is available to carry out target rotations and compute the factorial agreement.

The statistics developed for evaluating factorial agreement can have values between −1 and +1. No strict statistical criteria are available to decide whether two factors are the same, because sampling distributions of the statistics are unknown. Different lower limits have been proposed in the literature, such as .85 (Ten Berge 1986) and .90 (e.g., Van de Vijver and Leung 1997b). Caution is needed when interpreting the value of factor congruence coefficients. The main problem to consider is the "kindness" of target rotation procedures. There has been some debate about whether or not target rotations are 'too flexible' and always lead to high factor congruence coefficients, even when starting from random configurations (Bijnen, Van der Net, and Poortinga 1986; Barrett et al. 1998). Van de Vijver and Poortinga (1994), using simulated data, showed that if a factor with a fair number of items with equally high and low loadings in two cultural groups contains one or two items with substantially different loadings (e.g., poorly translated items), factor congruence coefficients tend to remain high (often well above even the more stringent criterion of .90).

Various computer programs are available for target rotations, not infrequently as part of larger statistical packages (e.g., the Procrustes-PC program, Dijksterhuis and Van Buuren 1989). Target rotations are sometimes part of larger factor analytic packages (e.g., SCA, Kiers 1990; CEFA, Tateneni, et al. 1998). Of the main statistical programs, target rotations are part of SYSTAT and SAS; SAS routines, for example can be downloaded from the Internet. McCrae et al. (1996) also provide a routine to carry out target rotations with SAS. An SPSS routine is included in Van de Vijver and Leung (1997b).

In cases of doubt about the similarity of the factors, at least two different approaches can be envisaged. The first is an inspection of the factor loadings in the two groups (after target rotation): what do the differences in factor loadings tell us? Are the differences in factor loadings large enough to be meaningful from a substantive perspective? What is the value of the factor congruence if the item with the largest difference in loadings is omitted from the analysis? Additional analyses such as these can help establish whether the low value of the factor congruence is due to anomalies in a few items or to major differences in factor composition of the questionnaire. A second approach involves using bootstrap procedures to estimate confidence intervals for the factor congruence coefficient. Chan and his colleagues (Chan et al. 1999) demonstrate this procedure; the article provides a SAS routine to carry out the bootstrapping procedure.

The comparison of factor congruence coefficients is usually based on pairwise comparisons. When the number of cultures or countries increases, the number of pairwise comparisons can quickly become prohibitively large; for example, when 20 countries are studied, the number of comparisons is 195. In research looking at many countries, attention usually focuses more on establishing an overall pattern than on countrywise comparisons. Three procedures are useful in dealing with a large number of countries (Welkenhuysen-Gybels and Van de Vijver 2002, in review). The first, a bottom-up clustering procedure, attempts to identify a homogenous set of two countries (defined as two countries with the highest factor congruence) and then searches for other countries which can be included in the cluster. The inclusion of countries continues until a point is reached when any further addition would lower the factor congruence coefficient below an a priori defined critical value. If not all the countries in the data set are included in the first cluster, a second one is formed with the next two countries with the highest factor congruence coefficient, and the whole procedure is repeated.

The second procedure, a top-down approach, starts by combining all the data in a single data set (all the covariance matrices of the various countries are combined in a weighted manner to form a single pooled matrix; Muthén 1991, 1994). The pooled matrix is then factor analyzed. After a target rotation of the country factors to the pooled factors, the factors as found in the separate countries are compared to the pooled solution. Countries are not compared with each other but with the pooled data, which can be seen as the overall 'averaged' data set. Van de Vijver and Poortinga (2002) used this procedure in an analysis of the Inglehart postmaterialism index from the World Values Survey (Inglehart 1993, 1997). They found that 39 of 40 countries in the study showed acceptable factor congruence values, the exception being Poland. The authors attributed the low value for Poland to the time of data collection (early 1980s) in which the newly formed trade union, Solidarity, had begun to challenge the communist regime. It was also found that the concept of postmaterialism became more salient with the affluence of a country; there was a significant correlation of .32 between the factor congruence and the country's GNP (per capita in 1990). Van de Vijver and Watkins (2002) also used this procedure to analyze responses from 5258 participants of the Adult Sources of Self-Esteem Inventory from college students and other adults from 19 countries. The two-factor solution, based on factors interpreted as representing the independent and interdependent self, was found to apply to most countries.

The third approach begins by computing pairwise factor congruence coefficients for all countries (per factor). This yields a country-by-country agreement matrix which can be subjected to a dimensionality-reducing procedure, such as multidimensional scaling or cluster analysis. Welkenhuysen-Gybels and Van de Vijver (2002, in review) used the postmaterialism data (Inglehart 1993, 1997) to compare these three methods. Important similarities were found across procedures, Poland was identified as an outlier in each procedure and the Western European countries were identified as a homogenous cluster. However, other clusters of countries differed somewhat across the three methods.

This discussion of procedures has focused on exploratory factor analysis, however, other exploratory multivariate techniques can be used. In principle, any statistical procedure that allows for a comparison of underlying dimensions or structures found in different countries can be used to examine structural equivalence. The technical aspects of the procedure vary with the statistical procedure used. Nonetheless, the rationale of comparing factors, clusters or dimensions across countries applies to all these (see Chapter 15 for an application of multidimensional scaling).

Confirmatory Factor Analysis

Another tool used to examine structural equivalence is confirmatory factor analysis, which is one of several statistical techniques that form part of structural equations modeling (Long 1983; Bollen 1989; Bollen and Long 1993; Byrne 1994, 1998, 2001; Marcoulides and Schumacker 1996; Tabachnik and Fidell 2001). Like its exploratory counterpart, confirmatory factor analysis decomposes correlations or covariances; it tests to what extent observed covariances can be reconstructed assuming a particular, prior specified factor constellation. However, confirmatory factor analysis does this in a more flexible way, allowing, among other things, for conditions to be imposed on the parameters to be estimated. In confirmatory factor analysis, specific parameters to be estimated (e.g., factor loadings in various countries) can be constrained to be equal. Compared to its exploratory counterpart, confirmatory factor analysis is more flexible, allows for a more fine-grained analysis of equivalence, and offers various tests of model fit.

As a hypothetical example to demonstrate the characteristics of confirmatory factor analysis, suppose that a questionnaire consisting of seven subtests, presumed to measure three correlated, underlying factors, has been administered to groups of individuals in two countries (see Figure 14.1). An analysis of the equivalence of the questionnaire amounts to an examination of which parameters of Figure 14.1 can be taken to be invariant across countries.

The invariance of various parameters can be examined: factor loadings (λ; in Figure 14.1; these indicate the regression of the subtests on the latent variable), factor covariances (φ; these indicate the associations of the three latent variables), and error variances (ε; these indicate the variance in the subtest unaccounted for by the latent variables). Whereas exploratory factor analysis focuses exclusively on the factor loadings in the analysis of structural equivalence, confirmatory analysis offers more scope for a detailed analysis of equivalence.

Two procedures are used to examine cross-cultural equivalence of confirmatory models, such as the one of Figure 14.1: a bottom-up and a top-down approach. Both procedures consist of a hierarchy of nested models; subsequent models that are tested in the series start from the previous and add or relax equality constraints. The top-down procedure starts from the most restricted model in which all parameters are constrained to be equal across countries and gradually and systematically allows parameters to vary across countries (top-down). The bottom-up procedure specifies the same structure across countries (such as specified in Figure 14.1) and then gradually and systematically imposes equality constraints.

In the bottom-up procedure, the first step consists of estimating the parameters per country without imposing any equality constraint across countries. If a poor fit is found in at least one country, it can safely be concluded that the postulated model does not hold and that at least some of the subtests and/or latent variables are not equivalent across countries. Item bias analyses discussed in the next section could be applied to the questions in the subtests for more precise identification of the sources of threats to structural equivalence. Exploratory or confirmatory analyses of the subtests might also provide clues.

Assuming that a reasonable model fit has been found in both countries, we conclude that a model postulating the theoretically expected structure was borne out. This finding provides evidence for the adequacy of the postulated model in both countries. The good fit indicates that in each country the indicators and latent variables are related as specified in Figure 14.1. However, the parameter values (e.g., factor loadings) may still differ across the countries. So, it remains a weak test of cross-cultural similarity.

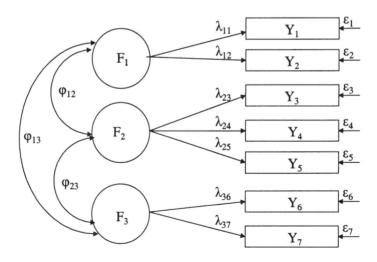

Figure 14.1. Example of a confirmatory factor analytic model (hypothetical).

In fit tests of confirmatory models it is common to see that some kinds of parameters are more central to the theory than others. The conceptual model will be less threatened when error variances differ across countries, whereas differences in factor loadings would be a more serious problem. (Allowing for differences in eigenvalues in exploratory factor analysis is related to allowing error variances to vary across countries.) Therefore, the next step starts with imposing equality constraints across countries. The lambda coefficients of Figure 14.1 are then constrained to be equal across countries. Indexes of model fit indicate how adequate the fit is, following the introduction of the equality constraints on the factor loadings. Fit tests of hierarchically nested models in confirmatory factor analysis have an attractive feature: tests of incremental fit are possible (commonly based on the overall chi square fit of the model). A statistical test is available to evaluate the loss of fit due to the introduction of the equality of factor loadings. If the loss of fit is not significant (meaning that the increase in chi square, as compared to the previous model, is not significant), this is taken as evidence that the factor loadings, usually the core of the conceptual model, are invariant across countries. If the fit statistic shows a significant decrement, the modification indices of the model could indicate which factor loadings are responsible for the loss of fit. Possibly only one or two items or scales are responsible for the lack of fit; lifting their equality constraints could then render the fit test nonsignificant. This would mean that equivalence could be established for some, but not all, of the subtests. Depending on whether the aims of the study are primarily substantive or more concerned with item properties and equivalence, the reduced data set can be further examined or an investigation begun of the reasons for the poor fit of the excluded items.

Once equality of factor loadings has been established, the next step is to examine factor covariances. Constraints on the equality of the factor loadings are maintained and restrictions on the factor covariances are added. The procedure of imposing equality constraints (on the factor loadings and covariances) and looking at the impact of this on the fit measures is repeated. In the last step of the analysis, error variances are examined in the same way. In sum, the procedure consists of beginning with a model without equality constraints, then setting parameters equal across countries, beginning with the parameters that are most important from a theoretical perspective and moving on to substantially less salient parameters. Incremental fit measures are then used to evaluate whether it was appropriate to impose the equality constraints.

The top-down procedure mirrors the above procedure. The series of analyses start with a model in which equality of parameters across countries is assumed. In subsequent steps, equality restrictions are relaxed. As before, the order should be decided on theoretical grounds, starting with allowing the conceptually least consequential parameters to vary. Incremental fit indices can be used to evaluate the effect of relaxing equality constraints.

A number of empirical studies have used confirmatory factor analysis to examine structural equivalence across instruments and populations. Windle, Iwawaki, and Lerner (1988) investigated the cross-cultural stability of the Revised Dimensions of

Temperament Survey among 234 Japanese and 114 American preschool children. Using confirmatory factor analysis, evidence was found for the stability of loadings for each of nine factors (i.e., temperament constructs) of the scale. De Groot, Koot, and Verhulst (1994) examined the cross-cultural stability of the Child Behavior Checklist, a measure of child pathology, in the United States and the Netherlands. Most syndromes (factors) were quite similar across these countries. Taylor and Boeyens (1991) applied confirmatory factor analysis to study the adequacy of the South African Personality Questionnaire (SAPQ) among blacks and whites in South Africa. Here only modest support was found for the equivalence of the constructs in the two samples. Ghorpade, Hattrup, and Lackritz (1999) demonstrated the structural equivalence of measures of locus of control and self-esteem in the United States and India. Finally, Cheung and Rensvold (2000) examined extreme scoring and acquiescence in cross-cultural research using confirmatory factor analysis. (Chapter 16 provides an elaborate example of the use of confirmatory factor analysis to study acquiescence.)

14.1.2 Measurement Unit and Full Score Equivalence

Cross-cultural survey research often deals with the question of the extent to which an instrument administered in different countries measures the same concepts. Structural equivalence is a necessary though insufficient condition for the higher forms of equivalence. Similarity of factor structures across countries does not yet imply comparability of scores across countries. Much research wants to make more direct cross-cultural score comparisons, which leads us on to measurement and full score equivalence.

Structural Equation Modeling
Analyses of measurement and full score equivalence (explained in Chapter 10) use information about the measurement unit of the answer scales. Analyses, based on correlations, such as exploratory factor analysis, cannot address measurement and full score equivalence. However, confirmatory factor analysis permits us to use information about the measurement unit by analyzing covariance matrices. Statistical procedures based on correlations and covariances can address structural equivalence, but the analysis of measurement and full score equivalence is restricted to covariance-based procedures.

Structural equation modeling can be used for examining a complicated issue in cross-cultural comparisons: the distinction between measurement unit and full score equivalence (see Chapter 10). The question to be considered is as follows: Is the cross-cultural difference in scores that was observed due to real cross-cultural differences or to method bias? (e.g., social desirability, acquiescence, previous instrument exposure). The establishment of full score equivalence is typically based on inductive arguments; cross-cultural differences are valid if rival explanations of score differences have been ruled out. The well-known induction problem in philosophy, according to which observation of countless white swans does not preclude that the next swan to appear will be black, applies equally here. Structural equation modeling enables the evaluation of the influence of external (background) variables on cross-cultural differences (assuming that these background variables have been measured). By including these background variables in the structural equation model, their influence can be tested. The more alternative explanations have been ruled out, the more confident we may feel to conclude that there are genuine cross-cultural differences in the target construct.

Item Response Theory

Item response theory is the generic name for a wide range of models, developed in psychometrics (e.g., Hambleton and Swaminathan 1985; Van der Linden and Hambleton 1997). In most models dichotomous answers are studied (correct–incorrect, agree–disagree). A respondent is represented by his or her latent trait and an item by a difficulty/popularity parameter (sometimes in addition to an item discrimination and a lower asymptote).

Item response theory has some attractive features for cross-cultural survey research. First, the mathematically assumed link of person and item parameters to produce responses can be empirically examined. The first part of the analysis always consists of a check of model fit. Most models assume a unidimensional scale, which may be highly restrictive. A model fit shows which items are appropriate in a particular culture. Second, after the fit of the data has been established (this may involve removing poorly fitting items), item response models allow for a comparison of item parameters across cultures. Statistical tests of differences of item parameters are available; these tests are often used to identify biased items. For example, Ellis, Becker, and Kimmel (1993) used item response theory to establish the equivalence of an English-language version of the Trier Personality Inventory and the original German version. Teresi (2001) gives an overview of applications of item response theory to detect item bias in the assessment of physical and mental health in multicultural populations. Third, item response theory is remarkably flexible in dealing with instruments that are not completely identical across cultures. For example, means of different cultural groups can be compared even if not all the items on which the comparison is based are identical. The models are equally flexible in comparing item parameters. As long as the underlying latent trait is identical, item parameters can be compared, irrespective of the size of the cross-cultural differences in mean scores.

The most important limitations of IRT are twofold (cf. Van de Vijver and Leung 1997b). First, the applicability of item response models may be reduced by the strict assumptions that have to be met. For example, responses in all cultural groups are assumed to be given independently and no transfer between item responses is allowed such as loss of motivation, fatigue, or learning from previous items. Goldstein (2000), commenting on a large-scale survey of adult literacy in four European countries, expresses a frequently heard complaint when he stated that the unidimensional model underlying item response theory is an oversimplification of the complex, multidimensional reality of adult literacy. Second, large sample sizes are required to obtain stable estimates, particularly in the three-parameter model.

Differential Item Functioning

Another frequently applied statistical procedure to examine full score equivalence is differential item functioning, also called item bias analysis. Two kinds of definitions of item bias have been proposed in the literature, each with its own statistical procedures. The first set of definitions focus on the domain of generalization; for example, according to Poortinga (1989), an item is biased if differences in scores on the item are not a reflection of differences in the domain of interest. In these terms, an item is biased if a difference in score of individuals or groups is not to be interpreted as a difference in the underlying construct of the questionnaire; if an item is biased, score differences are to be interpreted in terms of an ancillary construct, such as social desirability. More recently, another different set of definitions have been proposed; in these new definitions, an item is biased if respondents from different cultures with the same standing on the underlying construct (e.g., they are equally religious) do not have the same expected (mean) score on the item (e.g., Holland and Wainer 1993; Millsap and Everson 1993; Camilli and Shepard 1994; Van de Vijver and Leung 1997b). The score on the construct, which is unobservable, is usually operationalized as the total test score. The difference between the first and second type of definition is known as the distinction between unconditional and conditional bias definitions respectively (Mellenbergh 1982). The term 'conditional' here refers to using information about the respondent's standing on the latent construct to identify bias.

Dozens of statistical techniques have been proposed to study item bias. A selective list is presented here of item bias detection techniques which have either already been applied in cross-cultural survey research or have important potential for the field (see Table 14.1; a more comprehensive taxonomy of conditional bias techniques can be found in Millsap and Everson 1993). The list is based on the measurement level of the item score. For nominal-level variables an additional distinction is made between responses that are either dichotomous (e.g., yes/no) or polytomous (multiple categories, such as response options to questions about religious denomination or educational levels). A second criterion refers to the distinction between structure- and level-oriented techniques (see Chapter 10); whereas structure-level techniques examine patterns of correlation and covariance, level-oriented techniques compare patterns of mean scores across cultures. The third distinction refers to the use of statistics with known or unknown sampling distributions. For example, it is impossible to test country differences in item-total correlations because the sampling distribution of item-total correlations is unknown, which precludes a statistically rigorous test of item bias; on the other hand, in an analysis of variance the presence of item bias can be tested using *F ratios*, thereby allowing for a more rigorous test of item bias.

TABLE 14.1. Schematic Overview of Techniques to Identify Item Bias (Differential Item Functioning)

	Statistical Procedure	
	(a) Using Statistics with Unknown Sampling Distribution	
Measurement level	Level Oriented	Structure Oriented
Nominal		
Dichotomous	• Inspection of item means (Angoff 1982; see Figure 14.2)	• Inspection of discrimination indexes (Angoff 1982)
Dichotomous and polytomous	Infrequently used	Infrequently used
Ordinal	Infrequently used	Infrequently used
Interval/ratio	• Analysis of variance (Cleary and Hilton 1968)	• Exploratory factor analysis (McCrae et al. 1996)
	(b) Using Statistics with Known Sampling Distribution	
	Level Oriented	Structure Oriented
Nominal		
Dichotomous	• *Mantel-Haenszel* (Holland and Thayer 1988) • *Standardized p difference* (Dorans and Kulick 1986)	Infrequently used
Dichotomous and polytomous	• *Logistic regression* (Rogers and Swaminathan 1993) • *Loglinear modeling* (Kelderman and Macready 1990) • *Item response theory* (Hambleton and Swaminathan 1985; Van der Linden and Hambleton 1997)	Infrequently used
Ordinal	Infrequently used	Infrequently used
Interval/ratio	• *Analysis of variance* (Van de Vijver and Leung 1997b)	• Structural equation modeling (Little 1997)

NOTE: Italicized procedures are based on conditional definitions of item bias and use total test score as a proxy for the standing of an individual on the latent trait.

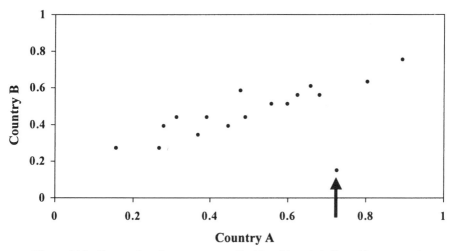

Figure 14.2. Scatter plot of country means (the biased item is indicated by an arrow).

The literature is currently dominated by applications of conditional definitions. As not all cross-cultural surveys have the large samples required for conditional techniques, unconditional techniques are also considered here. Techniques suitable for use with small samples include the comparison of item-total correlations and the inspection of a plot of item means. A constructed scatter plot is given in Figure 14.2. The item indicated by the arrow has an unexpectedly high score in Country A compared against its score in Country B. This item may be biased.

Such small-sample techniques are often simple to apply (Muñiz, Hambleton, and Xing 2001). The main problem is the lack of hard criteria for deciding whether an item is biased. In particular, it can be difficult to avoid making Type II errors (i.e., classifying biased items as unbiased). Bootstrapping and jackknifing procedures could conceivably help to overcome some of the small-sample problems of item bias detection procedures.

Only one of the conditional techniques, analysis of variance, is presented rather briefly here. This may nonetheless suffice as an introduction to various other procedures since most apply the same basic reasoning simply with different statistical procedures. Suppose that a questionnaire of 20 items dealing with attitudes toward health behavior (Likert-type, with response alternatives ranging from one to five) is administered to 1,000 respondents in two countries. The item bias analysis begins by determining score levels for splitting up the sample in homogeneous score groups (subgroups). It is assumed that all the individuals in a score group have the same attitude toward health behavior. Ideally, there are as many cutoff points as there are different scores; in our example scores can range from 20 to 100. Respondents with a score of one or a score of five on all items can be excluded from the analysis, as these extreme score groups have identical scores across countries on all items. In practice, samples are almost never large enough to allow for a split into the maximum number of score groups. Here, for example, this would amount to 79 groups (21 through 99). Thus scores are combined into groups. Within each score group, we assume that respondents have the same standing on the underlying concept. The cutoff points to split up the total sample in a number of groups have to be determined without considering culture. If five score groups are to be distinguished, the 20th, 40th, 60th and 80th percentile points of the total sample could be taken as cutoff points. After the cutoff points have been determined, each person gets a new score added: his or her score level; persons in the lowest group get a score of one, in the second a score of two, etc. The next step consists of a set of analyses of variance, one per item; score level and culture are taken as independent variables and the item score as the dependent variable.

The analysis of variance tests the significance of three effects: the main effects of score level and of culture as well as their interaction. The main effect of score level is not usually of interest for bias; it usually indicates that individuals scoring higher on the scale also score higher on the item. The other effects are more interesting from a bias perspective. The main effect of culture signifies that one culture scores systematically above another culture, even when score level has been taken into account. This means that a person from Culture A and a person from Culture B who have the same standing on the latent trait are not equally likely to endorse the item. For example, an item whether people bike (as part of a health program) may be more endorsed in a country in which biking is common than in a country in which biking is not common. An item that shows a main effect of culture is said to be uniformly biased (Mellenbergh 1982); a hypothetical example is presented in Figure 14.3. In this example, the last effect, the interaction between culture and score level is also interesting. A significant interaction between the two indicates that an item discriminates better in one culture than in the other (see lowest panel of Figure 14.3).

Applications of item bias techniques reveal a fairly consistent picture. First, uniform bias is more prevalent than nonuniform bias. Second, really poor items often show both kinds of bias. Third, conditional techniques usually flag more items as biased than do unconditional techniques. This is not surprising. Conditional techniques are often based on large samples and examine the data in more detail (as they also consider score level). Indeed a recurrent problem in particular for conditional techniques is that items that are found to show item bias do not seem to share any substantive characteristic. It is significant in this respect that expert judgments about bias in items and item bias statistics do not tend to converge (e.g., Huang, Church, and Katigbak 1995).

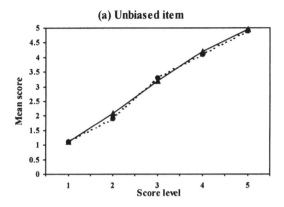

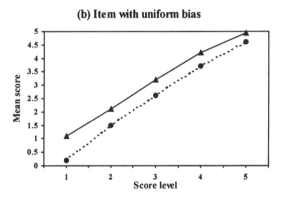

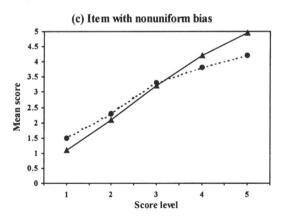

Figure 14.3. Graphical presentation of biased and unbiased items (solid and dotted graphs represent the scores in two different countries).

Finally, a word of caution is needed about claims often made in applications of item bias techniques. These are often applied to ensure full score equivalence. It is then taken for granted that once biased items have been removed, scores can be compared across countries. This ignores the fact that item bias is only one of a number threats to comparability. For instance, if a scale measures only some aspects of a construct in a specific culture (cf. Ho's example of filial piety, discussed in Chapter 10), item bias will not show that the scale is incomplete and fails to tap other relevant aspects. Item bias can reveal information about the inappropriateness of items, not about the inappropriateness of an instrument at a more global level. Similarly, item bias techniques will only identify response sets, such as social desirability, if these affect a small number of items. If most or all of the items are affected, item bias techniques are unable to differentiate between measurement and full-score equivalence. Used with these limitations in mind, item bias techniques are useful tools to identify bias.

14.1.3 Nonidentical Scale Items across Cultures

Statistical procedures are now available to analyze scales with partly dissimilar questions. In exploratory and confirmatory factor analysis, nonoverlapping items can be treated as missing values (for which no values are imputed). These then help to define the factors but are not used for evaluating cross-cultural similarity of the factors. Similarly, in item response theory nonoverlapping items which measure the same latent trait as the overlapping items do not have to be removed in the assessment of bias in the overlapping items. Other conditional item bias techniques can deal less easily with dissimilar stimuli. For instance, if we wanted to determine score levels in an analysis of variance, we would have to impute values for the nonoverlapping items for all the respondents in a cultural group. The validity of the imputation could remain doubtful.

When completely different scales have been used in each country, nomological networks can still be examined (by administering scales of related and unrelated constructs). Correlations between differently worded measures provide insight in the functional equivalence of a scale. In analyzing nomological networks across cultures, we need to assume equivalence of the measures used to validate the target measure. When no data are available to support this assumption, a vicious circle may arise. In order to assess the functional equivalence of a target measure, we need to establish the equivalence of the external measures. This would require another study, using new measures to provide evidence of functional equivalence, and so forth. Nomological networks also lack rich detail. Exploratory and confirmatory factor analyses and in particular item bias analyses often focus on item-level comparisons. In contrast, comparisons of nomological networks focus on correlations with external measures, which are by definition less fine grained. This lack of detail makes it less likely that subtle differences in measurement properties and related subtle cross-cultural differences are detected.

14.2   SUBSTANTIVE ANALYSIS

Before presenting the various statistical techniques, a brief description of more general topics in the analysis of cross-cultural survey data is given. The first topic concerns the *distinction between structure- and level-oriented analyses* (see Chapter 10). The former involves the comparison of constructs (e.g., can the same indicators be used to measure depression in the United States and Japan?), while the latter involves a comparison of mean scores (e.g., is there a difference in the average level of depression of people in the United States and Japan?). Level-oriented questions are more difficult to answer, as means are more susceptible to various sources of bias than are correlations and covariances.

The second topic involves *identity of measurement*: Were the scales used the same, partly the same (adapted), or entirely different and only functionally equivalent in the groups examined? The scope for statistical analysis is largest for identical instruments and smallest for completely different instruments. In the latter case, analyses are often restricted to the comparison of nomological networks (see previous section). Adapted instruments allow for more statistical tools to be employed, such as exploratory factor analysis, structural equation modeling, and item response theory (as described in the previous section). However, most statistical techniques, such as an analysis of variance or t tests assume that items which do not 'ask the same question' are excluded. When identical (usually translated) instruments have been used, these special conditions and precautions do not apply and all the statistical techniques discussed here can be used.

The unit of analysis relates to whether we deal with data at the level of the individual (information on patients or people in a firm), at an institutional level (information about hospitals or companies) or at a national or cultural level (health or work related information for a culture or country). Traditionally, studies deal with questions that focus on a single level (e.g., international comparison of health care), although the data may be derived from more than one level (e.g., patient satisfaction data at the individual level, organizational structures of hospitals at the institutional, and expenditure on health care *per capita* at the national level). In recent years, multilevel models have been used to examine data at different levels simultaneously. In multilevel models, discussed below, variables are studied at different levels of aggregation and these levels are linked. For example, in a cross-cultural study, math achievement of individual pupils can be studied as a function of intelligence (individual level), of a teacher's pedagogical skills (class level), of school quality (school level), and of educational expenditure per head at national level (country level). One of the strong attractions of these models is that they allow for the simultaneous analysis of data at different levels. However, two potential problems should be taken into account in multilevel applications (Achen and Shively 1995).

The first has to do with problems of cross-level inferences, such as the ecological fallacy (coming to wrong conclusions by applying aggregate-level phenomena to individual members of the group; Robinson 1950; see also Hofstede 1980). For example, there are good pupils in bad schools and bad pupils in good schools; application of school characteristics to these pupils would lead to an incorrect estimation of their scores. Conversely, there is the individual-differences fallacy (coming to wrong conclusions by applying individual-level phenomena to aggregate units; Richards, Gottfredson, and Gottfredson 1990, 1991). For example, from the observation that many people are poor in a society, it would be wrong to conclude that the country's Gross National Product is low.

The second problem is related to the meaning of scores after (dis)aggregation (Van de Vijver and Poortinga 2002). In a cross-cultural survey dealing with a topic that is sensitive in only some countries (e.g., abortion, divorce, the role of women), it may well be that individual differences within each of the countries can be interpreted in terms of the construct measured, but that cross-national differences are more a reflection of social desirability. In such cases, within and across culture scores do not have the same meaning. Confirmatory factor analysis (Muthén 1991, 1994) and exploratory factor analysis can be used to assess the similarity of meaning at different aggregation levels. The analysis compares the factor structure at individual and country level in much the same way as was described earlier.

In the past it was common to find reports of two-country studies in the literature; more recently, however, cross-cultural surveys involve many countries and there is a corresponding shift in study purpose. Multicountry surveys examine the role of one or more dimensions on which the countries differ, while two-country studies map differences of specific countries. In general, *cross-cultural survey research can compare specific countries* (e.g., a comparison of depression levels in the United States and Japan) *or can try to understand the role of an underlying dimension* (e.g., is there a relationship between a country's depression score and its Gross National Product?). While differences or similarities may be quickly found in two-country comparisons, explanations for differences will often be much more elusive. Observed score differences can be related to a host of differences between the countries. For example, individualism–collectivism (Hofstede 1980; Triandis 1995) is often used to explain differences in responses by Japanese and Americans. However, Japan and the United States differ in a great many respects and it is unlikely that every kind of difference is due to differences in individualism-collectivism. If cultures can indeed be ordered along a dimension, the number of rival explanations for observed difference reduces greatly when the number of cultures employed increases (Campbell 1986).

A final issue to be decided is *what to do with missing values (item nonresponse;* see Chapter 11 for a discussion of unit nonresponse). The traditional method of estimating missing values, based on either listwise or pairwise deletion, has well documented problems (e.g., Rubin 1987). Listwise deletion may reduce the sample size considerably and may yield statistics (means, standard deviations and covariances) that are poor approximations of their population values. If the missing values are not distributed randomly over the data matrix, pairwise deletion may produce the same problem. In addition, pairwise deletion can lead to problems in determining sample size. Covariance matrices based on pairwise deletion usually cannot be used for multivariate analyses like factor analysis. More sophisticated techniques are now available to impute missing values (e.g., Rubin 1987; Groves and Couper 1998).

14.2.1 Structure-Oriented Techniques

Simple Descriptive Measures, Graphical Data Representations
The first step in substantive data analysis, sometimes also combined with an analysis of the equivalence of the data, is often exploratory and consists of a visual inspection of various data descriptives. Since Tukey's (1977) book first appeared, exploratory data analysis has gained prominence; graphical data displays such as boxplots and stem-and-leaf diagrams have become standard tools. Visual inspection provides insight in various data set features, such as the quality of the coding, the occurrence of inadmissible scores, and the identification of outliers. It allows for a first rough scanning of cross-cultural differences and similarities and a comparison of the form of score distributions (such as means, standard deviations, floor and ceiling effects for questions, and multimodality of score distributions).

Internal consistencies are also often compared (for details, see Van de Vijver 1997). This comparison is quick and useful, though it may lack detail. If the test is significant, sources of the differences need to be sought. This may lead to an analysis of the structural equivalence of the instrument.

Exploratory Factor Analysis
Exploratory factor analysis is, as indicated, an important tool for cross-cultural survey research (for an introductory text, see Kim and Mueller 1978; more advanced textbooks are Harman 1976; Tabachnik and Fidell 2001). In Chapter 10 factor analysis was discussed as a means of analyzing structural equivalence. Here, it is described as a tool for examining models and theories about the structure of a scale in a cross-cultural framework.

In cross-cultural applications of factor analysis certain issues need to be considered that are usually irrelevant for monocultural research. One such issue relates to combining data from different cultural groups prior to analysis. Even if an analysis of structural equivalence shows that the factors are identical across the cultures studied, problems can arise when combining data. If groups differ in mean scores on the items in a scale, merging the data may affect the factor structure. Figure 14.4 illustrates aggregation effects for two items (constructed example). In two cultures the two items show a strong negative correlation (correlations of −.94 and −.97). Combining the data, however, yields a positive correlation (of .55).

There are at least three different ways to merge data prior to conducting a factor analysis. The first is to simply deny or ignore the existence of aggregation effects and to merge the data, without any form of standardization. In our example this would result in a positive correlation between the two items. In more general terms, the analysis combines individual- and culture-level differences. As a result, factor analysis may show a combination of individual and country factors which are difficult to interpret. If one country scores higher on various items than another country, a strong first factor may be found which gauges the overall country score differences.

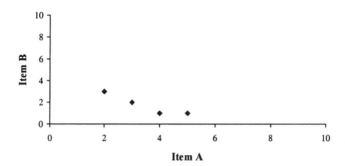

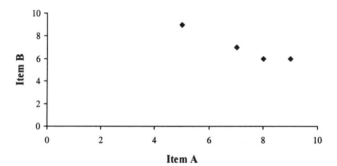

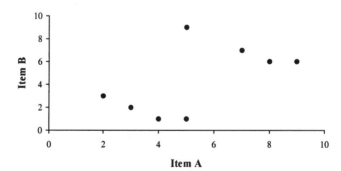

Figure 14.4. Example of aggregation effects.

The second way of combining the data uses one of a number of forms of standardization. For example, data could be standardized per country (so as to produce a mean of zero and a variance of 1 per country). In Figure 14.4 this would have given a correlation of –.96. A second way to standardize involves centering; deviance scores are computed (the country mean is deducted from each score, while country scores in standard deviation remain intact). The correlation for the centered data in Figure 14.4 is –.96. As can be seen, correlations of both standardizations come close to the correlations per country. If sample sizes were not equal across cultures, a weighting procedure could be applied in order to correct for sample size.

The third approach to combining data for a factor analysis uses pooled covariance matrices (e.g., Muthén 1991, 1994). Instead of pooling individual data, covariance matrices are first computed per culture. These are then averaged to make a pooled covariance matrix which is then factor analysed.

The psychological literature contains many examples of cross-cultural applications of exploratory factor analysis. An example of a personality measure studied in various cultural groups is the UCLA Loneliness Scale. Exploratory factor analyses have shown that the scale has a stable unifactorial structure among Anglo-American and Mexican-American adolescents (Higbee and Roberts 1994), South African students (Pretorius 1993), and Zimbabwean adolescents and adults (Wilson et al. 1992). Other cross-cultural applications of factor analyses can be found in the work on the translation of the five-factor model of personality, a currently very popular personality model. McCrae et al. (1998) compared the equivalence of Filipino and French translations of a Big Five questionnaire. All congruence coefficients were well above .90 (McCrae and Costa 1997). Cross-cultural work with the Eysenck Personality Questionnaire are further prominent examples (e.g., Eysenck and Eysenck 1983). The three Eysenck factors, psychoticism, extraversion, and neuroticism, are "strongly replicable" (Barrett et al. 1998, 805).

Studies of locus of control illustrate clearly that not all measures in cross-cultural comparisons yield stable factors. In a review of such studies, Dyal (1984) reports a lack of stability of factors in popular locus of control scales, even among closely related cultural groups. Van Haaften and Van de Vijver (1996) administered Paulhus's Locus of Control Scale (cf. Paulhus and van Selst 1990), measuring intrapersonal, interpersonal, and socio-political aspects among illiterate Sahel dwellers. The three scales could not be identified in a factor analysis.

Structural Equation Modeling: Confirmatory Factor Analysis and Path Analysis

The term 'structural equation modeling' encompasses a variety of statistical techniques used to model covariance and correlation matrices. Confirmatory factor analysis and path analysis (with or without including latent variables) are often associated with structural equation modeling (introductory texts can be found in Long 1983; Bollen 1989; Bollen and Long 1993; Byrne 1994, 1998, 2001; Marcoulides and Schumacker 1996; Tabachnik and Fidell 2001). Confirmatory factor analysis allows for a direct test of presumed relationships between variables (items or scales) and their underlying common latent variables. It can be seen as an extension of exploratory factor analysis with more control over the loadings and factor correlations and a more rigorous fit of model. Path analysis is an extension of regression analysis in that more dependent variables can be studied at once and that independent and dependent variables may be indirectly related (through mediating or moderating variables; Baron and Kenny 1986; Marcoulides and Schumacker 1996).

Structural equation modeling is a versatile tool for examining both static and dynamic constructs and their models. Thus, in addition to its use in exploring construct equivalence discussed earlier, confirmatory factor analysis can help to model cross-cultural data once equivalence has been established. Compared to exploratory factor analysis, structural equation modeling allows for a more direct 'translation' of theory into statistical terms; an extensive set of model fit tests is available to evaluate the adequacy of the model and, hence, the underlying theory.

In analyzing equivalence, a confirmatory factor analytic model is postulated in every cultural group and model tests are used to evaluate the hypothesis of identical parameters across cultures. In this instance culture is not part of the model parameters to be estimated. The observation that the same model holds in different cultures can be an important finding. A second way of employing culture as a variable is to use a dummy variable. Instead of analyzing two models, possibly with dissimilar parameters, culture can be taken as a dummy for the variable(s) showing different parameter values. Culture then functions as a moderator variable (Baron and Kenny 1986; Marcoulides and Schumacker 1996). If the moderator effect is present for a single independent variable and impacts on all variables linked by means of paths with this variable, introducing culture as moderator provides a succinct way of expressing cross-cultural differences. However, if the moderator effect is present for more variables, or does not affect all linked variables similarly, using culture as a moderator is not very efficient and the presentation of group-specific models may be easier to interpret.

Structural equation models have various attractive features for cross-cultural survey research. As discussed in Chapter 10, these models provide a flexible tool to compare structures across cultures. Moreover, the models allow for the estimation and comparison of mean scores across cultures, even when not all the relationships of a path model with latent variables are identical across groups; Byrne, Shavelson, and Muthén (1989) coined the term "partial measurement invariance" to describe this partial dissimilarity of paths and provide a procedure to compare the means of the latent variables. Little (1997) developed a "means and covariance structures" (MACS) model in which, once equivalence of a model is established, latent construct means of different cultural groups are compared. Malmberg et al. (2001) used this model to study the control beliefs of urban and rural Tanzanian children.

Cluster Analysis

The aim of cluster analysis is the classification of multivariate data in a limited set of nonoverlapping categories; each category has some common characteristic that is not shared by members of another category. (For an introductory text see Aldenderfer and Blashfield 1984; for more advanced treatment see Everitt 2001.) Its relevance for cross-cultural survey research is obvious, both for assessing equivalence and for substantive analyses. We are not aware of any published research which formally compares clusters obtained in different cultural groups in order to evaluate cross-cultural agreement. Espe (1985) applies cluster analysis cross-culturally in examining the adequacy of the Graphic Differential as a language-free alternative to the Semantic Differential (Osgood, Suci, and Tannenbaum 1957). The original scale uses bipolar adjectives such as "good-bad" and "strong-weak"; the Graphic Differential employs pictorial scales. Previous research using American students revealed the presence of three clusters of the bipolar adjectives in the Graphic Differential. Espe administered the pictorial scales to American and German students; they were asked to sort the figures into nonoverlapping groups. A matrix with the number of co-occurrences between pairs of figures was subjected to a cluster analysis. Three clusters were reported, more or less representing the three factors of the bipolar adjective version: evaluation, potency, and activity. The clusters were largely identical for the two groups, except for a few deviations. How much do we need? For example, whereas in the American sample an arrow pointing up was associated with evaluation, for the German subjects the arrow was more linked to activity. The author concluded that the graphic scales do not fully constitute a language-free, cross-culturally comparable alternative to the bipolar adjective scales.

Latent Class Analysis

If a group of respondents consists of different homogeneous subgroups, each having their own response characteristics (e.g., proponents and opponents of abortion), latent class analysis is a helpful tool to identify the homogeneous subgroups. In cross-cultural survey research the identification of latent classes can be very helpful. In particular, when the proportion of respondents belonging to different classes differ across cultural groups (e.g., more proponents of a group in country A than in country B), latent class analysis can help to clarify the nature of cross-cultural differences observed by identifying groups that are homogenous. In the last decade latent class analysis has been combined with various other statistical techniques, such as loglinear modeling (Hagenaars 1993) and has become a versatile tool for cross-cultural analysis of categorical data (Vermunt 1997). An example of a cross-cultural application of latent class analysis is described by Saris (see Chapter 18).

14.2.2 Level-Oriented Techniques

Analysis of Variance and Related Models
The most common techniques used to compare mean scores of cultural groups are t tests (in the case of two groups) and univariate or multivariate analysis of variance (in the case of at least two groups), with culture as the independent variable (Stevens 1996; Hinkle, Wiersma, and Jurs 1998; Frankfort-Nachmias and Leon-Guerrero 1999).

In cross-cultural surveys, the object is often not primarily to find significant differences, but to understand patterns of difference, such as which variables reveal large country difference and which point to only small differences. One way of addressing this question is to compare effect sizes for sets of variables (Cohen 1988; see also Chapter 21). Various effect sizes have been proposed. The first type refers to the proportion of variance accounted for by the factor culture (often designated as η^2 and ω^2). The second type refers to some standardized mean difference (such as the difference of two country means divided by their pooled standard deviation). Cohen (1988) suggested values of .2, .5, and .8 as minimum values indicating small, medium and large effect sizes, respectively. One advantage of the standardized mean difference measure is the ease of interpretation and directionality; the measure indicates how many units group means differ (expressed on the well known z scale) and it indicates which group scores higher (both positive and negative effect sizes are possible). On the other hand, the proportion of variance measure has the advantage of applying to any number of groups. Rosenthal, Rosnow, and Rubin (2000) developed correlational measures for analysis of variance designs to overcome the problem of nondirectionality of the omnibus effect size measures (such as η^2 and ω^2).

Profile analysis is a related procedure which is used to find the patterning of group differences in a set of related dependent variables. The profile of a culture is formed by its average scores on all the dependent variables. The analysis compares the form of the profiles. The first outcome is constituted by the complete coincidence of profiles and the absence of any cross-cultural differences. The second is parallelism of profiles: cultures differ in mean score but the effect size is relatively invariant across dependent variables (e.g., all dependent variables show a difference of about .25 standard deviation). This finding may point to real difference in scores between the cultures (full score equivalence) and/or to method bias (measurement unit equivalence). Finally, the profiles may differ in form and elevation, in which case effect sizes vary across dependent variables. An example can be found in work by Trentini and Muzio (1995), who applied profile analysis to examine work values of secondary school students, higher education students, and adult samples from various countries.

Finally, the analysis of covariance should be mentioned. This was originally developed for comparing groups that differ in some background characteristic which may impact on the dependent variable. For example, two groups compared on reading achievement do not have exactly the same level of schooling. If a proxy for schooling is available at individual level (e.g., number of years of education), this proxy variable can be used to statistically control for differences in level of schooling.

Two points need to be noted for applications of analysis of covariance in cross-cultural survey research. First, it is essential to carry out tests of homogeneity of regression lines across cultures. An analysis of covariance assumes parallelism of regression lines linking the covariate to the dependent variable across groups. Violations of the assumptions can lead to paradoxical results (Lord 1967). Second, the adequacy of covariance analysis for correcting for group differences can be easily overrated; it is questionable whether the analysis of covariance is able to correct for massive differences between cultures. For example, controlling for the number of years of schooling in an analysis of covariance will not work if the education levels of the populations studied differ substantially.

An interesting 'by-product' of the analysis of covariance is the test of the significance of the covariate; the analysis tells us whether the covariate is significantly related to the dependent variable in each cultural group. The relationship between the covariate and the dependent variable is often interesting in its own right. The greatest potential of covariance analysis in cross-cultural research may lie in the estimation of the impact of covariates on the size of cross-cultural differences. Poortinga and Van de Vijver (1987; see also Van de Vijver and Leung 1997b) describe a hierarchical regression procedure aimed at comparing cross-cultural differences before and after the introduction of the covariate in order to examine the role of the covariate in explaining cross-cultural score differences.

Regression Analysis

Regression analysis evaluates the association between one or more independent variables and a dependent variable (of interval or ratio level) in terms of the amount of variance of the dependent variable which is accounted for by the independent variable(s). Regression coefficients express the strength of the relationship between one or more independent variables and a dependent variable. The squared multiple correlation expresses the amount of variance explained by all the independent variables. This statistic gives an overall evaluation of the success of the independent variables in predicting variation in the dependent variable (an introductory text can be found in Fox 1997; more advanced textbooks are Cohen and Cohen 1975, and Pedhazur 1982).

Culture, as an independent variable, can be used in various ways in regression analysis. For example, the equality of regression lines in different cultures can be tested. Differences in both intercepts and regression coefficients may be relevant. The program P1R of the BMDP series and programs for structural equation modeling such as AMOS, EQS, and LISREL can be used to test the equality of regression lines. Another approach enters culture as a dummy-coded independent variable. The dummy variable often tests a 'main effect' of culture or an interaction of culture and another independent variable. If culture is coded as a main effect, its regression coefficient measures the association of culture with the dependent variable after controlling for all other independent variables. Culture can also be entered into the analysis as an interaction with another variable (in this case culture is a moderator variable). Poortinga and Van de Vijver (1987; cf. Van de Vijver and Leung 1997b, 117-118) proposed an approach in which predictors of cross-cultural score differences are examined. The analysis essentially consists of three steps. In the first step, the size of cross-cultural differences (in, say, religiosity) is evaluated in an analysis of variance, using culture as the independent variable. The size of these differences is expressed as a proportion of variance accounted for (Cohen 1988). In the second step, an explanatory variable is introduced (say, level of education). The standardized regression coefficient of education provides an indication of the strength of the association with religiosity. The third step then evaluates the size of the cross-cultural differences on the basis of an analysis of variance, with culture as the independent variable and the residual scores of the regression analysis as the dependent variable. A suitably chosen explanatory variable can be expected to reduce or even eliminate the size of the cross-cultural differences.

Earley (1989) uses regression analysis to investigate the role of individualism-collectivism in explaining cultural differences in social loafing. Social loafing is the tendency to exert less effort on work in a group than when working alone. It is widely documented that Americans show a tendency to show social loafing. Earley hypothesized that this tendency is due to the individualistic nature of American society. Chinese and American management trainees were asked to work as groups. The total output of the group was the dependent variable, and their individualism-collectivism level was measured as a covariate. Hierarchical regression analyses were used to test whether the effect of culture could be explained by the independent variables (i.e., experimental manipulations such as asking subjects to put their name on adhesive labels so as to create the impression of accountability of their individual performance) and the covariate. In the regression analysis, the independent variables, the covariates, and all the associated interactions were entered first; culture was entered as an independent variable in the last step. When the variables were entered in this sequence, culture was found to be nonsignificant. Thus, once effects associated with the independent variables and individual-collectivism were taken into account, cultural difference in social loafing was not significant. The independent variables and individualism-collectivism explained all the cultural differences in social loafing.

Multilevel Models

Multilevel models deal with hierarchical structures in data in a statistically more adequate way than do the previous procedures without losing information at aggregation level (introductory texts can be found in Goldstein 1987; Bryk and Raudenbush 1992; Lipsey and Wilson 2001; Hox forthcoming; see also Chapter 21). Suppose that we are interested in studying school performance on a standardized test that is administered nation-wide and that, in addition to the individual-level data, we also have data on teacher and school quality. If we want to examine test score variation as a function of teacher and school characteristics, we face several problems. Two rather obvious procedures have been somewhat discredited. The first is to disaggregate all the higher order variables to the individual level. The problem with this approach is that if we know that students are in the same class, then we also know that they have the same value on each of the class variables. Thus we cannot use the assumption of independence of observations that is basic for classic statistical techniques. The other alternative is to aggregate the individual-level variables to the higher level. Following this line, we would aggregate student characteristics over classes and do a class analysis, perhaps weighted with class size. The main problem here is that we throw away all the within-group information, which may be as much as 80% or 90% of the total variation, before we even start the analysis (de Leeuw 1992; in Bryk and Raudenbush 1992 xiv).

A good example of a cross-cultural multilevel study was conducted by Entwistle and Mason (1985). The authors studied the relationship between socio-economic status and fertility. In less affluent countries there tends to be a positive relationship between status and fertility, while in more affluent countries a negative relationship is often found.

14.3 CONCLUSION

The statistical toolbox of the cross-cultural survey researcher has become both larger and more sophisticated in the last few decades. Various tools such as multigroup structural equation modeling, multilevel modeling, item response theory, and various models of nominal and ordinal data are now available. It is important to point out that comparative techniques for instrument design have not enjoyed anything like the same degree of refinement. It would be fatal to neglect design and to expect analysis to produce a silk purse and the wherewithal to fill it. Statistical sophistication in data analysis cannot compensate for poor quality of study design nor for lack of cultural sophistication. Survey quality is the net result of many factors; statistical sophistication is just one of these. Both a sophisticated analysis of a poor instrument and a poor analysis of a good instrument yield low quality. Only through a combination of cultural awareness and statistical sophistication can we arrive at high quality survey research.

Chapter 15

MULTIDIMENSIONAL SCALING

JOHNNY FONTAINE

Multidimensional scaling models form relatively simple but versatile tools for the examination of dimensional structure in a broad range of data. The present chapter addresses multidimensional scaling in a cross-cultural survey context. The first part of the chapter presents the general techniques, while the second part describes how these techniques can be employed to deal with substantive questions in cross-cultural research. The chapter ends with some guidelines for the use of multidimensional scaling in cross-cultural research.

15.1 MULTIDIMENSIONAL SCALING MODELS

15.1.1 What Is Multidimensional Scaling?

Multidimensional scaling models (MDS) are suited for the structural representation of empirical interrelationships among a set of stimuli or items. These models both represent items as points in a multidimensional space in such a way that the distances between the points represent the observed dissimilarities between the items as well as possible (Kruskal and Wish 1978; Schiffman, Reynolds, and Young 1981; Davison 1983; Cox and Cox 1994; Borg and Groenen 1997).

The advantage of MDS is that it can be used for a broad range of data. All data that can directly or indirectly yield dissimilarities or similarities between pairs of items can be analyzed by MDS. The dissimilarities need only be measured at an ordinal level. Thus, unlike factor analysis, MDS can represent the structural information from direct similarity ratings and from similarity sorting. These two types of data are especially interesting for cross-cultural research. Similarity ratings and similarity sortings can be relatively easily explained to respondents who are not acquainted with Western research methods. Moreover, respondents can implicitly or explicitly choose the criteria according to which they judge the similarity between items. Consequently, such tasks reduce the threat of cultural imposition and can be used to investigate folk theories (lay theories) in domains such as emotion, personality, and theories of mind.

MDS can also be used as an alternative to factor analysis. Pearson correlations can be easily transformed into dissimilarities that can be used as input for MDS (namely,

$\sqrt{2 - 2r_{ij}}$ corresponds to Euclidean distances between standardized items). MDS is used like this if the theoretical model is explicitly formulated in geometrical terms, as in Holland's hexagonal model of vocational interests (Holland 1957) and Schwartz's circular order of the value domain (Schwartz 1992).

The geometrical configuration generated by MDS can be interpreted in two ways. First, the dimensions can be seen as organizing principles or properties of the construct domain. The coordinates on each of the dimensions indicate to which extent each item is characterized by each of the properties. A dimensional interpretation can be supported empirically by asking independent raters to rate each of the items on each of the underlying properties and to correlate these independent ratings with the coordinates of the items on the dimensions. An alternative approach is to look for bounded subregions within the space that contain qualitatively different types of items (Kruskal and Wish 1978; Davison 1983). This is called a regional interpretation. A regional interpretation can be strengthened by predicting the groups of items *a priori* that should be qualitatively similar and should thus be situated within a single bounded subregion, as is, for instance, the case in facet analysis (Borg and Groenen 1997).

15.1.2 Multidimensional Scaling and Structural Equivalence across Cultural Groups

In the MDS literature, a whole family of models has been developed for dissimilarity data from more than one observational unit. These are called three-way MDS models (Borg and Groenen 1997). Since cultural group is just one type of observational unit, these three-way models can be directly applied to cross-cultural research. In the present chapter, we restrict our attention to MDS models that generate a single configuration across cultural groups. These models can be easily applied in cross-cultural research and have already proved their value. Readers interested in more complex three-way MDS models are referred to more specialized literature, such as Borg and Groenen (1997) and Cox and Cox (1994).

We first present a conceptual framework that identifies four major levels of structural equivalence, then two possible approaches for empirically identifying the level of structural equivalence. Finally, we focus on the impact of unreliability on MDS representations.

Four Levels of Equivalence
Four hierarchically ordered levels of structural equivalence can be distinguished, based on the answers to two questions. The first question is whether the domain is organized according to the same underlying properties or dimensions in the different cultural groups. The second question is to what extent the items share the same position on each of the dimensions across cultural groups. The answers to the two

questions are interdependent. The dimensions are constructed based on dissimilarities between items, and the items derive their meaning from their position on the dimensions.

At the highest level of structural equivalence, the domain is organized according to the same underlying dimensions and all the items share the same positions on these dimensions. In this case, one can speak of concept identity, since both the organizing principles and the empirical manifestations of a domain are the same across cultural groups. This also means that a single geometrical representation is sufficient to represent the internal structure in all groups.

At the second level, only a representative subset of items shares the same position on the same underlying dimensions across cultural groups. While the domain is still organized according to the same underlying dimensions in the different cultural groups, there are culture-specific items or items that shift in meaning. However, a substantial subset of items that can be considered representative for the whole domain still shares the same positions on the underlying dimensions, called here representative item overlap. With representative item overlap, it is possible to study the domain based only on the overlapping items. Moreover, the overlapping items can be used as a common basis for the investigation of the structural characteristics of the culture-specific items. Based on the overlapping items, the structural information from all the cultural groups can be represented by means of a single configuration. In that configuration, overlapping items share the same positions across groups, and culture-specific items have a culture-specific position. In a common configuration, the meaning of culture-specific items can be conveyed easily to members of other cultural groups by relating them to the overlapping items.

At the third level, the underlying dimensions are shared across cultural groups, however, the items are not (or there is only non-representative partial overlap). For instance, if one is interested in differences between social roles from a cross-cultural perspective, no close translations might exist for social roles between cultural groups with a different social organization. However, the underlying organization of these roles in terms of dominance is probably universal. If there is no representative overlap between the items, it is not possible to represent the internal structure of a domain by means of a single structure. A specific representation is needed for each cultural group. At this level, the conclusion of concept equivalence has to be empirically supported by relating additional information about the items to the position on the dimensions. In the example of social roles, independent ratings of dominance within each cultural group could support the interpretation.

At a fourth level, the organizing principles are not shared across cultural groups. Since the structures are not comparable, a culture-specific representation is needed. At this level, there is concept bias. With different underlying dimensions, findings from one cultural group cannot be generalized to other cultural groups.

Classical MDS, Replicated MDS, and Generalized Target/Procrustes Analysis
The analyses start by applying multidimensional scaling within each cultural group in order to obtain culture-specific configurations. In line with the terminology of Schiffman, Reynolds, and Young (1981), we call this Classical MDS. After this, structural equivalence can be investigated in two different ways, either by means of a Replicated MDS or by means of Generalized Procrustes Analysis (GPA—similar to Target Rotation in Factor Analysis; see Chapter 14).

If there is concept identity, both the underlying dimensions and the position of the items should be identical across cultural groups. As indicated, this means that a single structure suffices to represent the internal structure of a domain across cultural groups. A single consensus structure can be computed across cultural groups by means of Replicated MDS (following the terminology of Schiffman, Reynolds, and Young 1981). With Replicated MDS, a consensus configuration is computed in such a way that it fits as well as possible the empirical dissimilarities from all groups. Concept identity can be empirically concluded if the consensus configuration generated by means of Replicated MDS fits the observed data as well as the culture-specific configurations generated by means of Classical MDS. A discrepancy in fit, however, cannot be interpreted unequivocally. It could mean that there is still concept identity but that the observed dissimilarities differ because of sampling fluctuations and unreliability. It could also mean that there is only a representative overlap in items and that some items—although seemingly equivalent—shift in meaning across cultural groups. Alternatively, it could mean that most items shift in meaning across groups, which could possibly point to inequivalence in the organizing principles of the domain. Which explanation deserves further exploration will depend upon the reliability of the empirical dissimilarities, the size of the differences in fit between the Classical and the Replicated MDS, and an interpretative scanning of the culture-specific configurations. The more unreliability there is, the smaller the sample size is. The smaller the discrepancy of fit and the more culture-specific configurations can be interpreted in some way, the more likely it is there will be a higher level of equivalence and the more interesting it becomes to look for structurally inequivalent items.

Structural item bias can be easily detected and modeled within an MDS framework by means of an iterative procedure. First, the fit of an item within the culture-specific configurations and within the consensus configuration is computed. Second, the item with the largest difference in fit is removed from the set of items. Third, all analyses are carried out again (computation of a culture-specific configuration via Classical MDS and the computation of a consensus configuration via Replicated MDS), and item fit measures are computed and compared between both levels. Again, the item with the highest drop in fit can be removed and the whole process can start again. This process can be reiterated until the fit of the culture-specific and the consensus configuration are minimal. In this way, it is possible to identify empirically a subset of items that have the same internal structure across cultural groups.

Since Replicated MDS can easily handle data sets with missing values, it is possible to represent the structurally unbiased and biased items within a single configuration. The structurally unbiased items receive a single position across cultural groups. The structurally biased items are treated as culture-specific items within each cultural group and each receive a culture-specific position within the consensus configuration.

In the last decade, the possibilities for investigating structural equivalence have been extended by the development of Generalized Procrustes Analysis (GPA) (Commandeur 1991; Borg and Groenen 1997). GPA is an interesting alternative to Replicated MDS. Since MDS models pairwise dissimilarities between items by distances, all operations on the dimensions that do not affect the relative distances lead to equivalent configurations. These operations are rotation (as in exploratory factor analysis), translation, reflection, and uniform rescaling (Borg and Groenen 1997). Thus direct comparisons of culture-specific configurations are only meaningful if these operations can be excluded as sources of differences. With GPA, the group-specific configurations are rotated, translated, reflected, and shrunk or dilated in such a way that the average pairwise correspondence between configurations is optimized. Moreover, this program generates a centroid configuration across all groups, and indices of correspondence between each of the separate configurations and the centroid configuration both at configuration and at item level. As with Replicated MDS, discrepancies between the culture-specific configurations and the centroid configuration cannot be unequivocally interpreted. If structural item bias forms a plausible explanation, GPA can be applied iteratively by treating the structurally biased items as culture-specific items, as was the case with Replicated MDS.

Reliability

As we have seen, unreliability of the observed dissimilarities forms an important possible explanation for observed structural differences. Only under very strict conditions, such as statistical independence and lognormal distribution of the random errors, do maximum likelihood MDS models exist that allow for statistical testing (Ramsey 1986). For most dissimilarity data, however, the assumptions of these models are violated and we can only work with descriptive fit measures. Fortunately, MDS representations are known for their robustness to unreliability in the dissimilarities (Borg and Groenen 1997). Still, a recent study using a bootstrap procedure on a large cross-cultural data set, demonstrated that up to more than half of the difference in fit between a Replicated MDS and a Classical MDS could be attributed to sampling fluctuation (Fontaine 1999). If possible, it is advisable to directly investigate the impact of unreliability on the data, for instance by means of a bootstrap procedure. Further research is needed to identify what the viable cut-off criteria are for distinguishing structural inequivalence due to unreliability and to sampling fluctuations.

15.2   APPLICATIONS IN CROSS-CULTURAL RESEARCH

The use of MDS to deal with substantive questions in cross-cultural research can be situated into three broad areas, as indicated in the previous section, namely, (1) in the domain of folk theories, as (2) an alternative for factor analysis, and (3) for modeling structural equivalence across cultural groups. We first present some cross-cultural studies on folk theories of emotions, of personality, and of theories of mind, where MDS is applied on similarity ratings or similarity sortings. Next, we present cross-cultural research where MDS is used as an alternative to factor analysis. Finally, we discuss a recent study where MDS was used to identify structural item bias in the emotion domain.

15.2.1 Folk Theories

Folk theories refer to the shared views within a cultural group on the workings of various domains of human functioning (D'Andrade 1987; Keesing 1987). An important aspect of these folk theories consists of the perceived interrelationships between the concepts of a specific domain of functioning and the major characteristics on the basis of which these concepts are related or separated from one another. As explained, folk theories have been investigated by similarity ratings or similarity sortings with indigenous concepts of the domain of interest.

Emotions

Research on the cognitive representation of emotions by means of similarity ratings between emotion terms has a long-standing tradition in emotion research (for an overview, see Russell 1991). In 1955, when the development of MDS models was still in its early stages, Gösta Ekman asked 168 Swedish respondents to rate the qualitative similarity between 23 Swedish emotion terms representing a wide variety of emotional states. By means of a split-half procedure, he found the similarity ratings across groups of 84 respondents to be highly reliable ($r = .99$). However, it was only after the landmark research by Osgood (e.g., Osgood, May, and Miron 1975) in which evaluation, activity, and potency were identified as three pancultural dimensions of affective meaning, that this type of research received a lot of attention. One of the landmark studies in this area by Russell and his colleagues was conducted, first with 34 English-speaking Canadians (Russell 1980), later with 35 monolingual Japanese, 33 monolingual Gujarati, 15 monolingual and 28 bilingual Croatian, and 18 monolingual, and 18 bilingual Chinese-speaking Canadian immigrants (Russell 1983), and finally with 50 Estonian, 30 Greek, and 50 Polish respondents (Russell, Lewicka, and Niit 1989). Most of these studies used the same procedure. First, respondents received 28 cards with emotion terms in their mother tongue (carefully translated from English). Then, they were asked to sort the cards into 4, 7, 10, and 13 groups on separate trials, with the simple instruction that the more similar feelings should be grouped together. Based on the times two emotion terms were sorted into the same category and taking into account in which trials this was done, similarities were computed for all pairs of emotion terms per subject. If two emotion terms were sorted into the same group in the four trials, there was a maximal similarity of 34 (4 + 7 + 10 + 13). According to Ward (1977), this multiple sorting procedure offers a good and economical alternative for the paired comparison procedure. Nonmetrical multidimensional scaling was then executed on the average pairwise similarities between the 28 emotion terms. In all the samples, two dimensions where systematically recovered, namely, Pleasure and Arousal. However, not all studies in this research tradition identify the same dimensions. For instance, Herrmann and Raybeck (1981) investigated the cognitive structure of emotions with 48 American (university students), 20 American (Mohawk college students), 17 Spanish, 15 Vietnamese, 21 Hong Kong, 17 Haitian, and 17 Greek respondents. They asked respondents from each cultural group to rate the pairwise similarity of 15 emotion terms on a 4-point scale (from similar to dissimilar meanings). Per cultural group, an average similarity matrix was constructed. MDS yielded comparable two-dimensional structures across cultural groups. The first dimension was clearly a Pleasure (evaluation) dimension, as in the studies of Russell. However, the second dimension was a Dominance dimension (separating anger from sadness and fear terms) and not an Arousal dimension (called projective versus introspective by the authors).

A major criticism of this research tradition from an anthropological perspective is that although respondents can use the criteria they see fit to sort the emotion terms, the terms themselves are generated in an Anglo-Saxon cultural group and then translated into the other languages (e.g., Lutz and White 1986). This could still imply cultural imposition. However, even cross-cultural studies with indigenous emotion lists reveal similar structures and similar differences in structures. For instance, Church et al. (1998) asked 49 Filipino college students to sort an indigenous set of 171 Filipino emotion terms into groups of similarity. They recovered a Pleasure and Arousal dimension, while Gehm and Scherer (1988) found a Pleasure and Dominance dimension with a set of 235 German emotion terms that were sorted by 10 German-speaking respondents in groups of similarity.

Thus, even with indigenous emotion lists, there is no convergence across studies about the second dimension of the emotion space. According to Russell (1991), this could be due to the selection of the emotion terms. If the list contains mainly emotion terms referring to interpersonal contexts (e.g., such as *affection* and *disdain*) Dominance could be expected as a second dimension, while if the list contains mainly emotion terms referring to non-interpersonal contexts (e.g., such as *tired* and *tense*) Arousal could be expected as a second dimension. However, it is also possible that the item selection effect interacted with unreliability of the similarities. For instance, in the studies of Gehm and Scherer (1988), and of Church et al. (1998) the number of respondents is rather small in comparison to the number of terms sorted into groups of similarity. Indeed, a three-dimensional structure is revealed in research where both a large and representative set of emotion terms is used and a large sample of respondents is asked for the similarity judgments. Instead of working with a theoretically-selected list of emotion words, Shaver et al. (1987) empirically constructed a list of 135 emotion terms based on a free recall task and a prototypicality-rating task. The selected emotion terms were sorted by 100 English-speaking Canadian respondents into categories of similarity. The number of times two emotion terms were sorted into the same category was used as a similarity measure. The three dimensions of Pleasantness, Arousal, and Dominance clearly emerged from the similarity sorting task. Following the same procedures as Shaver et al. (1987), Fontaine et al. (1996, 2002) independently constructed an indigenous Dutch and an indigenous Indonesian emotion list of each 120 emotion terms and asked 109 Dutch and 105 Indonesian respondents to sort these terms into categories of similarity. In each cultural group, they replicated the expected three-dimensional structure.

Personality

In a review article, White (1980) compared the results of two anthropological studies on folk theories of personality in two very different non-Western cultural groups. During fieldwork in the state of Orissa in India, Shweder (1972) constructed a list with 81 indigenous personality descriptors in the Oriya language. Twenty-five adult males sorted these terms into groups of similarity. An MDS on the pairwise similarities revealed a two-dimensional structure. The dimensions could be interpreted as Dominance (e.g., *tough-minded*) versus Submission (e.g., *weak*) and Solidarity (e.g., *beneficent*) versus Conflict (e.g., *rude*). White (1978) replicated this structure in a very different cultural group in Santa Isabel on the Solomon Islands. He collected 37 personality descriptors in the A'ara language, and asked 25 local informants to sort each term with the five other terms of the remaining list that most closely matched in meaning. Based on this information, an average similarity between personality traits was computed, and MDS was executed. Two dimensions represented most of the similarity information and could also be interpreted in terms of Dominance (e.g., *strong*) versus Submission (e.g., *fearful*) and Solidarity (e.g., *kind*) versus Conflict (e.g., *recalcitrant*). Moreover, the structures closely related to what is found in empirical research in the United States (e.g., Rosenberg, Nelson, and Vivekenanthan 1968; Wiggins 1979). The dimensions of Dominance and Solidarity can be interpreted as instances of the general dimensions of Potency and Evaluation revealed by Osgood and colleagues (Osgood, May, and Miron 1975) for the domain of interpersonal trait descriptives.

Folk Theories on Knowing

In the United States, two dimensions underlie the folk theory of knowing, namely, an Information-processing dimension (from input/perceptual to output/conceptual) and a Certainty dimension (Schwanenflugel et al. 1994). Recently, Schwanenflugel, Martin, and Takahashi (1999) have studied the folk theory of knowing in Germany and Japan

(Schwanenflugel et al. 1994). The authors first translated a set of 30 English verbs of knowing into German and Japanese. This set was constructed in the USA based on extensive empirical research and frequency ratings (Schwanenflugel et al. 1994). Then, the researchers conducted two preliminary studies in order to check whether the English list could indeed be considered as representative for the domain of knowing in Germany and in Japan. Then, 30 German and 29 Japanese respondents had to rate all 435 pairs of the 30 verbs in terms of the similarity in "the way you use your mind when you are in the process of doing the described mental activity" on a 7-point scale. In addition, 15 German and 13 Japanese respondents were asked to rate directly the extent to which each of the 30 verbs referred to the Information-processing dimension and the Certainty dimension. As in the United States, the Information-processing dimension emerged consistently across the two cultural groups. The Certainty dimension also emerged in the German sample. In the Japanese sample, however, this dimension was less obvious. Especially verbs referring to the uncertainty pole of the dimension seemed to vary in position. The interpretation of the dimensions was supported by correlating the coordinates of the verbs in the geometrical representation with the direct ratings of verbs on the Information-processing dimension and on the Certainty dimension.

15.2.2 Representing Geometrical Models

Vocational Interests and Minorities

A domain of research where MDS is regularly applied as an alternative for factor analysis is the domain of vocational interest. John Holland (1957, 1997) has constructed a hexagonal model for vocational interests. He distinguishes six general interest types; realistic, investigative, artistic, social, enterprising, and conventional. A two-dimensional graphical representation with the six interest types ordered into a hexagon in the specified order can represent all hypotheses with respect to the relationships between the interest types. Interest types that share compatible interests are situated adjacent to one another, while conflicting interest types are situated opposite one another. The two dimensions underlying the hexagonal model have been interpreted as Interest in People versus Things (social versus realistic) and Interest in Data versus Ideas (enterprising and conventional versus investigative and artistic) (Prediger 1982). The generalization of this model from European-Americans to ethnic minorities in the United States has been debated in recent years. For instance, Rounds and Tracey (1996) found that Holland's model fitted the data of 73 U.S. European-American majority samples better than the data of 20 U.S. ethnic minority samples. However, Day and Rounds (1998) came to a different conclusion. They had unusually large samples at their disposal from different cultural groups: African-Americans (6759 men and 11400 women), Asian-Americans (2970 men and 3553 women), Mexican-Americans (2555 men and 3464 women), Native Americans (1113 men and 1530 women), and European-Americans (6637 men and 9469 women). All respondents took the test as a part of the nationally administered

American College Testing Assessment Program (ACT). MDS analyses on the intercorrelation matrices of the 90 items for each of the 10 samples indicated highly similar structures in all samples with a clear differentiation of the six interest types within a two-dimensional representation. Apparently, the combination of large and representative samples and a formal assessment context which motivated respondents to fill out the questionnaire seriously offered clear support for the model for each of the ethnic minorities.

Values across Cultural Groups

Another area where MDS is regularly used as an alternative for factor analysis is in value research within the Schwartz value tradition. Schwartz (1992) developed a pancultural value theory about the content and the structure of the human value domain. He identified 10 value types which he took to form the content of the value domain; hedonism, stimulation, self-direction, universalism, benevolence, tradition, conformity, security, power, and achievement. All hypothesized structural relationships among the 10 value types, which are based on an analysis of the mutual compatibilities and conflicts among the transsituational goals to which each of the value types refer, can be represented in a two-dimensional circumplex structure. In that structure, compatible value types are situated adjacent to one another, while conflicting value types are situated opposite to one another. Moreover, based on the same analysis, he distinguished four higher-order value types; Self-Enhancement (power, achievement, and hedonism) which is in conflict with Self-Transcendence (benevolence and universalism), and Stability (tradition, conformity, and security) which is in conflict with Openness to change (self-direction and stimulation). Schwartz has investigated the validity of this theoretical framework in many countries around the world. The Schwartz value survey contains in total 57 value items. For each value item respondents have to indicate the extent to which they consider this to be a guiding principle in their personal life. Per sample, the intercorrelation matrix is computed between all value items and used as input for MDS. Typical for Schwartz's use of MDS is that he works with a regional interpretation instead of a dimensional interpretation. As already explained, a regional interpretation investigates whether bounded subregions of items can be identified within the geometrical representation that contains only items belonging to a single value type. Thus first it is established whether the 10 *a priori* value types can be identified as bounded regions in a two-dimensional representation, then whether the 10 regions are organized in the predicted circular fashion (Schwartz 1992; Schwartz and Sagiv 1995). The results from numerous studies confirm the Schwartz hypothesis, if an average correlation matrix is computed across samples and an MDS is applied on that average correlation matrix. However, considerable deviations from the theoretical structure are observed when the MDS is applied to individual samples. In a study with 88 samples from 40 countries, Schwartz and Sagiv (1995) found that on average 16% of the value items (about 9) did not emerge in the predicted region. The authors demonstrated by means of random split-halves that most of the deviations in the individual samples are caused by sampling fluctuations. Bootstrap analyses offered further support for this conclusion (Fontaine 1999). It remains a matter of further research to disentangle random sampling fluctuations from genuine differences in the structure of the value domain.

15.2.3 Structural Equivalence and Structural Item Bias in the Emotion Domain

Fontaine et al. (2002) applied three-way MDS models to investigate structural equivalence in the cognitive representation of emotions in the Netherlands and in Indonesia. In order to reduce the danger of Western imposition, a list of 120 culturally relevant emotion terms was selected within each cultural group, based on a free recall and a prototypicality-rating task. In Indonesia, a group of Indonesian respondents was asked to sort the Indonesian terms in categories of similarity. In the Netherlands, Dutch respondents were asked to do the same for the Dutch emotion terms. Thus the data consisted of similarities between pairs of 120 emotion terms in each cultural group. Structural equivalence was investigated in five different steps.

First, a Classical MDS was performed on both data sets separately. A three-dimensional representation fitted the data very well in both cultural groups. The Indonesian representation accounted for 90% of the variance in the Indonesian dissimilarities and the Dutch representation accounted for 88% of the Dutch dissimilarities. The dimensions could be interpreted in terms of Evaluation (positive versus negative), Arousal (active versus passive), and Dominance (strong versus weak).

A second phase investigated the extent to which it was possible to reach translation equivalence between the 120 Dutch and the 120 Indonesian emotion terms. Translation equivalence could be established for 50 pairs of emotion terms. With the 50 pairs, all parts of the culture-specific emotion spaces were represented: They formed a representative subsample of emotion terms within each cultural group.

A third phase investigated whether these 50 pairs of emotion terms took the same position on the same underlying characteristics. A three-dimensional Classical MDS and a three-dimensional Replicated MDS were conducted on 50 Indonesian and 50 Dutch emotion terms, with 92% (Classical) and 85% (Replicated) variance accounted for in the Indonesian data, and 90% (Classical) and 81% (Replicated) variance accounted for in the Dutch data. Thus the fit of the Replicated MDS was good and the fit of the Classical MDS even better. Since the culture-specific configurations indicated that the underlying dimensions could be interpreted in the same way in the two cultural groups, the authors attributed the decrease in fit to translation-equivalent emotion terms that did not share the same position in the emotion space.

In the fourth phase, an iterative bias procedure was performed at item level. The authors used a cut-off of maximally 15% per item, meaning that the proportion of variance accounted for by the common configuration could be maximally 15% lower than the proportion of variance accounted for by a culture-specific configuration.

With that cut-off, eight pairs of emotion terms were identified as biased, including *guilt*, *shame*, and *sinfulness*. For the remaining 42 emotion terms, the fit of the common configuration was 88% for the Indonesian data and 90% for the Dutch data, compared with 92% for the Indonesian culture-specific configuration and 94% for the Dutch culture-specific configuration. Thus for these 42 terms structural equivalence could be established both at dimensional and at item level.

A fifth phase investigated whether all Indonesian and Dutch emotion terms could be represented within a single emotion space with the 42 unbiased terms used as common base. For the Dutch and the Indonesian data, the fit was 87% for the common configuration with partial overlap, compared to 90% for the Indonesian culture-specific configuration and 88% for the Dutch culture-specific configuration. Since we could not depend on perfectly reliable dissimilarity data, the structural item bias could both have resulted from genuine differences in meaning across cultural groups, or from unreliability and random sampling fluctuation. Based on the analysis procedure, we can be confident that the cognitive structure of emotions is organized according to the same principles in the two cultural groups and that at least 42 pairs of emotion terms—a representative subsample—share the same position within the common emotion space.

15.3 SOME GUIDELINES

Based on a brief overview of areas where MDS has been used in cross-cultural research, we can give some guidelines for further use of MDS. First, as in factor analysis, the resulting structure depends on the items that have been included in the research. Thus it is very important to pay sufficient attention to the selection of a representative set of items within each cultural group. The lack of convergence about the second dimension in the emotion domain (Russell, 1991) forms a point in case here. Second, before concluding that there are substantial cultural differences in structure or deviations from a theoretical structure, the reliability of the dissimilarity data should be checked. MDS is known for its robustness (Borg and Groenen 1997), however, it seems that there are limits to this robustness. The findings on emotions and on vocational interests are revealing in this respect. With a substantial improvement in the reliability of the similarities, most of the so-called substantial deviations from the expected structure disappeared. Third, direct ratings of the items on the supposedly underlying characteristics of the domain in each of the cultural groups can validate findings to an important extent, as found in the Schwanenflugel et al. (1999) study. They can form an independent source for the interpretation of the dimensions. Finally, cross-cultural researchers should not limit themselves to culture-specific MDS analyses. Both older MDS models, such as Replicated MDS, and recent developments, such as Generalized Procrustes Analysis, provide the cross-cultural researcher with powerful tools to investigate structural equivalence for a broad range of data. They can support claims of structural equivalence and offer an empirical possibility to identify structurally biased items.

Chapter 16

CROSS-CULTURAL EQUIVALENCE WITH STRUCTURAL EQUATION MODELING

JAAK BILLIET

This chapter demonstrates how a multigroup test of a measurement model for several attitudinal latent variables and a structural model of the inter-relations can be carried out using structural equation modeling (Bollen 1989; Jöreskog and Sörbom 1993). Data from a study of national identity in two culturally different populations, the Flemish and the Walloons, are used to illustrate the procedure. Three substantive concepts are expected to be completely equivalent in the two populations, while a fourth concept is expected to differ between the two populations. The invariance of latent variables (see Rensvold and Cheung 1998) in the two cultural groups is examined by determining whether or not the structural relationships between these variables are identical or different. It will be useful to begin with some remarks about the Belgian context of 'national identity,' the central concept in the study.

16.1 (SUB)NATIONAL IDENTITIES IN BELGIUM

In some countries, citizens do not have a single and self-evident national identity but can choose between more or less competing different identities: on the one hand an official nationality (e.g., Spanish or British), and on the other a subnational or peripheral identity (e.g., Catalonian or Welsh). The political autonomy acquired by numerous subnational entities in recent decades has tended to encourage subnational authorities to begin nation-building projects of their own. Even where citizens have little interest in the political issues involved in this process, they still have to come to terms with this double identity.

National identity can be conceptualized as the choice that citizens have to make between a national and a subnational identity. It is assumed here that citizens as a rule do not consistently adopt either the national or the subnational point of view and hence cannot be categorically divided into two different groups. Instead, the majority constantly oscillates between the two identities. The likelihood of some taking the national or the subnational point of view will probably vary: some people will tend to take the national position more frequently, while others will be more inclined to adopt the subnational position. However, only a minority will consistently prefer one identity over the other. The variable national identity can thus be conceived of as a bipolar scale, with people who consistently favor the national and subnational positions located at the two extremes (Maddens, Billiet, and Beerten 2000, 45).

In Belgium, several nationalities compete for public support. The most obvious identity is the official Belgian nationality, as recorded in passports. However, various reforms of the state (1970, 1980, 1988–89, 1993) have given the two main subnational regions, Flanders and Wallonia, a considerable degree of autonomy, and both have engaged in establishing their own 'nation.' (The situation in the Brussels region is ignored here because it is too complex with regard to national identity; Maddens, Billiet, and Beerten 2000, 47.) Wallonia and, to an even greater degree, Flanders do not consider themselves as mere administrative units, but as nations in the making. In Flanders, the regional government actively promotes a Flemish national identity. Although the political elite likes to portray the Belgian and regional identities as complementary, tension clearly exists between the two. For example, the aspirations of the regions toward a national status obviously pose a threat to the Belgian nation-state. As a result, the Belgian establishment takes every opportunity to reinforce citizens' loyalty toward the unitary state, often referring to the Belgian monarchy as a symbol of that unity. Following the first electoral success of the extreme-right and separatist party Vlaams Blok at the end of the nineteen eighties, the Belgian authorities attempted to combine their standard rhetoric, emphasizing the unity of Belgium, with a discourse defending values such as solidarity, tolerance, and multiculturalism.

In this discourse, the Belgians are bound together by their common interests and by their acceptance of the basic tenets of the constitution, not by common descent. This newly created social representation of the Belgian state serves as a model of peaceful co-existence for different cultural groups (i.e., Flemings and Walloons, as well as immigrants). Being 'Belgian' then stands for being open toward other cultures and showing solidarity with different ethnic groups. In these terms, identifying with the Belgian nationality can be seen as rejecting xenophobia and narrow-minded nationalism (Maddens, Billiet, and Beerten 2000, 47).

The social representation of a nation (Moscovici 1984) contains shared images and beliefs about the national ingroup and its relationship to other groups or 'foreigners' (van Dijk 1993, 39). A distinction is also usually made between an *ethnic representation* and a *republican representation* of the nation (Fennema and Tillie 1994). The ethnic representation sees national identity as a static cultural heritage that should be safeguarded and passed on to future generations. Foreigners are seen as an outside threat to this cultural identity. In the republican representation, the nation is seen as a dynamic contract between the different components of society. Membership of the nation does not depend on descent, but on a willingness to accept the basic rules of the society. From this perspective, cultural identity neither needs to be nor should be protected against foreign influences. On the contrary, cultural identity is seen as the result of a constant dialogue and blending with other cultures (Maddens, Billiet, and Beerten 2000, 46).

The republican representation of identity has now found wide acceptance throughout Belgium. It contrasts sharply with the dominant representation of the Flemish identity, which tends toward the ethnic representation. Van Dam's (1996) detailed analysis of the elite discourse on Flemish identity shows that 'Flanders' is associated with protection of the Flemish cultural heritage — in particular the Dutch language — and hence with a defensive attitude toward other cultures. Keywords in the elite discourse on Flemish identity are: suffering, subordination to a dominant power, threat, strife, uncertainty, culture, finance and nation-state. A similar analysis of self-representation by the Walloon elite shows that the Walloon identity is primarily associated with the social-economic emancipation of the Walloon region. From the Walloon perspective, regional autonomy is necessary to defend their common social-economic interests in the Belgian state, not to preserve a Walloon cultural heritage. In addition, Walloons strongly emphasize the open and non-racist nature of their identity. Van Dam characterizes the self-representation of the Walloon identity with the following keywords: trust, tolerance, creativity, diversity, the social dimension, the *terroir* or local community, the *francité* ('Frenchness'), self-evident character and region. This representation comes close to the republican ideal type (Maddens, Billiet, and Beerten 2000, 47).

It is evident, therefore, that the representation of Belgian identity is not uniform across the two regions. In Flanders, the political extreme right, with its roots in the Flemish nationalist movement, has always been opposed to the Belgian nation state. In Wallonia, the extreme right, although more marginal than in Flanders, has continued a tradition of ethnic Belgian nationalism and is generally averse to Walloon regionalism (de Witte and Verbeeck 1998, 73–79). The representation of the Belgian identity is thus less homogeneous in Wallonia and often contains aspects of both the official republican representation and the ethnic representation of the extreme right.

These differences in the historical roots of subnationalism and the social representations of the nation lead us to expect differences in the relationships between (sub)national identity and attitudes toward immigrants and multiculturalism. In the remaining discussion we set out to replicate findings using data from 1991 about the relationship between (sub)national identity and ethnocentrism (Maddens, Billiet, and Beerten 2000) with data from the 1995 Belgian General Election Survey. The structural model is extended to include two other variables, authoritarianism and distrust of politics (output alienation). These are closely related to attitudes toward immigrants and, perhaps, to (sub)national identity (Gamson 1968).

16.2   DATA, MEASUREMENTS, AND EXPECTED RELATIONSHIPS

The two-step random samples from the 1995 data set from the national register have 2,099 (Flanders) and 1,258 (Wallonia) cases respectively. The Flemish sample was split up in two samples, one used for exploration of the model, the other for the confirmation of the two-group model. Two samples of approximately equal size were used (963 Flemish and 932 Walloon). The English, Dutch (Flanders) and French (Wallonia) wording of items is presented in the Appendix.

The first concept is *attitude toward foreigners* (Moroccans and Turks) in the sense of 'feeling threatened' by them in the economic and cultural spheres. The concept is measured by a balanced set of six Likert items with response categories from strongly disagree (score 1) to strongly agree (score 5). This allows us to control for a style factor (acquiescence) in the measurement model (see Billiet and McClendon 2000). The two items on cultural enrichment and distrust deal explicitly with acceptance or rejection of a multicultural society. Two other items deal implicitly with these.

The second concept is the political aspect of *(sub)national identity*. This is related to political options for more or complete autonomy for the different regions in Belgium. It is measured by five indicators. The first item used a region-specific wording. In Flanders, the (Dutch) wording roughly translates as "striving for the autonomy of Flanders" and the Walloon item in French translates roughly as "stop the division of Belgium". In the analysis, the item scores were reversed (see Appendix).

The third concept, *political distrust*, is measured by a balanced set of items which can be controlled for acquiescence (see Appendix). The scale consists of two negatively and two positively worded items.

Political distrust is closely related to 'output alienation' (Gamson 1968) or political powerlessness and can be seen as one of the five components of Srole's (1996) concept of anomie, which is both empirically and theoretically linked with ethnocentrism (Tajfel 1981; Turner 1982; Scheepers, Felling, and Peters 1990, 1992). Previous research has revealed that political distrust and the rejection of multiculturalism go hand in hand (Billiet 1995; Fennema 1997). At the same time, the direction of the relationship between (sub)national identity and distrust of politics is less clear (Maddens, Beerten, and Billiet 1994, 87-89).

The fourth concept is *authoritarianism*. The classic concept of authoritarianism refers to a cluster of nine 'subsyndromes' (Adorno et al. 1950). Key indicators are a strict adherence to conventional values and norms, uncritical subjection to and an uncritical attitude toward moral authorities, and authoritarian aggression toward norm violators. These last two describe the tendency of authoritarian people to define themselves in sharp contrast to other social or ethnic groups (contra-identification).

Authoritarianism was measured in the study with four items covering some of these aspects: authority, conformism, and repression (see Appendix). Some of the subsyndromes of authoritarianism (Adorno et al. 1950), such as strict conformity to traditional norms and values, an uncritical submission to moral authorities, and acceptance of aggressive oppression of deviant people, can contribute to explaining the negative attitude of authoritarian people to ethnic minorities and multiculturalism (Turner 1982; Altemeyer 1988; Billiet, Eisinga, and Scheepers 1996). Scheepers, Felling, and Peters (1990, 1992) provide a sociological explanation for the strong relationship between authoritarianism and distrust of politics; authoritarianism is seen as a reaction to political alienation (compensation) which is, in turn, caused by unfavorable social circumstances.

As mentioned, two of the sets of indicators are balanced, allowing a style factor (acquiescence) to be specified. This factor cannot be specified for authoritarianism, however, since the item set is not balanced with positive and negative question wordings. As a result, the tendency to agree cannot be separated from the substantial concept of authoritarianism. Thus a positive relation between authoritarianism and style can be expected in both regions.

The analysis focuses on the relationship between (sub)national identity and ethnocentrism in two cultural groups (populations) in Belgium. In earlier research two conflicting hypotheses about this relationship were explored (Maddens, Billiet, and Beerten 2000). One hypothesis states that national identity is intrinsically related to a negative attitude on the part of the national in-group toward the relevant out-groups (i.e. foreigners). The more consistently someone identifies with a single nation, the more he/she will tend to contra-identify with foreigners by maximizing the perceived difference between members of the nation and non-members. Conversely, people who do not have a strong preference for either nationality are likely to have a weaker national identity; they will be less inclined to dissociate themselves from foreigners and be more inclined to accept multiculturalism.

This 'intrinsic' hypothesis can be traced back to Sumner (1906), who considered out-group aggression and in-group identification as basically correlative. However, this view is now generally considered to be extreme, since social-psychological research has shown that social identification does not require a contra-identification with a salient out-group (Brewer and Miller 1996, 47–48). The implications of this for national identity are that citizens can also be expected to identify strongly with the nation without adopting a hostile attitude toward foreigners.

The other hypothesis states that the degree to which both attitudes coincide is likely to depend, among other things, on the social representation (i.e., ethnic or republican) of the national identity (Moscovici 1984). If the ethnic representation is dominant, a strong and consistent identification with the nation will presumably imply both a negative attitude toward foreigners and opposition to multiculturalism. Conversely, if a republican representation is dominant, a strong identification with the nation should normally coincide with a positive attitude toward foreigners. Because of the differences between Flanders and Wallonia in the social representations of the nation, a linear but reversed relationship between (sub)national identity and ethnocentrism was expected in the two cultural groups.

In line with the social representation hypothesis, previous research found that the relationship was indeed linear and took a different direction in Flanders and Wallonia. Flemish xenophobes tended to favor the Flemish identity, while Flemish non-xenophobes tended to favor the Belgian nationality. Conversely, Walloon xenophobes were somewhat more inclined toward the Belgian identity, while Walloon non-xenophobes preferred the Walloon identity. The relationship was, however, not very strong in the two populations. However, when age and education were controlled for, it was somewhat stronger in Flanders than in Wallonia. This is may be because the ethnic nature of the Flemish identity representation in Flanders is much more pronounced than is the Belgian identity representation in Wallonia (Maddens, Billiet, and Beerten 2000).

16.3   ANALYSIS

The first analysis examines whether the measurement model and the relationships between the concepts are invariant in the two populations. Next, two relevant social background variables are included as covariates in the structural model. These variables are Educational level, a four-point ordinal variable ranging from lower education (1) to higher education (4), and Age, coded as an ordinal variable consisting of six classes (1 to 6) corresponding with the following age categories: 18–24, 25–34, 35–44, 45–54, 55–64, and over 64 years of age.

16.3.1   Measurement Model

The test starts with a model of complete invariance across the groups of the measurements and with complete invariance of the relationships between the concepts, followed by steps in which equality constraints across groups are relaxed (see Table 16.1). As can be seen in the table, equality constraints were relaxed in the Walloon group. The removal of an equality constraint implies that the parameter can vary freely across Flanders and Wallonia.

TABLE 16.1. Fit Measures of the Measurement Model and the Relationships Between Four Attitudinal Variables and a Style Factor in Two Cultural Groups

Model	Chisq	Df	RMSEA	p-Value of Close Fit	NFI
1a Basis four concepts + style	1549.93	307	.065	.970	.972
1b Free $\lambda^W_{8,3}$	1525.50	306	.065	.985	.973
1c Free $\lambda^W_{8,1}$	1467.65	305	.064	.995	.973
1d Free $\lambda^W_{19,1}$	1443.27	304	.096	.997	.973
1e Free $\psi^W_{3,3}$	1394.74	303	.062	.999	.974
1f Free $\psi^W_{1,1}$	1355.04	302	.061	1.0	.975
1g Free $\psi^W_{1,3}$	1318.15	301	.060	1.0	.976
1h Free $\psi^W_{4,4}$	1275.88	300	.059	1.0	.976
1i Free $\psi^W_{3,5}$	1254.04	299	.058	1.0	.977
1j Free $\psi^W_{3,4}$	1158.92	298	.055	1.0	.979
1k Free $\lambda^W_{11,3}$	1119.81	297	.054	1.0	.979
1l Free $\psi^W_{2,5}$	1097.12	296	.053	1.0	.989
1m Free $\psi^W_{5,5}$	1070.75	295	.053	1.0	.980
1n Free $\lambda^W_{17,3}$	1054.26	294	.052	1.0	.981
1o Free $\psi^W_{4,5}$	1049.00	293	.052	1.0	.981

W = Wallonia; λ = regression coefficient (factor loading); ψ = (co)variance of latent variable.
The first subscript after the Greek symbols in the first column represents the indicator (item), the second the latent variables (the attitudinal factors and style factor) in the following way:
1 = Feeling threatened by foreigners; 2 = Response style; 3 = (Sub)national identity; 4 = Distrust of politics; 5 = Authoritarianism.

Several criteria are used (Bollen and Long 1987) for the evaluation of the models. The first is the drop in chi-square value (Chisq) for one degree of freedom (df). Provided the drop in value is substantially more than three units, one can conclude that relaxing the equality constraint of the parameter examined has led to a better fit and that the parameter has different values in the two groups. Furthermore, the p-value of close fit should be near to 1.0 and the Normed Fit indices (NFI) should be close to 1. Ideally, the root mean square error of approximation (RMSEA) should be lower than 0.05 (Bollen and Long 1987).

Several observations can be made concerning the invariance of the factor loadings. The factor loadings of the indicators 1 to 6 of the attitude toward immigrants were completely invariant in the two groups. However, in Wallonia, two indicators expected to measure other concepts ($\lambda^w_{19,1}$ and $\lambda^w_{8,1}$) also loaded on 'feeling threatened.' Two indicators of national identity ($\lambda^w_{8,3}$ and $\lambda^w_{11,3}$) had different loadings in the two cultural groups. An indicator of political distrust ($\lambda^w_{17,3}$) also loaded on (sub)national identity in Wallonia. All other factor loadings were invariant, including the loadings of the style factor (see Table 16.2).

The loadings ($\lambda_{2,j}$) on the acquiescence style factor are low (.15); this is not surprising, since the indicators were not primarily intended to measure a style factor. The variances of this factor are significantly different from zero ($t = 5.42$), and equal in the two cultural groups. Thus agree–disagree items about attitudes toward immigrants are equally affected by acquiescence in the two populations. As expected, the response style does not correlate with substantial variables, with the exception of authoritarianism ($\psi_{2,5}$). Because the item set for measuring acquiescence was not balanced, the latent variables were expected to correlate somewhat (although probably not as strongly as found in Flanders). People who are culturally conservative and who want to conform with group norms can be expected to tend to acquiesce.

Table 16.2 shows that, as expected, the relationship between feeling threatened by immigrants and (sub)national identity is reversed across the two cultural groups. National identity is, however, not factorially invariant. There are substantial differences in one of the indicators, namely, 'splitting social security' ($\lambda_{8,3}$), which has a stronger relationship in Flanders with the latent variable, national identity, than in Wallonia. The reason for this is that public opinion in Flanders is convinced that the Flemish will benefit from splitting social security, whereas Walloons tend to be less outspoken about this issue. This probably explains the ambiguous attitude of the Walloon regionalists toward splitting the social security system. Moreover, this item loaded weakly in Wallonia 'on feeling threatened' by immigrants. The other item with a weaker loading in Wallonia dealt with exclusive identity ($\lambda_{11,3}$). It makes sense that this item does not correlate as strongly with the concept, since most citizens in Wallonia have a mixed identity (Maddens, Beerten, and Billiet 1994).

In sum, it can be concluded that the latent concept national identity is not completely equivalent and differs somewhat in meaning between the two populations. How serious is this difference? This question is important in view of the reversal in the correlation between the attitude toward immigrants and the measured dimension of national identity. The question is, then, whether two equivalent subjective states are inversely correlated or whether the subjective states are not comparable. Since three of the five indicators are invariant and since the most important indicator is about the center of political decisions ($\lambda_{7,3}$), the results suggest that national identity is neither factorially invariant nor highly different.

TABLE 16.2.   **Measurement Model for Attitude toward Immigrants, National Identity, Political Distrust, Authoritarianism, and the Tendency to Agree (Style) among Flemish and Walloons**

Indicators	Λ^1: **Flanders** Feeling Threatened	Response Style*	(Sub)national Identity (all constrained)	Distrust of Politics	Authoritarianism
1. *Distrust***	.83 (fixed)	.15			
2. *Employment*	.82 (51.69)	.15			
3. *Culture*	.78 (47.79)	.15			
4. *Welfare*	−.76 (−52.81)	.15			
5. *Enrichment*	−.80 (−55.23)	.15			
6. *Welcome*	−.75 (−49.19)	.15			
7. *Independence*			.81 (fixed)		
8. *Soc_sec*			**.73 (24.37)**		
9. *Decide*			.61 (20.65)		
10. *First_id*			.51 (15.28)		
11. *Exclus_id*			**.62 (18.22)**		
12. *Unable*		.15		.71 (23.66)	
13. *Feel too good*		.15		.86 (23.48)	
14. *Take. Account*		.15		−.57 (fixed)	
15. *Able*		.15		−.54 (−18.74)	
16. *Adaptation*					.66 (fixed)
17. *Strong/weak*					.70 (27.91)
18. *Repression*					.69 (27.26)
19. *Strong leaders*					.68 (22.10)

Standardized Covariances of the Latent Variables

Threatened	**1.07 (37.91)**				
Response style	—	1.0 (5.39)			
(Sub)national	**.28 (7.13)**	—	**1.1 (18.24)**		
Distrust	.52 (14.44)	—	—	**1.1 (12.02)**	
Authoritarianism	.61 (19.50)	**.65 (8.67)**	—	**.60 (12.42)**	**1.1 (16.55)**

(Table continues on next page)

TABLE 16.2. (continued)

Indicators	Λ^2: **Wallonia** Feeling Threatened	Response Style*	(Sub)national Identity (all constrained)	Distrust of Politics	Authorita-rianism
1. Distrust**	.83 (fixed)	.15			
2. Employment	.82 (51.69)	.15			
3. Culture	.78 (47.79)	.15			
4. Welfare	−.76 (−52.81)	.15			
5. Enrichment	−.80 (−55.23)	.15			
6. Welcome	−.75 (−49.19)	.15			
7. Independence			.81 (fixed)		
8. Soc_sec	**.17 (4.26)**		**.51 (12.97)**		
9. Decide			.61 (20.65)		
10. First_id			.51 (15.28)		
11. Exclus_id			**.36 (9.41)**		
12. Unable		.15		.71 (23.66)	
13. Feel too good		.15		.86 (23.48)	
14. Take. Account		.15		−.57 (fixed)	
15. Able		.15		−.54 (−18.74)	
16. Adaptation					.66 (fixed)
17. Strong/weak			**.11 (2.64)**		.70 (27.91)
18. Repression					.69 (27.26)
19. Strong leaders	−.11 (−3.05)				.68 (22.10)

Standardized Covariances of the Latent Variables

Threatened	**.92 (33.61)**				
Response style	—	1.0 (5.40)			
(Sub)national	**−.18 (−3.53)**	—	**.87 (14.26)**		
Distrust	.52 (14.44)	—	**−.29 (−5.18)**	**.88 (11.42)**	
Authoritarianism	.61 (19.50)	.24 (3.26)	**−.28 (−4.86)**	**.50 (10.64)**	**.88 (14.93)**

The factor loadings in the table are the loadings of the common metric completely standardized solution. Critical ratios are given between brackets. Both samples have about equal sizes (963 and 832 observations). Non-invariant parameters are in bold.

*All loadings constrained to be equal across indicators and groups.

**See the Appendix for an explanation of the indicator labels.

Two other correlations of attitudes with (sub)national identity differ strongly in the two cultural groups. In Wallonia, both political distrust and authoritarianism are negatively correlated with national identity. This means that the more people consider themselves to be Walloon rather than Belgian, the less they distrust politicians (and the political system) and the less they support authoritarian ideas. In contrast to Flanders, Walloon regionalism is left wing, hence the correlation between feeling Walloon and lower support for authoritarian ideas. In Flanders, these correlations are not significant, thus the analyses again demonstrate that the two concepts cover different realities in the two cultural groups.

16.3.2 A Causal Substantive Model

In the next analysis a causal model with the relations between the latent variables and social background variables is tested to establish whether some of these relations change after controlling for education and age. Prior to the two-group comparison, a basic model was tested in the Flemish exploratory sample. The relations between the variables are specified in Figure 1. Since the values of the parameter values in one group are of no interest here, the relations are identified by Greek symbols. This may be helpful in understanding Table 16.3. The selected Model (2) fitted well in the Flemish sample (Chisq = 926.48; df = 175; RMSEA = .036). This model was therefore taken as the initial model in the multigroup analysis (see Model 3a). On the basis of our earlier findings on the structural relationships (in Model 1o; see bottom line of Table 16.1), however, some changes were carried out. Undirected relations between the attitudes were translated into directed (causal) relations (ψ's were replaced by corresponding β's). The measurement model that was selected in Model 1o remained unchanged in the subsequent causal analysis, even when improvements of the factor structures were suggested by the modification indices. An exclusive reliance on statistical modification criteria might have led to a long series of model changes in the measurement and structural models.

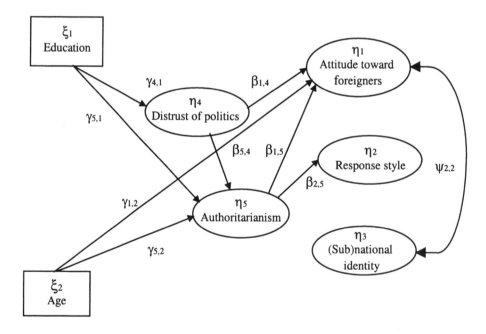

Figure 16.1. Model of the Causal Relationships Between the Social Background Variables and the Attitudinal Variables (Starting Model 2: Flanders).

TABLE 16.3. Test Information of the Structural (Causal) Model in Two Cultural Groups

Model		Chisq	Df	RMSEA	p-Value of Close Fit	NFI
3a	Model 1o and free $\beta_{1,5}$, $\beta_{2,5}$ and $\beta_{5,4}$ and $\beta_{1,4}$, instead of corresponding ψ's. Fix: $\gamma_{1,1}$, $\gamma_{2,1}$, $\gamma_{3,1}$, $\gamma_{2,2}$, $\gamma_{3,2}$ and $\gamma_{4,2}$ (= Model 2) Model 2 and adjustment for two group comparison: Free $\beta^F_{2,5}$ and $\beta^F_{3,5}$ and $\beta^F_{5,4}$ instead of $\psi^F_{2,5}$, $\psi^F_{3,5}$ and $\psi^F_{5,4}$	1562.92	367	.059	1.0	.975
3b	Free $\gamma^W_{5,2}$	1533.91	366	.058	1.0	.976
3c	Free $\gamma^W_{4,2}$	1524.73	365	.058	1.0	.976
3d	Free $\gamma^F_{3,1}$ and fix $\gamma^W_{3,1}$	1485.59	364	.057	1.0	.976
3e	Free $\gamma^F_{2,2}$	1459.91	363	.056	1.0	.977
3f	Free $\gamma^W_{2,2}$	1450.47	362	.056	1.0	.977
3g	Free $\gamma^W_{4,1}$	1431.36	361	.056	1.0	.977
3i	Free $\beta^F_{2,4}$ and fix $\beta^W_{2,4}$	1395.42	360	.055	1.0	.978

From Table 16.3 one can conclude that in the two groups, the relations between feeling threatened by foreigners ($\gamma_{1,1}$) and response style ($\gamma_{2,1}$) with education, and between (sub)national identity ($\gamma_{3,2}$) and age are fixed at a value of zero (Model 3a). The effect of education on national identity was also fixed to zero, but it turned out to be significantly different from zero in Flanders ($\gamma^F_{3,1}$ in Model 3d). Other differences between the two groups dealt with the effect of age on both authoritarianism ($\gamma^W_{5,2}$) and on distrust of politics ($\gamma^W_{4,2}$) in Wallonia (Models 3b and 3c). Age also affected response style, but the effect was different in the two populations ($\gamma^F_{2,2}$ and $\gamma^W_{2,2}$ in Models 3e and 3f). The effect of age on distrust of politics also differed; in Flanders this relation was zero, but not in Wallonia ($\gamma^W_{4,1}$ in Model 3g). Finally, the relation between the response style and distrust of politics was not zero in Flanders ($\beta^F_{2,4}$ and $\beta^W_{2,4}$ in Model 3i).

The causal relations between the latent variables in the structural model are reported in Table 16.4. There are several differences in the relationships between the groups. The predictor variables are presented in the rows in the upper part of each subtable while the explained variables are presented in the headings of the columns. We begin with the invariant part. The causal structure of the relationships between the predictors and the explained variables is almost the same in the two populations. The effects of age, distrust of politics, and authoritarianism on attitudes toward immigrants are exactly the same in the two populations. The negative effect of education on authoritarianism is also equal in the two populations.

TABLE 16.4. Structural Relationships between the Variables in Two Cultural Groups in Model 3i

| | Explained Variables | | | | |
Predictors	Distrust of Politics	Authorita- rianism	Feeling Threatened	Response Style	(Sub)national identity
Flanders					
Education ($\gamma_{i,1}$)	**-.49 (-13.6)**	**-.22 (-8.4)**	—	—	**.25 (6.9)**
Age ($\gamma_{i,2}$)	—	**.39 (11.9)**	-.10 (-4.0)	**.20 (3.4)**	—
Distrust of politics ($\beta_{i,4}$)	—	**.42 (10.9)**	.29 (7.8)	**-.44 (-5.0)**	—
Authoritarianism ($\beta_{i,5}$)	—	—	.54 (14.3)	**.74 (7.7)**	—
Covariances (ψ's)					
Feeling threatened					.25 (7.1)
Response style					
(Sub)national identity				—	
Wallonia					
Education ($\gamma_{i,1}$)	**-.31 (-8.6)**	**-.22 (-8.4)**	—	—	—
Age ($\gamma_{i,2}$)	**.11 (5.9)**	**.20 (2.9)**	-.10 (-4.0)	**.11 (2.2)**	—
Distrust of politics ($\beta_{i,4}$)	—	**.39 (6.8)**	.29 (7.8)	—	—
Authoritarianism ($\beta_{i,5}$)	—	—	.54 (14.3)	**.46 (4.6)**	**-.31 (-5.0)**
Covariances (ψ's)					
Feeling threatened					.05 (1.3)
Response style					
(Sub)national identity				**-.12 (-2.4)**	

Chi-square = 1395.42 (df = 360); RMSEA = .055; p-value of close fit = 1.0 (contribution of Flanders: 42%). Non-invariant parameters are in bold. Critical ratios are given between brackets.

In Flanders, there is a direct effect of education on (sub)national identity, indicating that the more educated are more inclined to be Flemish nationalists ($\gamma^F_{3,1}$ = .25). In Wallonia, distrust of politics is positively affected by age: older people are more likely to distrust politicians ($\gamma^W_{4,2}$ = .11).

Another structural difference deals with the effect of distrust of politics on the response style in Flanders ($\gamma^F_{3,1}$ = -.44). Those who score higher on distrust seem less inclined to endorse agree–disagree items. This relation may have a substantive meaning (negativism), since the response effect was excluded from the measurement of distrust of politics (therefore, acquiescence could not explain the effect). Bear in mind that there are two reasons for the strong effect of authoritarianism on the response style, especially in Flanders $(\beta^F_{5,4} = .74; \beta^W_{5,4} = .46)$. First, the measurements of authoritarianism (agree–disagree items) are not a balanced set and could not be tested for acquiescence. Thus the substantive relationships with other variables measured in the agree–disagree format are artificially inflated. Second, authoritarian people may be substantively more likely to acquiesce.

Surprisingly, the correlation between feeling threatened and national identity is no longer significant in Wallonia $(\psi^W_{1,3} = .05; p > .05)$. The negative relation found in the first analysis was spurious because of the negative relationship between education and both national identity and the attitude toward immigrants in Wallonia. This relation is, however, still completely different from Flanders.

16.4   CONCLUSION

The chapter investigated the relationship between (sub)national identity and the attitudes toward immigrants in two different cultural groups within one nation. The social and political context, in particular the differences in social representations of the nation in different parts of Belgium, led us to hypothesize that the relationships between the two core variables are unequal in the two cultural groups. This hypothesis was explored in the context of two other relevant attitudinal variables, distrust of politics and authoritarianism, and two single-indicator social background variables, education and age.

The analysis was undertaken in two steps, beginning with a systematic test of a measurement model, followed by a systematic test of the substantive (causal) model in the two groups. Three substantive concepts were found to be almost completely equivalent in the two populations. The fourth, (sub)national identity, differed in meaning somewhat between the two groups. Two out of the five indicators for subnational identity seemed to have a somewhat different weight in the composite scale for the two cultural groups. A third indicator, which used different wording in the two cultures, has nevertheless the same weight. This suggests that factorial invariance (Rensvold and Cheung 1998) can be obtained by indicators that are different. We also demonstrated how a response style could be modeled in the two cultural groups for two of the latent variables. Both populations proved to be subject to acquiescence; however, relations of acquiescence with other variables were somewhat different in the two groups.

The causal relations between the latent variables and two background variables were analyzed in the second step. The relations between attitudes toward immigrants and their antecedents were completely identical in the two populations. Two types of difference were found. First, two effects were observed in one group but not in the other. Second, there were some differences in the strength of the relationships. After including the social background variables, there was no relationship in Wallonia between feeling threatened by immigrants and (sub)nationalism. In Flanders, those who scored higher on Flemish autonomy were somewhat more inclined to reject multiculturalism. In Wallonia, there was a direct effect between education and (sub)national identity. In Flanders there was no such effect, but in the path model, there was a path from education via authoritarianism to (sub)national identity. The substantive relations between the core variables differed, but (sub)national identity may have a somewhat different meaning in the two populations.

Establishing the equivalence of measurements in the two cultural groups is crucial in this kind of analysis, since we want to be sure that the same concepts are measured in the two cultural groups and comparisons of national identity and attitudes toward migrants across the two populations presuppose identity of concepts across the groups. This chapter has shown how structural equation modeling can be employed to investigate structural equivalence in a cross-cultural framework.

APPENDIX: QUESTION WORDINGS OF THE MEASURED ATTITUDINAL VARIABLES

1. English Translation of the Items

Labels	Indicators of the attitude toward immigrants (balanced set of Likert items)
Distrust	"In general, immigrants are not to be trusted."
Employment	"Guest workers endanger the employment of the Belgians."
Culture	"Muslims are a threat for our culture and customs."
Welfare	"The immigrants contribute to the prosperity of our country."
Enrichment	"The presence of different cultures enriches our society."
Welcome	"We should kindly welcome the foreigners who come to live here."

Labels	Indicators of National identity (mixed scoring of items)
Independence	"Strive for the independence of Flanders" (*strongly agree = 5—strongly disagree = 1*) . (*In Wallonia*: "Stop the division in Belgium" *with reversed scores in the analysis*).
Soc_sec	"Split up (federalize) social security" (*strongly agree = 5—strongly disagree = 1*).
Decide	"The form of the country is still being discussed. Some think that Flanders (Wallonia) must be able to decide everything itself. Others think that Belgium, that is, the Flemish and the Walloons together, must be able to decide about everything. Where would you place yourself on the scale?" (*an 11-point scale with Belgium = 0 and Flanders/Wallonia = 10*).
First_id	"Which group do you consider yourself to be a member of in the first place? And in the second place?" (*Belgium = 1; Flemish/Walloons = 3, other = 2*).
Exclus_id	"Which one of the following statements applies most to you? I consider myself only as Flemish/Walloon, I feel more Flemish/Walloon than Belgian, I feel as much Flemish/Walloon as Belgian, I feel more Belgian than Flemish/Walloon, I consider myself only as Belgian" (*scores 5 to 1*).

Table continues on next page

1. English Translation of the Items (continued)

Labels	Indicators of political distrust (balanced set of Likert items)
Unable	"The politicians have lost the ability to listen to ordinary people like me."
Feel too good	"Once they are elected, most politicians feel themselves too good for people like me."
Take. Account	"If people like me make their views known, the politicians generally take them into account."
Able	"Most of our politicians are able people who know what they are doing."

Labels	Indicators of authoritarianism
Adaptation	"Young people are often rebellious, but they will have to adapt as they grow older. "
Strong/weak	"People can be divided into two distinct classes: the weak and the strong. "
Repression	"Most of our social problems would be solved, if we could somehow get rid of the immoral, crooked people."
Strong leaders	"What we need are strong leaders who tell us what to do."

2. Original Dutch and French Question Wording

THREAT: Attitude toward Immigrants and Multiculturalism

	Dutch (Flemish sample)	Français (Walloon sample)
Distrust	1. Migranten zijn over het algemeen niet te vertrouwen	1. En général, on ne peut pas se fier aux immigrés
Employment	2. Gastarbeiders zijn een gevaar voor de tewerkstelling van de Belgen	2. Les travailleurs immigrés sont une menace pour l'emploi des Belges
Culture	3. Moslims zijn een bedreiging voor onze cultuur en gebruiken	3. Les musulmans sont une menace pour notre culture et nos usages
Welfare	4. De migranten dragen bij tot de welvaart van ons land	4. Les immigrés contribuent à la prospérité de notre pays
Enrichment	5. De aanwezigheid van verschillende culturen verrijkt onze samenleving	5. La présence de différentes cultures enrichit notre société
Welcome	6. Wij zouden de buitenlanders die zich in België willen vestigen hartelijk welkom moeten heten	6. Nous devrions souhaiter de tout coeur la bienvenue aux étrangers qui veulent s'établir en Belgique

NAT_ID: Belgian-Flemish (Walloon) nationalism

Dutch	Français

Independence 7. De onafhankelijkheid van Vlaanderen nastreven (1 = helemaal oneens—5 = helemaal eens).

7. Arrêter la division de la Belgique (5 = pas du tout d'accord—1 = tout à fait d'accord).

Soc_sec 8. De sociale zekerheid splitsen (1 = helemaal oneens—5 = helemaal eens).

8. Scinder (fédéraliser) la sécurité sociale (1 = pas du tout d'accord—5 = tout à fait d'accord).

Decide 9. Over de staatsvorm die het land moet hebben wordt nog steeds gediscussieerd. Sommigen vingen dat Vlaanderen over alles moet kunnen beslissen. Anderen vinden dat België, Vlamingen en Walen samen dus, over alles moeten kunnen beslissen. Waar zou U zichzelf plaatsen op de schaal? (Schaal: België = 0; Vlaanderen = 10) schaal).

9. En Belgique, on discute souvent des problèmes de la forme de l'Etat. A ce sujet, certains trouvent que les nouvelles unités - les régions et les communautés - devraient décider de tout, alors que d'autres au contraire trouvent que c'est la Belgique qui devrait décider de tout. Vous personnellement, où situeriez-vous entre ces deux avis? (Echelle: Belgique - 10 = Nouvelles unités).

First_id 10. Tot welk geheel rekent U zichzelf op de eerste plaats? (Vlaanderen = 3; België = 1, andere = 2).

10. A quoi avez-vous le sentiment d'appartenir en premier lieu? Et en deuxième lieu? (Flandres = 3; Belgique = 1, Autre = 2).

Exclus_id 11. Welke van de volgende uitspraken is voor U meest van toepassing? Ik voel me enkel Vlaming (5), meer Vlaming dan Belg (4), evenveel Vlaming als Belg (3), meer Belg dan Vlaming (2), enkel Belg (1).

11. Laquelle des affirmations figurant correspond le plus à votre vision de vous-même? Je me sens uniquement belge (1), je me sens plus belge que wallon(ne) (2), je me sens aussi bien wallon(ne) que belge (3), je me sens plus wallon(ne) que belge (4), je me sens uniquement wallon(ne) (5).

DISTRUST: Political distrust

	Dutch	Français
Unable	12. De politici hebben nooit geleerd om te luisteren naar mensen zoals ik.	12. Les politiciens n'ont jamais appris à écouter des gens comme moi.
Feel too good	13. Van zodra ze gekozen zijn, voelen de meeste politici zich te goed voor mensen zoals ik.	13. Dès qu'ils sont élus, la plupart des politiciens ne s'intéressent plus aux gens comme moi.
Take. Account	14. Als mensen zoals ik aan politici hun opvattingen laten weten, dan wordt daar rekening mee gehouden.	14. Si des gens comme moi font connaître leurs opinions aux hommes politiques, ils en tiendront compte.
Able	15. De meeste politici zijn bekwame mensen die weten wat ze doen.	15. La plupart de nos hommes politiques sont des gens compétents qui savent ce qu'ils font.

AUTHOR: Authoritarianism

	Dutch	Français
Adaptation	16. Jonge mensen zijn vaak opstandig, maar als zij ouder worden moeten zij zich aanpassen.	16. Les jeunes sont souvent révoltés mais ils doivent bien s'adapter quand ils vieillissent.
Strong/weak	17. Er zijn twee soorten mensen, sterken en zwakkelingen.	17. Il y a deux types d'hommes: les forts et les faibles.
Repression	18. Onze sociale problemen zouden grotendeels opgelost zijn als we ons op één of andere manier konden ontdoen van immorele en oneerlijke mensen.	18. Nos problèmes sociaux pourraient être résolus si d'une manière ou d'une autre on pouvait se débarrasser des gens malhonnêtes et sans moralité.
Strong leaders	19. We hebben nood aan sterke leiders die ons voorschrijven wat we moeten doen.	19. Nous avons besoin de dirigeants forts qui nous disent ce que nous devons faire.

Chapter 17

MULTITRAIT – MULTIMETHOD STUDIES

WILLEM E. SARIS

17.1   A MODEL OF MULTITRAIT – MULTIMETHOD STUDIES

Many features of a research project are dictated by the research goals and purpose of the study. This determines what entities questions ask about (e.g., politicians, policies, institutions, or events or facts) and whether, for example, evaluations, behaviors, or factual reports are required from respondents. Other features of the study, however, allow for greater freedom of choice, notably question wording, the choice of response scales, using additional components such as introductions or bridges to questions, as well as the position of the question in the questionnaire and the mode of data collection.

It is a well-established fact that these choices can have an effect on the results of a study. Most attention has been given to effects on the distribution of the variables (Sudman and Bradburn 1974; Belson 1981; Schuman and Presser 1981; Dijkstra and van der Zouwen 1982; Billiet, Loosveldt, and Waterplas 1986; Molenaar 1986; Groves 1989; Saris et al. 1996; Saris and Kaase 1997). More recently, attention has also been paid to effects on the relationships between variables (Andrews 1984; Alwin and Krosnick 1991; Költringer 1995; Saris et al. 1996; Scherpenzeel and Saris 1997). This chapter focuses on the effects of choices such as mentioned above on the relationship between variables. An illustration of such effects is provided in Table 17.1, which presents findings for two countries for the variables General Life Satisfaction (GLS), Satisfaction with One's Home (SH), Satisfaction with the Financial Situation of One's Household (SF) and Satisfaction with Social Contacts (SC). The data were collected from random samples of the population in the Netherlands ($n = 1599$) and in Hungary ($n = 312$). In each country the same people were asked these questions twice. The question wording was kept the same but the response scale differed. In one instance, a 10-point scale was used, and in the other, a 5-point scale. The effect of this arbitrary choice is illustrated in Table 17.1, which shows that the method used can strongly affect results. The correlation between GLS with all three other variables is smaller in the Netherlands than in Hungary using the 10-point scale; however, this pattern is not found for two of the three correlations on the 5-point scale.

TABLE 17.1. Polychoric Correlations of Four Satisfaction Variables Measured with Two Different Methods Obtained from the Same Respondents

The Netherlands	10-point scale				5-point scale			
	GLS	SH	SF	SC	GLS	SH	SF	SC
GLS	1.000				1.000			
SH	.458	1.000			.381	1.000		
SF	.456	.434	1.000		.445	.349	1.000	
SC	.491	.325	.333	1.000	.462	.232	.270	1.000
Hungary	10-point scale				5-point scale			
	GLS	SH	SF	SC	GLS	SH	SF	SC
GLS	1.000				1.000			
SH	.490	1.000			.341	1.000		
SF	.637	.468	1.000		.664	.380	1.000	
SC	.519	.254	.308	1.000	.296	.182	.247	1.000

GLS = general life satisfaction, SH = satisfaction with housing, SF = satisfaction with financial situation, SC = satisfaction with social contacts.

This difference cannot be explained by sampling fluctuations, because in each country only one sample is involved and the same people provided answers on both scales. Instead, the differences are due to a specific combination of random errors and method effects (for more details on the study, see Saris et al. 1996).

We adopt here an approach suggested by Andrews (1984) for estimating the size of the effects of errors and the procedure to correct for them. This approach is the most explicit and general of a number of different procedures available. Once a special methodological study has been completed, it provides researchers with the information necessary to be able to make different measurement instruments comparable within a study and across studies.

To be able to describe this approach, we first have to formulate the problem of measurement error more formally. We use a formulation from Saris and Andrews (1991) to do so. The path model presented in Figure 17.1 from Saris and Andrews (1991) summarizes their idea of the relationship between methods and error.

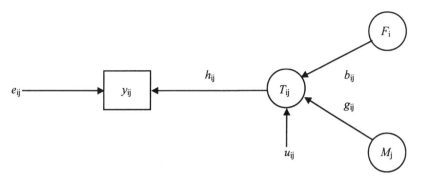

Figure 17.1. A Model for the Response on a Question Incorporating Method Effects, Unique Components, and Random Error.

This idea can be formulated more formally as follows: The responses y_{ij} on item i using method j, can be decomposed into a stable component T_{ij}, which is called the 'true score' on a trait in classical test theory (Lord and Novick 1968; Heise and Bohrnstedt 1970) and a random error component e_{ij}. If the response variable and the variable representing the stable component are standardized, equation (1) results:

$$y_{ij} = h_{ij} \ T_{ij} + e_{ij}, \tag{17.1}$$

where h_{ij} represents the strength of the relationship between the stable component, or true score, and the response. The formula indicates that an observed score is a sum of an unobservable true score on the trait, weighted by the relationship between the true and observed score, and a random error component.

The true score can be further decomposed into a component representing the score on the variable of interest, F_i, a component due to the method used, M_j, and a unique component due to the combination of method and trait, u_{ij}. After standardization, this leads to the following formulation:

$$T_{ij} = b_{ij} \ F_i + g_{ij} \ M_j + u_{ij}, \tag{17.2}$$

where b_{ij} represents the strength of the relationship between the latent variable of interest and the true score and g_{ij} represents the method effect on the true score. All variables are standardized, except for the disturbance variables, which are usually not standardized. The correlations between the disturbance variables and the explanatory variables in each equation and across equations are assumed to be zero, and we assume that the method and trait factors are uncorrelated.

If all variables except the disturbance terms are standardized, the coefficients h_{ij}, b_{ij}, and g_{ij} represent the strength of the relationships between the variables in the model. These coefficients are defined in the literature as follows:

- h_{ij} is called the 'reliability coefficient.' The square of this coefficient is an estimate of the test–retest reliability in the sense of classical test theory (Heise and Bohrnstedt 1970; Lord and Novick 1968).
- b_{ij} is called the 'true score validity coefficient' because the square of this coefficient is the explained variance in the true score due to the variable of interest.
- g_{ij} is called the 'method effect' because the square of this coefficient is the explained variance in the true score due to the method used.
- The variance u_{ij} plus g_{ij}^2 is sometimes called 'invalidity' because it is the variance explained in the true score that is not due to the variable of interest (Heise and Bohrnstedt 1970).

With this information, the total measurement error in the responses (Y_{ij}) can be decomposed into a random component $(\mathrm{Var}(e_{ij}))$ and a (systematic) bias component $(\mathrm{Var}(u_{ij}) + g_{ij}^2)$.

With this notation and simple path analysis we can examine all the possible effects of measurement error on the correlations. In order to estimate the measurement error as defined here, measures of several different indicators of the same construct are needed. In the present study, as in Andrews (1984), repeated measures of the same constructs with different measurement instruments are used. This design is called a 'multitrait-multimethod' (MTMM) design and was first introduced by Campbell and Fiske (1959). As the name suggests, a number of different traits (constructs) are measured with a number of different methods.

Figure 17.2 presents a structural equation model for analyzing MTMM data. Latent factors are introduced for the trait (construct) and method factors, and a distinction is made between true scores and observed scores as in Figure 17.1.

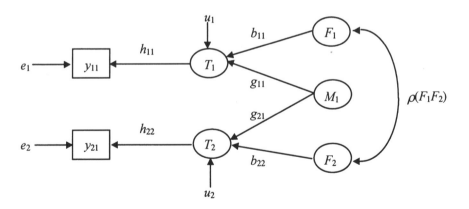

Figure 17.2. A Model for Two Correlated Variables Incorporating Method Effects, Unique Components, and Random Error.

The only difference between Figure 17.1 and Figure 17.2 is that two variables are studied at the same time and that these are assumed to be correlated. This correlation is denoted by $\rho(F_1\ F_2)$. It is assumed that the method factors are uncorrelated both with each other and with the trait factors. The disturbance variables are assumed to be uncorrelated with each other and the factors. All other assumptions made for the model in Figure 17.1 also hold, and the parameters have the same meaning as before.

In statistical terms this model is 'not identified,' since it is impossible to distinguish the unique component u_{ij} (the interaction between the trait and the method) from the random error e_{ij}. The distinction between these two terms is only possible if repeated observations of the same trait with the same method are available (test–retest designs, panel data). However, in this model we have only one observation of each trait with each method at one point in time; this makes an additional assumption necessary. If the latent variable F_j is measured by all the questions and these questions differ only in the method M_j, then the (presumably realistic) assumption can be made that the unique component (u_{ij}) is zero (Saris 1990). The assumption makes the model identifiable and enables the estimation of the error variance $\mathrm{Var}(e_{ij})$.

The correlation between the latent variables of interest, $\rho(F_1\ F_2)$, can be derived from Figure 17.2. Path analysis suggests that the correlation between the observed variables, denoted by $r(y_1,y_{21})$, is equal to the correlation produced by F_1 and F_2 and the spurious relationship due to the method specific variation in the observed variables:

$$r(y_1,y_{21}) = h_{11}\,b_{11}\ \rho(F_1\ F_2)\ b_{22}\,h_{22} + h_{11}\,g_{11}\,g_{21}\,h_{22'} \tag{17.3}$$

Since the validity coefficients and the reliability coefficients are maximally 1, it follows from (17.3) that $r(y_1,y_{21}) = \rho(F_1\ F_2)$ only if the reliability and validity are maximal and the method effect is 0. Such a situation is extremely unlikely, therefore the two correlations will usually be different. Since the reliabilities, validities, and method effects differ from method to method, this could be the explanation for the differences in correlations between the different methods presented in Table 17.1. Any correlation between the factors of interest can produce very different correlations between the observed variables, depending on the size of the validity and reliability coefficients and method effects.

In order to derive the correlation between the latent factors, we have to express the correlation between the factors in the observed correlations and the different validity and reliability coefficients and method effects. This expression follows immediately from equation (17.3):

$$\rho(F_1\ F_2) = [r(y_1,y_{21}) - (h_{11}\,g_{11}\,g_{21}\,h_{22})] / (h_{11}\,b_{11}\,b_{22}\,h_{22}). \tag{17.4}$$

This result suggests that the correlation between the factors can be estimated from the observed correlation if estimates for the validity and reliability coefficients and the method effects are available.

17.2   EMPIRICAL INFORMATION CONCERNING RELIABILITY, VALIDITY, AND METHOD EFFECTS

Saris and Andrews (1991) describe how the estimates of the reliability and validity coefficients and the method effects can be obtained. Scherpenzeel (1995) applies this procedure to life satisfaction research. The estimation procedure can only be carried out if a systematic study of the possible effects of measurement procedures on the construct of interest is carried out. One such study was conducted by participants of a large international network (Saris et al. 1996). For each of the language areas described in the lower part of Table 17.2, a study was conducted to obtain estimates of reliability, validity, and method effects. This was followed by a meta-analysis to investigate the effects of the different characteristics of the instruments used (for details of the study and discussion of related issues, see Saris and Münnich 1995).

Table 17.2 summarizes the findings from Scherpenzeel (1995). The overall mean validity and reliability coefficients for satisfaction measures are at the top of the table. In the other rows of the table, the adjustments of the expected value are specified for different choices possible with respect to the design of the measurement instrument. In each row, the adjustment for a different specific study characteristic is indicated (for further discussion, see Scherpenzeel 1995). A restricted number of characteristics were taken into account, including the specific trait studied, the answer scale, the data collection mode, the position of the question in the questionnaire, as well as other factors related to the study design (such as whether an instrument was used in combination with others, the position of the instrument in the multimethod sequence, and the country in which the data was collected).

Table 17.2 can be used to predict the validity and reliability of any measure of satisfaction. We demonstrate how this can be done, using one variable from the Dutch satisfaction study described earlier. The data for the study were collected by tele-interviewing. The variable discussed is the measure of General Life Satisfaction using a 5-point answer scale, which was asked at the beginning of the interview and the repetition of the same question, using a 10-point scale, some 30 to 60 minutes later in the interview. Using the information from Table 17.2, we can estimate the true score validity and reliability coefficients for this measure, as shown in Table 17.3. By adding all adjustments to the mean value, we obtain an estimate of the validity and reliability coefficient for this variable.

TABLE 17.2. Meta-Analysis of Life Satisfaction Data Across Countries

	N Measures	Validity Coefficient *Mean = .940* Multivariate Deviations	Reliability Coefficient *Mean = .911* Multivariate Deviations
Satisfaction domain			
Life in General	54	−.006	−.038
House	54	.005	.029
Finance	54	.003	.020
Contacts	54	−.001	−.011
Response scale			
100 p. number scale	64	−.021	−.027
10 p. number scale	72	.011	.051
5/4 p. category scale	72	−.022	−.026
Graphical line scale	8	.058	−.007
Data collection			
Face-to-face interview	96	.011	.012
Telephone interview	52	.002	−.051
Mail questionnaire	40	−.014	−.011
Tele-interview	28	−.022	.067
Position			
1 – 5	48	.011	.026
6 – 45	68	.017	−.001
> 50	100	−.017	−.012
Time between repetitions			
Alone in interview	32	.010	−.071
First/last 5–20 minutes	64	.017	.063
First/last 30–60 minutes	80	−.021	−.023
Middle 5–20 minutes	16	.043	.028
Middle 30–60 minutes	24	−.017	−.016
Order of presentation			
First measurement	60	−.015	−.025
Repetition	156	.006	.010
Country			
Slovenia	12	.020	−.013
Germany	16	.007	.028
Catalonia (Spain)	12	−.039	−.022
Italy	12	.013	.043
Flanders (Belg) + the Netherlands	64	−.028	−.039
Wallonia (Belgium)	12	−.026	−.028
Brussels (Belgium)	12	.006	.000
Sweden	12	.023	.099
Hungary	12	.050	.046
Norway	16	−.018	.031
Russians (Russia)	12	.043	.004
Tatarians (Russia)	12	.033	.003
Other ethnicities in Russia	12	.039	.000

TABLE 17.3. Prediction of the Validity and Reliability of a Measure in the Dutch
Example Study, on the Basis of the Instrument Characteristics

	Validity Coefficient	Reliability Coefficient
	Mean = .940	Mean = .911
Adjustments for:		
Domain: General life satisfaction	−.006	−.038
Response scale: 5-point scale	−.022	−.026
Data collection: Tele-interview	−.022	+.067
Position: 2nd question in interview	+.011	+.026
Time between repetitions: 30–60 min.	−.021	−.023
Order of presentation: Repetition	+.006	+.010
Country: The Netherlands	−.028	−.039
Sum	.858	.886

17.3   COMPARISON BETWEEN HUNGARY AND THE NETHERLANDS

Coefficients for the other traits and countries of Table 17.2 can be estimated in the same way as illustrated above. The results of this kind of calculations for all satisfaction traits and methods are presented in Table 17.4. In this table we have added the method effect. This effect can be calculated from the information on the validity coefficient because the method variance should be $1 - b_{ij}^2$ if the unique variance is zero. Thus the estimate of the method effect parameter is the square root of the method variance.

Table 17.4 shows that the questions in these two countries/languages do not differ much in reliability but do differ considerably in validity and method effect. This has serious consequences for the observed correlations. For the 10-point scales, the correlations with GLS were all larger in Hungary than in the Netherlands (see Table 17.1), while this was only the case for one correlation when a 5-point scale was used. The extra correlation produced in the Netherlands might be due to a method effect. On the basis of equation (17.4) the influence of a correction for the random and systematic method effect can be estimated. In the Netherlands the correlation between GLS and SH was .381 and in Hungary .341. Correction for measurement error in the Dutch data yields a value of .279.[1] As said before, in the Dutch study the observed correlation overestimated the corrected correlation because of systematic method effects. The corrected correlation in the Hungarian data is .357. In the Hungarian study the validity and the reliability are rather high. Therefore the small increase due to method effects is compensated for by the lack of reliability, so that

TABLE 17.4. Quality Estimates of the Indicators in the Dutch MTMM Study, Predicted on the Basis of the Meta-Analysis

	Validity	Reliability	Method Effect
Dutch Questions			
GLS	.86	.89	.51
SH	.91	.99	.42
SF	.91	.98	.42
SC	.90	.93	.44
Hungarian Questions			
GLS	.96	.90	.28
SH	.99	1.00	.14
SF	.99	.99	.14
SC	.99	.94	.14

GLS = general life satisfaction, SH = satisfaction with home, SF = satisfaction with financial situation, SC = satisfaction with social contacts.

the observed correlation is smaller, not larger, than the correlation corrected for measurement error.

This example illustrates that one cannot simply compare the correlations across countries. The reliability and validity of the measures used need to be taken into account. Only after correction for the relevant random and systematic errors can the relationships in the different countries be compared.

17.4 CONCLUSION

Observed correlations from different countries are susceptible to various errors, both nonsystematic and systematic. Corrections for these errors increase the adequacy of conclusions. Whenever possible, comparison of observed (uncorrected) correlations is to be avoided and correlation coefficients should only be compared after correction for random and systematic error (reliability and validity). This enhances the accuracy of comparisons, not only across different methods within one study, but also across studies and countries. When tables like Table 17.2 are constructed for other research topics (see for example Andrews 1984; Rodgers, Andrews, and Herzog 1992; Költringer 1995; Scherpenzeel 1995; Scherpenzeel and Saris 1997), the procedure described in this chapter can be used for any correlation matrix and structural equation model. This versatility is an attractive feature of the procedure described here. Using the quality measures obtained with these calculations, equation (17.4) can be used to obtain the correlations corrected for measurement error. This could be computed by hand, as in this chapter, but programs such as LISREL (Jöreskog and Sörbom 1989) can also be used to estimate the corrected correlations, using the model specified in Figure 17.2, or a larger model for all traits for which data have been collected.

Until now, researchers had only two options: to use multiple indicator models, which are more expensive and require the analyst to have sophisticated technical statistical knowledge, or to ignore measurement errors completely. Ignoring measurement error is hardly an acceptable option. In our view it is always better to correct for measurement error, even if correction cannot be perfect, than to ignore the measurement error. Ignoring the effect of random error implies that the random measurement error and the method effects are assumed to be zero. This assumption is often untenable and leads to errors in analyses and to wrong conclusions. The approach described here does not require multiple indicators in substantive studies, provided a methodological project has been carried out to estimate the relevant method effects (Saris et al. 1996). We have shown here that there are good alternatives to assuming that there is no measurement error and that reasonable estimates of the size of the errors are often available, especially for studies in the field of social indicators. For other variables we refer to the work by Andrews (1984), Költringer (1995), and Scherpenzeel and Saris (1997).

Note

[1] The Netherlands: $\rho(GLS, HS) = (.381 - .89 \times .51 \times .42 \times .99)/(.89 \times .86 \times .91 \times .99) = .279.$
Hungary: $\rho(GLS, HS) = (.341 - .90 \times .28 \times .14 \times 1.0)/(.90 \times .96 \times .99 \times 1.0) = .357.$

Chapter 18

RESPONSE FUNCTION EQUALITY

WILLEM E. SARIS

This chapter deals with the process by which respondents select answers from among answer categories offered, a process we call here the response function (cf. van der Zouwen 1976). Survey respondents have to interpret each question, retrieve an opinion or other stored information, render a judgment, and report an answer (Tourangeau and Rasinski 1988). Researchers are interested in respondents' judgments, but what they have to work with are the reported responses. If the response function is the same for all the respondents in a study, the fact that our data are answers and not judgments does not matter so much. However, if response functions vary across respondents and, in comparative research, across cultures, then comparing responses instead of judgments introduces a bias into the comparison. The question then is whether or not comparisons are possible across cultures without correction for these differences in response functions.

We know from everyday experience that people differ in the way they express themselves. The British, for example, are stereotyped as masters of understatement, while Southern Europeans are held to prefer extreme and exaggerated statements (cf. Hui and Triandis 1989). Saris (1986) showed that variation in response functions exists and is a problem for survey research. However, there are other reasons for differences in response functions, and one and the same question might produce more measurement error in one country than in another. In this case responses are not a good indication of the judgments of respondents and comparison across countries will be difficult. Scherpenzeel (1995) and Holleman (1999) provide ample evidence for differences in the response functions stemming from different errors in the different countries investigated.

Below, we examine differences in response functions across countries and modes of data collection for categorical variables that are used frequently in cross-cultural research. An outline of the problem is followed by a discussion of how differences in response functions can be detected. The problem is illustrated in detail using a Eurobarometer question on voting behavior, followed by a discussion of mode effects in other parts of the Eurobarometer. The chapter closes with a discussion of the findings.

18.1   WHAT IS THE PROBLEM?

The latent class model developed by Lazarsfeld (1950a, 1950b) helps illustrate the problem. Imagine the simplest case of a variable (x) with two categories, for example, people who believe that their own country has benefited from the EU $(x = 1)$ or not $(x = 2)$. The percentage of people of the population in each category π_1^x and π_2^x is unknown. The only information that can be obtained is the percentage of respondents in a sample who answer a question on the topic positively or negatively. But questions can be interpreted in many different ways. If either the interpretation of the question or the way people express themselves varies across countries, different response distributions can result.

Using a slightly different notation from Goodman (1974a, 1974b) and Hagenaars (1990), the conditional probabilities of reacting positively or negatively, given the score on the variable x, can be presented in a matrix (as below) for respondents from, say, France (i) and Spain (j). In Table 18.1 π_{11}^i is the probability in France (i) of saying 'yes' $(= 1)$ if the judgment is also 'yes' $(x = 1)$. In the same way π_{21}^i is the probability that French respondents will say 'no' if their judgment is in fact 'yes' $(x = 1)$; π_{12}^i is the probability that French respondents (i) will say 'yes' $(= 1)$ if their judgement is 'no' $(x = 2)$; and π_{22}^i is the probability that French people will say 'no' if their judgment is also 'no' $(x = 2)$. In the other table, the same probabilities are provided for respondents from Spain (j).

The proportion of French people (i) who answer 1 (yes) is denoted by π_1^i and the proportion of French people who answer 2 (no) by π_2^i. The proportion of people who say 'yes' in the French study is then equal to

$$\pi_1^i = \pi_{11}^i \pi_1^x + \pi_{12}^i \pi_2^x \qquad (18.1)$$

and the proportion of people who say 'no' is equal to

$$\pi_2^i = \pi_{21}^i \pi_1^x + \pi_{22}^i \pi_2^x . \qquad (18.2)$$

The same formulation can be used for Spain, using j instead of i in the equations. If we denote the matrix with response probabilities in France by $\mathbf{\Pi}^i$ and for Spain by $\mathbf{\Pi}^j$, the response distribution in France by $\boldsymbol{\pi}^i$ and for Spain by $\boldsymbol{\pi}^j$, and the distribution of the judgments in France and Spain by $\boldsymbol{\pi}^x$, assuming that they are identical, then the equation indicated above can be presented in two simple matrix multiplications presented in equations (18.3a/b):

$$\boldsymbol{\pi}^i = \mathbf{\Pi}^i \boldsymbol{\pi}^x \qquad (18.3a)$$

and

$$\boldsymbol{\pi}^j = \mathbf{\Pi}^j \boldsymbol{\pi}^x . \qquad (18.3b)$$

TABLE 18.1. Probabilities of Observed Responses per Value of the Latent Variable (Judgments) for Two Countries

Observed Response	Country i			Country j		
	Latent Variable			Latent Variable		
	$x = 1$	$x = 2$	Marginal	$x = 1$	$x = 2$	Marginal
1	π_{11}^i	π_{12}^i	π_1^i	π_{11}^j	π_{12}^j	π_1^j
2	π_{21}^i	π_{22}^i	π_2^i	π_{21}^j	π_{22}^j	π_2^j
Sum	1.0	1.0	1.0	1.0	1.0	1.0

If π^x is the same in both countries and if the probabilities are the same that respondents will answer a question positively in France (π_{11}^i, π_{12}^i) and in Spain (π_{11}^j, π_{12}^j), given the score on the latent variable x, then the distribution of the observed distributions (π_1^i, π_2^i) and (π_1^j, π_2^j) will also be the same, apart from random fluctuations. If, however, the probabilities are unequal, then the distribution for the different variables will also be different. Thus if the matrices with the response probabilities are the same ($\mathbf{\Pi}^i = \mathbf{\Pi}^j$), then the response distributions will also be the same. If, however, these matrices differ from each other, the resulting response distributions will also differ.

This point can be illustrated with a simple constructed example. Suppose that $\pi_1^x = .9$ and $\pi_2^x = .1$, while the response probabilities are as given in Table 18.2. This hypothetical difference in the tendency to say 'yes' to the same question in France and Spain will lead to a difference in the distributions for the two variables. The distribution in France would be $\pi_1^i = .73$, which is $.8 \times .9 + .1 \times .1$, and $\pi_2^i = .27$. In Spain, the distribution will be $\pi_1^j = .85$ and $\pi_2^j = .15$. This difference in distribution would not have occurred if the response probabilities had been the same.

TABLE 18.2. Response Probabilities

France	$x = 1$	$x = 2$	Spain	$x = 1$	$x = 2$
1	.8	.1	1	.9	.4
2	.2	.9	2	.1	.6

Various factors can lead to a preference for certain response categories in one country but not in another. For example, different response category labels may have been used in the different translations, or the category labels may have different meanings for respondents in the different countries, even if the translated labels seem to be functionally equivalent. Both can realistically be expected to occur, therefore it is essential to detect whether the response functions or response probabilities are the same in the different countries. If that is not the case, the responses collected cannot be compared directly.

However, in practice we do not know the response probabilities, we only have the distributions for the two response variables. These do not permit us to estimate the response probabilities. In panel studies, however, it is possible to estimate these response probabilities, as illustrated in the next section.

18.2   ESTIMATION OF RESPONSE PROBABILITIES

In panel survey research, the same respondents answer the same questions on at least two different occasions. The relationship between the (two) observations of the same variable can be presented in a table (the so-called 'turnover table'). To estimate the response probability, it is important that this table presents the relationship between the responses observed at approximately the same time. The mode of data collection may differ for the two observations. Panel respondents are, for example, often interviewed face-to-face for the first contact and interviewed by telephone thereafter. Our example includes a change of data collection mode. Differences in mode create the problem that the response probabilities for the first method can be different from those of the second method. This is the same problem of comparability as previously discussed across cultures. One important difference, however, is that in this case we have data for the same respondents collected in two different ways. This allows us to produce a table such as Table 18.3.

This table shows the distributions of the two variables in the marginals while the combinations of the response variables can be found in the cells of the matrix. For the marginals of this table we can write as before:

$$\pi^f = \Pi^f \pi^x \qquad\qquad (18.4a)$$

and

$$\pi^t = \Pi^t \pi^x, \qquad\qquad (18.4b)$$

where π^f is the vector with the marginal distribution obtained by face-to-face interview;
π^t is the vector with the marginal distribution obtained by telephone interview;
π^x is the vector with the marginal distribution of x;
Π^f is the response probability matrix in face-to-face interview given the score on x;
Π^t is the response probability matrix in telephone interview given the score on x.

TABLE 18.3. **Relationship Between the Observation of the Same Variables with Two Different Modes**

Variables	Telephone		
Face-to-face	1	2	total
1	.652	.078	.730
2	.198	.072	.270
Total	.850	.150	1.00

From this model it follows that the table denoted by T^{fi} can be written as a function of the matrices with the response probabilities and the values of the latent variable, if it can be assumed that the modes are independent of each other, given the value of x, and that x is stable over time. We first create a diagonal matrix (X) which contains the values of the latent variable in our example on the diagonal, that is, the number or proportion of people in the classes x_1 and x_2:

$$X = \begin{vmatrix} .9 & .0 \\ .0 & .1 \end{vmatrix}. \tag{18.5}$$

Using this matrix, T^{fi} can be shown to be:

$$T^{fi} = \Pi^f . X . \Pi'. \tag{18.6}$$

Thus, if the matrix with the proportions of people in the latent classes is pre- and post-multiplied by the two matrices representing the response probabilities, we obtain table T^{fi}. This formulation is attractive because it makes the connection between the table obtained from the panel study and the model characteristics in which we are really interested. As we do not know the values of the probabilities in the two matrices Π^f, Π', and the matrix (X) the estimation of these values has to be made on the basis of the information in the table which can be obtained from the data. The ML estimation procedure (Haberman 1979) using the EM algorithm (Goodman 1974a, 1974b; Hagenaars 1993) can be used for this. The program LEM used here was written by Vermunt (1995). The program uses turnover tables like the ones presented above as input. The user has to specify some mild restrictions on the probability matrices because otherwise the models of interest are not statistically identified, due to the fact that the number of unknown parameters is larger than the number of independent cells in the table. The program (LEM) provides a goodness of fit test for the whole model. The procedure is illustrated below.

In cross-cultural research the response probabilities can be estimated for both countries simultaneously from the tables presenting the relationship between the responses in the face-to-face mode and the telephone study. Crucial in this statistical procedure are the relations between the response probabilities and the frequencies in the table presented in equation (18.6). Various restrictions can be imposed which lead to different models:

Model 1   The response probabilities are the same for telephone and face-to-face studies in all countries.

Model 2   The response probabilities are the same across countries but different for telephone and face-to-face interview modes.

Model 3   The response probabilities for face-to-face and telephone modes are the same but the probabilities across countries are not the same.

Model 4   The response probabilities differ between countries and between the data collection methods.

The first restriction is the more attractive one because it allows for comparisons across studies and countries. If this restriction does not hold but the second one does, then comparisons across countries are possible, but not across the mode of data collection without correction. If the third restriction holds, comparisons across modes within each country are possible, but comparisons across countries are not possible without further consideration. If the fourth restriction holds, then no comparisons can be made across either countries or modes. Below we demonstrate how the data were analyzed to test these restrictions.

18.3   RESEARCH DESIGN

To test the equality of the response probabilities data have to be collected from the same respondents using two different data collection methods, so that the turnover tables described in the previous section can be constructed. Leaving sufficient time between the two observations ensures that respondents are not affected by the fact of replication. The analysis here is based on a panel experiment described in Saris and Kaase (1997). A face-to-face study was followed (after more than two weeks) by a telephone interview with subsamples of respondents from the first study in different countries. France, Belgium, and Spain were selected for the panel experiment. In France approximately 350 respondents completed a personal interview followed by a telephone interview, in Belgium, approximately 250 respondents, and in Spain, 320. Although these subsamples are much smaller than the original face-to-face samples, the distribution of responses for most variables did not deviate significantly from the responses in the original samples. This suggests that respondents in the subsample (panel) did not differ in their views from respondents only interviewed face-to-face. To be able to continue the analysis, an important assumption needs to be made, namely, that the respondents who participated in both the face-to-face and the telephone interview did not differ in their response behavior from the people who participated only in the face-to-face interview.

18.4   DATA ANALYSIS

The equality of the response probabilities can be estimated and tested by analyzing the turnover tables from the Eurobarometer panel, using the program LEM (Vermunt 1999). First we formulate a model by specifying specific restrictions on the response probability matrices for each country and for each mode. The program then estimates the response probabilities and provides a goodness of fit test for the whole model, as illustrated below. Our example item is the third from a battery of questions:

> What is your opinion on each of the following proposals? Please tell me for each proposal, whether you are for it or against it. Any citizen of another EU country who resides in (our country) should have the right to vote in local elections.
> Pro / Against / DK, No answer

The three subtables representing the relationship between the face-to-face and telephone responses for this item in the three countries are presented in Table 18.4, which has been used as data input to test different models in the three countries.

TABLE 18.4.   Relationships Between the Face-to-Face and Telephone Responses for the Variable 'Right to Vote in Local Elections' in Three Countries

Face-to-face	Telephone			
	Pro	Against	DK	Total
France				
Pro	159	5	9	173
Against	7	128	4	139
DK	16	5	1	22
Total	182	138	14	334
Belgium				
Pro	103	5	1	109
Against	3	80	1	84
DK	21	10	1	32
Total	127	95	3	225
Spain				
Pro	175	4	14	193
Against	6	63	5	74
DK	25	8	5	38
Total	206	75	24	305

TABLE 18.5. Matrix with Identical Response Probabilities across Countries and Modes

	Face-to-Face	Telephone
Country 1	$\Pi^{f1} = \begin{vmatrix} \text{free} & \pi^{fl}_{12} & \pi^{fl}_{13} \\ \pi^{fl}_{12} & \text{free} & \pi^{fl}_{23} \\ \pi^{fl}_{31} & \pi^{fl}_{32} & \text{free} \end{vmatrix}$	$\Pi^{t1} = \begin{vmatrix} \text{free} & \pi^{fl}_{12} & \pi^{fl}_{13} \\ \pi^{fl}_{12} & \text{free} & \pi^{fl}_{23} \\ \pi^{fl}_{31} & \pi^{fl}_{32} & \text{free} \end{vmatrix}$
Country 2	$\Pi^{f2} = \begin{vmatrix} \text{free} & \pi^{fl}_{12} & \pi^{fl}_{13} \\ \pi^{fl}_{12} & \text{free} & \pi^{fl}_{23} \\ \pi^{fl}_{31} & \pi^{fl}_{32} & \text{free} \end{vmatrix}$	$\Pi^{t2} = \begin{vmatrix} \text{free} & \pi^{fl}_{12} & \pi^{fl}_{13} \\ \pi^{fl}_{12} & \text{free} & \pi^{fl}_{23} \\ \pi^{fl}_{31} & \pi^{fl}_{32} & \text{free} \end{vmatrix}$
Country 3	$\Pi^{f3} = \begin{vmatrix} \text{free} & \pi^{fl}_{12} & \pi^{fl}_{13} \\ \pi^{fl}_{12} & \text{free} & \pi^{fl}_{23} \\ \pi^{fl}_{31} & \pi^{fl}_{32} & \text{free} \end{vmatrix}$	$\Pi^{t3} = \begin{vmatrix} \text{free} & \pi^{fl}_{12} & \pi^{fl}_{13} \\ \pi^{fl}_{12} & \text{free} & \pi^{fl}_{23} \\ \pi^{fl}_{31} & \pi^{fl}_{32} & \text{free} \end{vmatrix}$

According to the simplest model, all the response probabilities are the same across countries and modes (Model 1). Model 1 is specified by the following patterns for the matrices Π^{fi} and Π^{ti} (see Table 18.5, in which 'free' means that this parameter is free and should be equal to 1 minus the other probabilities in the column). As can be seen, the response probabilities are the same for the different countries. In addition, we see that the face-to-face and the telephone response probabilities are assumed to be the same. Finally the elements π_{12} and π_{21} are assumed to be identical for reasons of identification. For the specific example, Model 1, assuming all probabilities across modes and countries to be identical, gave a χ^2 statistic of 41.12 with 13 degrees of freedom. A test on the 5% level would lead to rejection of this model.

Having observed that this model does not fit, we have two options: we can either relax the assumption of the equality of the probabilities across countries or relax the assumption of the equality of the probabilities across modes.

The bad fit of Model 1 came as no surprise; in all three tables cell 31 is much larger than cell 13, suggesting that respondents react differently in face-to-face than in telephone interviewing. Thus Model 2 specifies that all response probabilities are the same across countries but that one or more probabilities can be different for the different modes. We hypothesize that respondents who are at least weakly in favor of the 'right to vote' issue will say "Don't know" more quickly or give no answer in a face-to-face interview than in a telephone interview. Consequently, π_{31} is expected to be different for the telephone and the face-to-face interviews. Since in each column of the response matrix one parameter is free to make the probabilities add up to 1, π_{11} can also now vary between the face-to-face and the telephone mode, but not across countries. Thus only one extra parameter needs to be added (π^{tl}_{31}). In all other countries, the coefficients are assumed to be the same. The fit of this model is acceptable because the χ^2 fit statistic was 19.3 with 12 degrees of freedom. The introduction of only one additional parameter, the difference between the telephone and face-to-face interviews, is sufficient to obtain a model with a good fit for these data. No differences between the countries needed to be allowed for.

For illustrative purposes, we present the analysis for Model 3 as well. Model 3 allows for differences between countries but not between interviewing modes. The model specification in each country is the same as before, with the requirement that the face-to-face and telephone response probabilities are the same. On the other hand, these coefficients do not have to be the same from country to country. Therefore, 15 parameters have to be estimated now instead of five as in Model 1 or six as in Model 2. In this specific case, the fit of the model with ten additional parameters is scarcely better that the fit of Model 1. The χ^2 statistic has a value of 38.2 (with 3 degrees of freedom), which leads to rejection of the model. For ten extra parameters an improvement in χ^2 of only three points was achieved. Moving from Model 1 to Model 2, only one parameter more was introduced, yet the reduction in χ^2 was 22 points, a large improvement. In order to obtain a fitting model, therefore, a difference in response probabilities across modes is necessary, but not a difference in response probabilities across countries.

If Model 2 and Model 3 had not fitted the data, then the last alternative would have been to allow for differences between the countries and the modes (Model 4).

The advantage of testing the models in this sequence is that time can be saved since the process can stop as soon as a model fits the data. Furthermore, these models are hierarchical, so that the χ^2 statistics for the different models can be subtracted from each other, and a test can be carried out on the improvement of the model by the additionally introduced parameters.

18.5   RESULTS OF THE ANALYSIS OF EUROBAROMETER ITEMS

Table 18.6 summarizes results from applying this procedure to various different question types and response categories used in Eurobarometer questions (see Appendix). The table indicates that the *media involvement* questions were not affected by the mode of data collection. The response categories are relatively detailed and ask for estimates of frequencies. There are also no differences in response probabilities across countries, so the responses to these questions can be used directly for comparison across modes and across countries.

The second category, two questions on *political involvement*, has one question with the same response probabilities for the two modes and for the three countries, while the other question produces differences between modes and across countries. The explanation for this is that the question on political discussion has the response categories: "frequently, occasionally, never" and "DK, No answer." The other question on persuading people has the response categories: "often, from time to time, rarely, never, DK, No answer," that is, both different wording and one further response category. Matching up the assumed degrees of frequency of the categories in the two questions, the extra category seems to be "rarely." This is precisely the category that causes differences between the modes and the countries. This suggests that if the categories "rarely" and "never" were collapsed, the problems might disappear. This hypothesis was tested and the model with equal probabilities across modes and across countries now fitted the data. This suggests that for purposes of comparison these two categories should be combined. Another option, either at the design or implementation stage, is to omit the category "rarely."

The third pair of questions ask about *satisfaction*. Both questions have unequal response probabilities across modes and countries. Looking at the response categories, however, the problem comes as no surprise. The response categories for both questions in English are: "very satisfied, fairly satisfied, not very satisfied, not at all satisfied" and "DK, No answer." The French and Dutch translations of the labels were, however, more like: "very satisfied, rather satisfied, rather dissatisfied, not at all satisfied." Although the Dutch and French categories treat the third category differently, I was assured several times that these categories are linguistically equivalent. Nevertheless, the statistical analyses showed that significant differences in the reactions of the respondents across countries are found in the response probabilities.

The questions on *involvement in the EU* are interesting in this connection. The question on knowledge uses a response formulation comparable to the satisfaction questions: "very well informed, quite well informed, not very well informed, not at all well informed" and "DK, No answer." If the formulation explains the differences discussed above, this question should have the same problems as the satisfaction question and in Table 18.6 we see that this is indeed the case. The question on interest in EU matters has different response categories: "a great deal", "to some extent", "not much," "not at all" and "DK, No answer"), but the problem is comparable. The term 'not much' is formally a negation of 'much'. So if one is less than 'much interested' in the EU, one could choose "not much". But then the meaning of the category "to some extent" is not clear. It could be seen as a part of the category "not much", but that would lead to confusion. On this basis, the same problems as for the other questions could be expected and, as Table 18.6 indicates, were indeed found. The translations in French and Dutch again treated the categories differently.

For the next set of questions on *opinions on EU membership*, no mode effect was found for 'benefit' and comparison across countries is also possible. For the membership question, on the other hand, respondents had a significantly higher probability to say 'good' if they were in the category "good" on the latent variable in face-to-face interviews than in telephone interviews. The difference was .92 against .80. The hypothesis that this difference existed in all three countries was not rejected. Thus this phenomenon seems to be a cross-cultural difference between telephone and face-to-face interviews.

Finally, seven *opinion* questions, all with the same response categories; "pro," "against," and "no opinion" were analyzed. As Table 18.6 shows, the first two questions concerning the introduction of the European Monetary Union and a common defense policy for the EU did not point to either mode effects or differences

TABLE 18.6. The Evaluation of Difference between Modes and Countries for the Panel Data of the Eurobarometer Study

	Equality of Parameters			
	Across Countries		Not Across Countries	
	Across Modes (Model 1)	Not Across Modes (Model 2)	Across Modes (Model 3)	Not Across Modes (Model 4)
Media involvement				
Radio	+			
Newspaper	+			
TV	+			
Political involvement				
Political discussion	+			
Persuade others	−	−	−	+
Satisfaction				
Life in general	−	−	−	+
Democracy in country	−	−	−	+
Involvement in EU				
Interest	−	−	−	+
Knowledge	−	−	−	+
Opinion on EU membership				
Benefit of EU membership for country	+			
Evaluation of EU membership for country	−	+		
Opinion on EU policies				
European Monetary Union	+			
EU defense	+			
Participation local elections	−	+		
Participation EU elections	−	+		
Candidacy local elections	−	+		
Candidacy EU elections	−	+		
Subsidiarity	−	+		

− means that the model is rejected; + means that the model is not rejected.

across countries. On the other hand, the questions on elections indicated a mode effect. Respondents with a score of "DK, No answer" on the latent variable in the telephone interview have a higher probability to say "pro" than in the face-to-face interview. The last opinion question on subsidiarity between local, national, and EU governments produced the same effect.

It is difficult to explain these effects; a general acquiescence bias (Schuman and Presser 1981) is unconvincing since this would affect all the questions. It is also not an effect of the topic because then one would have to find a different explanation for the last question. An explanation as a learning effect is also difficult to uphold, because then one would also expect this outcome for all questions and not only for a limited number. There seems to be no good reason why respondents would learn that they have to respond "yes" to this question instead of "no." We will have to wait for further research to clarify the matter.

18.6   CONCLUSION

The comparability of responses across modes of data collection and across countries was studied by testing the equality of the response probabilities for different questions and response categories across modes and countries. Table 18.6 shows that the response probabilities across countries were different for five questions. In four questions the category labels were not unambiguous in the English source version, which led to differences in the translations and consequently to differences in the response functions.

It is interesting that these differences were found although native speakers of English, all survey experts, considered the response categories in the different languages as optimally comparable. This shows that such judgments of experts are not always sufficient. In order to be sure about the equality of response functions, tests are necessary.

Table 18.6 indicates that for 11 out of the 18 questions the responses are affected by the mode of data collection. This is a rather large number and indicates that switching between modes leads to differences in the results (see also Saris and Kaase 1997 with a different approach; Silberstein and Scott 1991; Kalfs 1994; Scherpenzeel and Saris 1997).

The present findings contradict the standard literature on mode effects (van der Zouwen 1976; Groves and Kahn 1979; de Leeuw 1992). Our findings are most comparable to those of Scherpenzeel and Saris (1997), who also used panel data. In panel studies the confounding factors are better controlled although one faces the additional problem of memory effect.

The findings reported here suggest that testing the equality of response functions is an essential step in the development of measurement instruments for cross-cultural research. Without certainty about the equality of the response functions, we cannot tell whether differences in responses and correlations between countries are due to substantive differences or to differences in the response functions.

APPENDIX: THE QUESTIONS USED

The questions presented below are all standard questions from the Eurobarometer.

Topic	Question
Evaluation of membership of the EU	
Membership	Generally speaking, do you think that (our country's) membership of the EU is
	Response categories : a good thing / bad thing / neither good nor bad / DK, No answer
Benefit	Taking everything into consideration, would you say that (our country) has on balance benefited or not from being a member of the (EU and EC)?
	Response categories: benefited / not benefited / DK, No answer
Satisfaction	
Life satisfaction	On the whole, are you very satisfied/ fairly satisfied/ not very satisfied/ not at all satisfied with the life you lead? Would you say you are?
	Response categories: very satisfied / fairly satisfied / not very satisfied/ not at all satisfied / DK, No answer
Satisfaction with democracy	On the whole, are you very satisfied/ fairly satisfied/ not very satisfied/ not at all satisfied with the way democracy works in (our country)? Would you say you are?
	Response categories: very satisfied / fairly satisfied / not very satisfied/ not at all satisfied / DK, No answer
Political interest	
Political discussion	When you get together with friends, would you say you discuss political matters frequently, occasionally, or never?
	Response categories: frequently / occasionally / never / DK, No answer
Persuade others	When you hold a strong opinion, do you find yourself persuading your friends, relatives, or fellow workers to share your views? Does this happen?
	Response categories: frequently / occasionally / never / DK, No answer
Media involvement	
Newspaper	About how often do you read the news in daily newspapers?
	Response categories: every day / several times a week / once or twice a week / less often / never / DK, No answer
Radio	About how often do you listen to the news on the radio?
	Response categories: every day / several times a week / once or twice a week / less often / never / DK, No answer
TV	About how often do you watch the news on television?
	Response categories: every day / several times a week / once or twice a week / less often / never / DK, No answer

The wording of some questions differs somewhat in different Barometers. These questions concern specific opinions and knowledge. Questions of this type examined here are:

Topic	Question (Please tell me for each proposal, whether you are for it or against it.)
Opinions	
European Monetary Union	There should be a European Monetary Union with one single currency replacing by 1999 the (national currency) and all other national currencies of the member states of the EC or EU
EU defense	The (EC and EU (EC)) member states should work towards a common defense policy
Participation local elections	Any citizen of another (EC and EU (EC)) country who resides in (your country) should have the right to vote in local elections
Participation European elections	Any citizen of another EC or EU country who resides in (our country) should have the right to vote in European elections
Candidacy local elections	Any citizen of another EC or EU country who resides in (our country) should have the right to be a candidate in local elections
Candidacy EU elections	Any citizen of another EC or EU country who resides in (our country) should have the right to be a candidate in European elections
Subsidiarity	The EC or EU should be responsible only for matters that cannot be effectively handled by national, regional, and local governments

PART FIVE

DOCUMENTATION AND SECONDARY ANALYSIS

Chapter 19

USING PUBLISHED SURVEY DATA

JAN W. VAN DETH

19.1 WHY USE PUBLISHED DATA?

Virtually every researcher starts with the idea that collecting his or her own data is necessary and preferable. With the obvious exception of research questions addressing comparisons with past phenomena and developments, original research is usually based on the collection of fresh data. From this viewpoint, only poor and unimaginative scholars would rely on data collected by more ambitious colleagues. After all, who would want to spend his or her life analyzing data collected for other purposes by other people who have already published the most interesting results?

The idea that collecting your own data is the ideal situation for researchers is based on a clear misunderstanding of the role of empirical testing and exploration in research and also on an overestimation of the need for newly collected data. In fact, using existing data is the rule rather than the exception in social research. This was certainly the case in the early days of empirical research, but it is also true for the more recent past (cf. von Alemann and Tönnesmann 1995; Gehring and Weins 1998). With the rapid expansion of social science research, a large amount of information has been stored in data archives and on Web sites, waiting for other uses. If, for instance, one wanted to test the proposition that men and women show clear differences in political and social involvement, there are plenty of surveys available in many countries that include both the gender of the respondents and some measure of political activities. It would thus seem unnecessary to collect fresh data in this case. Alternatively, if your research hypothesized a possible decline in differences between the political activities of men and women over the last few decades, basing your analyses at least partly on existing data would be the only way to obtain empirically well-founded answers. For these types of questions, you would need to search for available studies and thus confront the difficult problems of using data collected for quite different purposes. In this chapter, the advantages and problems of reanalyzing existing survey data are discussed. Patently, each and every analysis is by definition based on 'existing data' (sometimes even called 'preexisting data' (Kiecolt and Nathan 1985, 75). The terms 'published data' and 'existing data' are used here as synonyms, although not all existing data are published. They are preferred here to the distinction between 'desk research' and 'field research', which refers more to a division and location of labor. The term 'archival data' (Elder, Pavalko, and Clipp 1993) is also not used here, since data available for reanalysis do not necessarily have to be archived

In his seminal work on "secondary analysis of survey data" Hyman defined these activities as ". . . the *extraction of knowledge on topics other than those which were the focus of the original surveys*" (1972, 1; italics from original). Yet this useful demarcation is too limited for the broad field of using existing data and too much emphasis is laid on the goal or focus of researchers involved in collecting the original data. Glass simply speaks of secondary analysis as ". . . answering new questions with old data" (1976, 3). More recent definitions refer to ". . . the further analysis of data by anyone other than those responsible for its original commissioning or collection" (Dale, Arber, and Procter 1988, 4) or as ". . . any further analysis of one or more datasets which yields findings or knowledge additional to those presented in the original report" (Hakim 1982). The different modes and purposes of these "further analyses of data" can be summarized more precisely in the following way:

> A *reanalysis* studies the same problem as that investigated by the initial investigator; the same data base as that used by the initial investigator may or may not be used. If different, independently collected data are used to study the same problem, the reanalysis is called a *replication*. If the same data are used, the reanalysis is called a *verification*. In a *secondary analysis*, data collected to study one set of problems are used to study a different problem. (Committee on National Statistics 1993, 9; italics from original; as cited by Herrnson 1995, 452)

The use of existing data simply implies that you do not collect your own data (verification strategies and secondary analysis) or that you enlarge available data with additional findings (replication strategies). Unfortunately, many data cannot be used for 'further analysis' in a straightforward manner and answering 'new questions with old data' can be much more complicated than it looks at first sight.

19.2   WHY USING EXISTING DATA?

Empirical analyses of, for example, a decline in gender-biased political involvement, the spread of individualistic value orientations, or of attitudes toward inequality and poverty call for the use of previously collected data. Fortunate researchers might find excellent data on these topics gathered for the purpose of longitudinal analysis. For instance, Ronald Inglehart (1977) collected data on value orientations among citizens in Western Europe with the explicit goal of building time series which would permit the study of social and political change. Today, researchers interested in the dynamics of value change can rely on an extensive set of data collected for this purpose in many countries over a relatively long period of time. Verification of earlier findings has become a minor branch of the study of cultural changes (cf. Abramson and Inglehart 1995; van Deth and Scarborough 1995). The same applies—perhaps even more strongly—to the development of research on age-graded life patterns in society using the growing body of archival data available for 'studying lives' (Elder, Pavalko, and Clipp 1993). Obviously, a researcher might replicate part of the work and add his or her own new data collection in order to study the differences between the present situation and the past. In this way, verification or replication means that available resources are utilized in a very efficient fashion and that the opportunities for innovative research are clearly expanded.

Anyone studying social change and dynamic processes obviously needs to reanalyze existing data, since only comparisons between measures at two or more different points in time are relevant for such research. In some cases, the researcher is able to start a new time series and then wait another ten or twenty years, but often the goal is to study specific developments, such as the rise of unemployment in the 1980s, the dissolution of authoritarian regimes in Eastern Europe, or the consequences of a speech by the American president. For such unique events, research strategies based on verification or secondary analysis are the only way to obtain empirical information, irrespective of how well-funded a project might be. The first reason for relying on existing data, then, is based on the need for *comparability*. In order to obtain theoretically relevant information about the impact or development of specific phenomena, comparisons with previously collected data are required and the analyses have to take into account the complexities of studying social change and dynamic processes. This can be done by replicating an earlier study or by secondary analyses. Obviously, verifying original data can be part of this design, too. Replication studies can also be used as a substitute for significance tests (cf. Hyman 1972, 136) and for estimating the reliability of measurement instruments in some form of test–retest procedures (cf. Saris and Münnich 1995).

Apart from the fairly clear needs in studying social change and dynamic processes in a specific period, several other considerations and circumstances provide a strong case for relying on existing material instead of collecting new data. If the research question or hypothesis is clearly spelled out and existing data sets include acceptable measurements for the major theoretical constructs, why would one repeat the whole painstaking process of collecting fresh data? Instead, the researcher can use resources in a much more efficient way by concentrating on analyzing the data and interpreting the results. In addition, a lack of resources or limited competencies might make using existing data the only option. Collecting high quality survey data is a costly affair. Modern interview techniques such as CAPI or CATI have reduced the time it takes to have data available, but the costs are still significant and vary substantially across countries. For this reason, even the most ambitious researcher is often driven to investigate the opportunities of verification or secondary analysis.

A last reason to support the use of existing data is of a more fundamental nature. The guiding principle of science is *skepticism*, implying that each and every scientific proposition should be the object of public discussion and should undergo rigorous critical appraisal. It implies, too, that all results obtained must be able to be replicated by other researchers. From this it follows that even for industrious and well-funded researchers not forced to make comparisons, analysis of existing data is a key and core activity. In the permanent exchange of critique and recommendations among scientists, the same data are scrutinized from different perspectives, analyzed with different techniques, and (re-)interpreted on the basis of additional information.

In order to underline this crucial aspect of scientific research, several professional organizations have discussed the need for access to existing data (e.g., the American Political Science Association) or have already set up clear policy rules (e.g., the American Psychological Association). 'Principle 6.25 Sharing Data' of the 'Ethical Principles of Psychologists and Code of Conduct,' for example, declares that "... psychologists do not withhold the data on which their conclusions are based from other competent researchers who seek to verify the substantive claims through reanalysis..." (APA 1994, 293). Herrnson (1995) reflects the discussion among political scientists. Obviously, skepticism will stimulate verification and replication strategies, but secondary analyses are also usually the outcome of this skepticism.

Assessment of the reliability or trustworthiness of published results (or of the researchers involved), is not the main goal of these activities. But this openness does have important consequences for participants: "The ability to replicate research provides a mechanism that should ensure that survey data is not systematically misinterpreted or used against the interests of any one group" (Dale, Arber, and Procter 1988, 55). Notice, furthermore, that one of the strongest appeals for founding data archives started with the observation: "In the past few years there has been increasing evidence of fraud in scientific research" (Bryant and Wortman 1978). Reanalyzing existing data is "extremely important to the further development of the discipline":

> The most common and scientifically productive method of building on existing research is to replicate an existing finding—to follow the precise path taken by a previous researcher, and then improve on the data or methodology in one way or another (King 1995, 445).

With the expansion of social science data archives in the last decades, an enormous amount of data is just waiting to be reanalyzed by skeptical and creative researchers. Glover presents a brief summary of the purposes and potentials of secondary analyses: for 'originality,' 'testing of theoretical perspectives,' 'creativity,' to 'break monopolies in social research.' They are useful for 'carrying out analyses of subpopulations,' and 'valuable in making comparisons over time', while being 'time-saving' and 'inexpensive' (Hakim 1982, 169–171; see also Glover 1996, 28). The main reasons for using existing data may be pragmatic and everyday or based on ambition and the search for comparability in studies of social change and dynamic processes, but it is evident that science can only flourish when existing material is reanalyzed. Using existing data can also lead to the discovery of entirely new interpretations. Dale, Arber, and Procter (1988, 54), for example, mention the relationship between smoking and lung cancer detected in reanalysis studies of medical records collected for very different purposes. Since by definition *serendipity* is not directed at some known goal, this important advantage is not listed here as a reason for using published data.

19.3   CORE PROBLEMS WITH PUBLISHED DATA

Public access to all previously collected data might also seem a self-evident requirement. However, it means that researchers collecting data have to document details of their study extensively and make both the data and documentation available to the scientific community. Although these two requirements are accepted in theory everywhere, things still look rather different in practice. The flow of information from principal investigators of studies to researchers looking for data for their own purposes is not without complications.

From the perspective of a principal investigator, it is almost impossible to foresee the exact documentation needs of future researchers. However, some minimum requirements can be easily formulated for survey research. Systematic overviews of standards required are provided, for example, by Sommer, Unholzer, and Wiegand (1999) and Kaase (1999). Survey study documentation should include information on the following points:

- Exact definition of the population and precise description of sample procedures;
- Detailed description of the fieldwork and the timetable used (including a report on the quality of the realized sample with respect to the population);
- A copy of the original questionnaire and all other materials used, including showcards, together with a report on the design of the instruments, pilot studies, and pretests of the questionnaire;
- A copy of interviewer instructions or the interviewer manual, together with an overview of major interviewer characteristics;
- A detailed description of the final data set (including information on nonresponse, missing values and recoding of the original data).

Ideally, principal investigators deliver this documentation along with their data to a public data archive—often an archive in their own country (see Appendix 1). These archives store the information in a systematic way and disclose their holdings to the research community. Many professional journals and publishers now require that publications unambiguously identify data used in order to facilitate reanalysis.

Simple as these recommendations may seem, the world looks quite different from the standpoint of researchers trying to find existing data. Locating data for reanalysis presents problems because many principal investigators do not spend much time documenting and archiving data. Frequently, important information is lost after the results have been published or the researcher moves on to another field of research. This neglect of minimum standards among principal investigators hampers and handicaps the present use of existing data in survey research. Complications range from the lack of a population definition to being confronted with an unlabeled machine-readable data set with no further information. Despite efforts of data archives in recent decades to improve this situation, existing data are often not documented according to even the minimum standards mentioned above.

Insufficient or absent documentation constitute serious barriers to the reanalysis of existing data. Researchers searching for existing data will frequently discover that several data sets include more or less appropriate measures, but that none of these measures meet their needs precisely. When using existing data, the ordinary choice is usually not between good and excellent empirical research, but between debatable research and no research at all. The skills and imagination of the researcher often determine what information can be extracted from existing data. No specific recommendations can be formulated here, but it is clear that little will be gained by starting with a rigid idea of what data are needed. Theoretical concepts and not specific questions or operationalizations should be the main concern when searching for existing data. (Later sections deal with constructing equivalent measures for similar concepts on the basis of different operationalizations.)

Inadequate or faulty documentation, different operationalizations, and poor or unavailable measures of key constructs are serious hurdles for every attempt to use existing data. Even more serious than these practical problems are a number of conceptual complications. The example of studying value change already mentioned indicates some of the hazards and difficulties encountered. Inglehart developed a post materialism instrument in the early 1970s; one of the items formulated to tap a feeling of insecurity was 'fighting rising prices'—an issue of then pressing importance during a time of double-digit inflation in many countries. Price stability has completely disappeared from the political agenda in the last three decades and the intelligent researcher is now confronted with a dilemma. The widely used Inglehart instrument still includes the item 'fighting rising prices.' A straightforward analysis of social changes can thus be performed using and/or replicating these existing data. But wouldn't it be better to replace this item now with an item referring to, say, unemployment or social security payments? Doing so would probably increase the validity of the instrument (cf. Clarke et al. 1999). However, the price to be paid would be that the huge amount of data collected since the early 1970s could no longer be used. In other words, unless the equivalence of formally different measures can be established, replication or verification strategies are effectively blocked. If the item were replaced, reanalysis studies would then either be based on data of disputable validity or the whole idea of trend analyses of cross-national value change would have to be abandoned for the next twenty years or so.

While replicating the Inglehart instrument raises doubts about the meaning of the results, another example illustrates an even more fundamental problem. Since the 1970s, the *American General Social Survey* (GSS) has used a four-item battery to measure changing attitudes toward the division of labor between men and women. These items are all phrased in terms of the position of women, for example, 'A preschool child is likely to suffer if his or her mother works'. Precisely because rapid social change has taken place in this area, it is virtually impossible to use these items as they stand today. Respondents are less willing to accept questions of this kind that suggest only women need to change (Braun 1998). The fact that social change cannot be measured by simply repeating old questions does not mean that we cannot or should not try to measure social change. Instead, this example shows that change can be evaluated using more sophisticated research strategies. Braun (1998) extended the German version of a set of the GSS questions and established equivalence between the old measures and the new (different) measures.

The use of existing data is often avoided because of the types of problems encountered by analysts of changing value orientations or by Braun in his analyses of changing gender roles. Verification strategies seem to be the least problematic in terms of validity and reliability, but they call for well-documented data. For replication strategies and secondary analyses, on the other hand, validity and reliability are of supreme importance. Here very interesting results can be obtained by reanalyzing even poorly documented data. Overviews of the basic aspects of measurement error, validity, and reliability problems when using published data can be found, for example, in Jacob (1984), Elder, Pavalko, and Clipp (1993), and Hyman (1972).

19.4   *EX POST* HANDLING OF MAJOR PROBLEMS

Reanalyzing existing data is crucial from both practical and theoretical perspectives. After the principal investigator makes the data accessible, other researchers start from the situation they encounter and begin by assessing the appropriateness of the data available. Dale, Arber, and Procter (1988, 20-31) discuss a similar approach under the headings 'Preliminary considerations' and 'Preliminary questions relating to the proposed analyses.' Although obviously correct, these headings do not give sufficient emphasis to the need to assess how appropriate existing data are for the research questions at hand. Is the population definition appropriate for the present research aim? Do sampling procedures and design provide the required degree of representativity and have basic indicators (such as the distributions of sex, age, and education) been examined? Do rates of nonresponse and missing values indicate systematic errors? What evidence is provided on the reliability and validity of key constructs? Have the data been cleaned for filter and coding errors? Are coding schemes available for responses to open or partially open questions (such as occupation or party preference)? Only after the *apparent appropriateness* of existing data has been established, should attempts be made to increase the *potential usefulness* of the data with *ex post* strategies. Jacob's unambiguous warning that "Without exception, all published statistics should be treated with suspicion" (1984, 51) is good advice for every researcher planning to use existing survey data.

19.4.1 Harmonization

A first way to improve the potential usefulness of existing data for purposes of cross-national comparisons involves the harmonization of instruments from the different surveys selected for replication or secondary analyses. Harmonization means that for different studies a common scheme is used to code interview questions and answer categories and/or for analyzing variables derived from those inputs (Office for National Statistics 1996). Classical examples here are the coding schemes for occupation and education used in comparative research. For instance, a British replication of a French study on employment cannot simply be based on the regular French occupational scheme (which relies heavily on the distinction between private and public sector occupations) or on the scheme used in British government surveys (which does not make a systematic distinction between private and public sector). The data can be harmonized by either recoding to one of these schemes or by referring to some other scheme used in cross-national research, such as the scheme used for the European Labor Force Survey (Eurostat 1998). In each case important information will be lost because culture-specific details are sacrificed to make reanalysis possible (Glover 1996, 33–34). Similar challenges exist in American surveys. Each decennial census conducted in the United States over the past 200 years has used different race and ethnicity categories. Obviously, this problem is also encountered when fresh data are collected. Successful harmonization strategies, however, considerably increase the amount of existing data available for reanalysis by offering *ex post* strategies to deal with manifest differences between coding schemes used.

The following illustration of the problems and potential of using harmonization strategies is based on Eurobarometer studies conducted for the European Commission. These form a long time series (since the early 1970s) and are conducted at least twice yearly in EU member countries. Most Eurobarometer surveys record respondent occupation. Since the Eurobarometer studies are centrally coordinated and occupation is a widely used standard socio-demographic background variable, we might expect not to encounter the kind of culturally determined problems sometimes involved in comparisons based on country-specific data. Figure 19.1, however, tells a different story.[1]

Before this graph could be constructed, the valid scores had to be modified, since identical codes appeared to have different meanings in different waves. In the set of Eurobarometers selected here, no less than ten different coding schemes were used, varying from seven different categories (ECS-71) to a total of 18 different categories (starting with EB-37). In one case, the coding scheme used even varied between countries within one cross-sectional study (ECS-71). These problems in the coding schemes make it impossible to assume, for example, that a code of '2' represents 'fishermen' in each study and in each wave. The first step in using such data, therefore, is to make a careful comparison of the coding schemes and codes used in each individual survey or wave and in every individual country. Only after each category has been assigned a unique code, can a first depiction of the results be undertaken. Figure 19.1 gives the distribution of occupations for each Eurobarometer study after the original categories (as found in the available data sets) were assigned unique codes. Figure 19.1 would seem to indicate that the occupational structure in Europe is characterized by several massive and very rapid shifts. Apart from relatively minor fluctuations in 1972, 1981, and 1995, we would have to

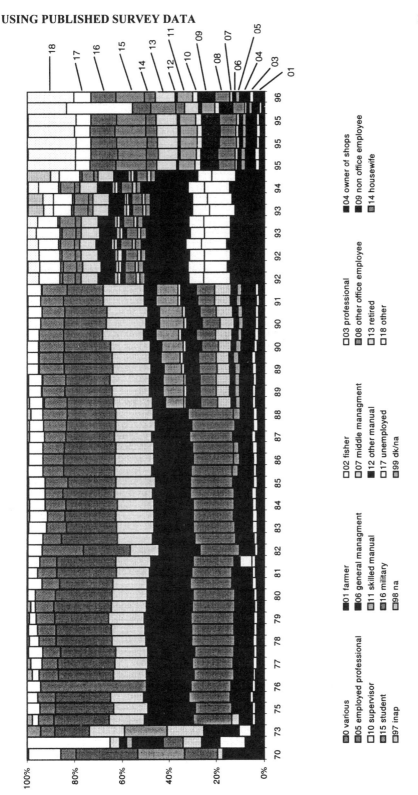

Figure 1: Respondents' Occupation: ECS-70 (1970) to EB-44.4 (1996); Raw Data/Unique Codes

■0 various   ■01 farmer   □02 fisher   □03 professional   ■04 owner of shops
■05 employed professional   ■06 general managment   ■07 middle managment   ■08 other office employee   ■09 non office employee
□10 supervisor   □11 skilled manual   ■12 other manual   □13 retired   ■14 housewife
■15 student   ■16 military   □17 unemployed   □18 other
■97 inap   ■98 na   ■99 dk/na

conclude that in 1974, 1988, 1992, and 1996 very large sections of the working population changed their occupation. This is, of course, not the case and points to very serious problems with the data. The very time-consuming task of assigning unique codes to each original category obviously did not result in meaningful results.

How can the information on the occupational structure in Europe in the last few decades be used in a more intelligent way? Although the ten or so coding schemes showed clear variations, these differences are easy to trace and mainly concerned with various degrees of specification. In these circumstances, an elaborated and encompassing coding scheme can be developed, which covers all the unique codes in the individual studies. This procedure resulted in a new coding scheme with a total of 34 unique categories and the occupation of each respondent was recoded according to this scheme. However, this general coding scheme was not designed as a simple compilation of distinct categories. On the contrary, from the very beginning it was clear that meaningful results could only be obtained when the new codes were assigned systematically. For instance, different categories of office employees were assigned different, but consecutive codes. Using two and three digit codes allows groups to be identified without losing the information about specific categories. So the broad category 'manual worker general' was assigned code '50,' while a more specific category like 'skilled manual' was coded '52.' Some studies have only code '50;' for other studies a very similar category can be constructed by aggregating all the codes between '50' and '59.' In this way, it is possible to maximize comparisons of similar groups without giving up more detailed information. The first digits of the new codes distinguish between the major categories: farmers/fishermen, self-employed, professionals, employees, manual workers, other occupations, and those not gainfully employed. Several complicated categories from the first three waves (ECS-70, ECS-71, and ECS-73) appear to overlap partially with these major categories and cannot be recoded in a straightforward way. Distinct codes (200 and higher) were unavoidable for these categories.

The results of this extensive recoding into a harmonized coding scheme are presented in Figure 19.2. On the basis of this graph, substantive conclusions about the evolution of the occupational structure in the last three decades can be reached. These changes appear to be much less dramatic and more likely than those suggested by Figure 19.1. The fact that more than ten different coding schemes have been used in the original data sets does not prohibit the use of these data in an intelligent way. However, it does require time and patience to arrive at this point—a point at which the real analyses have yet to begin.

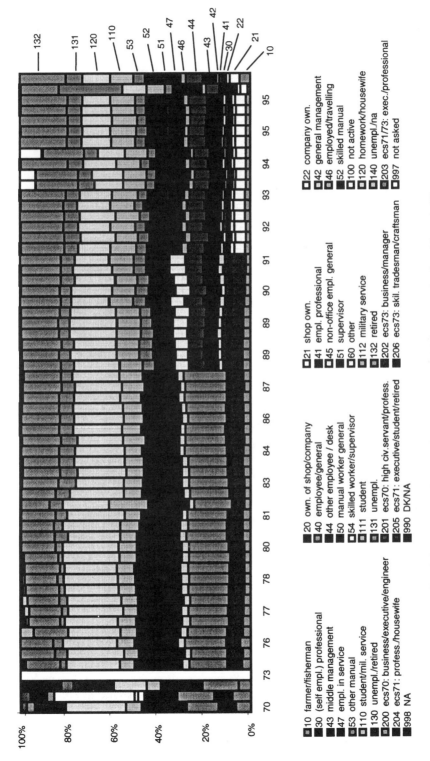

Figure 2: Respondents' Occupation: ECS-70 (1970) to EB-44.4 (1996); Harmonised and Recoded Data

10 farmer/fisherman
30 (self empl.) professional
43 middle management
47 empl. in service
53 other manual
110 student/mil. service
130 unempl./retired
200 ecs70: business/executive/engineer
204 ecs71: profess./housewife
998 NA

20 own. of shop/company
40 employee/general
44 other employee / desk
50 manual worker general
54 skilled worker/supervisor
111 student
131 unempl.
201 ecs70: high civ.servant/profess.
205 ecs71: executive/student/retired
990 DK/NA

21 shop own.
41 empl. professional
45 non-office empl. general
51 supervisor
60 other
112 military service
132 retired
202 ecs73: business/manager
206 ecs73: skil. tradesman/craftsman

22 company own.
42 general management
46 employed/travelling
52 skilled manual
100 not active
120 homework/housewife
140 unempl./na
203 ecs71/73: exec./professional
997 not asked

19.4.2 Equivalence

Different occupational coding schemes can be the result of miscommunication, organizational problems, or simply a lack of interest in careful research documentation. The Eurobarometer example just discussed illustrates problems with different coding schemes in cross-national data. However, different occupational coding schemes can also be the result of cultural and historical differences about the socio-political meaning and relevance of specific concepts such as what is 'private' and what is 'public.' Frequently, this is also the outcome of conflicts and compromise among the people who commissioned the study and performed the primary analyses (Glover 1996, 30). Consequently—and in particular for comparative research—these differences should not be ironed out by rigid harmonization strategies, but should be preserved and nourished as key assets to be obtained. This brings us to a second *ex post* strategy to deal with problems encountered in using existing data: attempts to establish *equivalence* instead of *identity*.

Similar phenomena can have different meanings in different contexts, while different phenomena can have similar meanings in different contexts. Meaning depends on context. The rather naive request for identical measures (especially in replication studies and secondary analyses in comparative designs) simply neglects this truism. Thus, not much is gained with a literal translation of, for instance, the German concept *Bürgerinitiative* into *citizen's initiatives*, since these terms do not refer to similar political phenomena in Britain, the United States, and Germany. Other examples are problems mentioned earlier of measuring changing value orientations or changes in the field of gender roles, and cross-national measures of education and occupation. In these cases, perfect duplications of the original instruments contribute significantly to the problems instead of enabling us to reach meaningful conclusions. What is required are equivalent, not necessarily identical measures (cf. Harkness 1998; van Deth 1998b).

Broadly speaking, we refer to the equivalence of two phenomena if they have the same value, importance, use, function, or result. The important aspect of this concept is the restriction of similarity to one or more specifically defined properties. We should be willing to accept conclusions about coalition behavior of clearly different Christian Democratic parties in different countries, if the parties manage to hold a center position in various coalitions for many decades. It is the similarity of the relevant properties of different phenomena that lies at the center of the idea of equivalence in comparative research. From this line of argument it follows that equivalence cannot be presumed or established in some *a priori* fashion; equivalence depends on the position of some concept in a specific context. Only by studying this position can information be obtained to assess the degree of equivalence. Developing this strategy means that the ultimate test of instruments is not to be found in a meticulous search for identical measures, but in ". . . *the similarity of the structure of indicators . . . Equivalence is a matter of inference, not of direct observations*" (Prezeworski and Teune 1970, 117–118; emphasis from original). As Van de Vijver and Leung state: "*Equivalence should be established and cannot be assumed*"

(1997b, 144; emphasis from original). Though the information required can be derived from simple correlation coefficients, more sophisticated approaches and techniques such as confirmatory factor analysis (cf. Parker 1983) or (full) structural equation models (cf. Saris and Münnich 1995; Van de Vijver and Leung 1997b, 99–106) are available. The fact that equivalence cannot be considered an intrinsic property of some measure or indicator widens the scope for using existing data notably.

From this emphasis on inference in assessing the degree of equivalence, an important conclusion can be drawn for the opportunities of using existing data. If we establish equivalence as the functional equivalence of some concept by analyzing relationships in terms of underlying patterns, we do not have to restrict our analyses to inferences based on a common set of stimuli, items, or indicators. Inference can be based on different stimuli, items, or indicators in various settings, since equivalence is a question of relationships. This means that the search for relevant data has not to be restricted to material including identical measures only. From the perspective of constructing equivalent measures, it can be very useful to rely on data sets which contain different measures for similar concepts.

If inference instead of identity is stressed, internal and external ways to reach conclusions about relational aspects of our observations or indicators can be established. Internal consistency means that the stimuli or items used should show more or less the same structure in different environments; external consistency means that indicators are related in the same way to an element not belonging to the initial set of indicators. Van de Vijver and Leung distinguish between 'levels' of equivalence: 'construct (or structural) equivalence', 'measurement unit equivalence', and 'scalar equivalence' (or 'full score comparability') (1997b, see also Chapter 10). Although this distinction is useful for their discussion of research methods, it is not very helpful in the present context, which focuses on pragmatic strategies to deal with equivalence problems. The strategies presented here can be applied at each level distinguished by Van de Vijver and Leung. The question whether a concept such as 'party identification' is equivalent in various settings has, then, no meaning. Only by introducing relationships between two or more items to measure party identification or linkages between party identification and another concept can we discuss the degree of equivalence. These relationships can be conceptualized among a common set of stimuli or indicators, but different measures can be used in different settings (cf. van Deth 1998b). Whether auxiliary information is required or not is irrelevant for the distinction between external and internal consistency. The intelligent use of auxiliary information is a necessary condition for the application of every *ex post* strategy in research based on existing data.

19.4.3 Mixed Strategies

Due to the complexities of harmonizing data or establishing equivalence among different measures, a mixed research strategy is recommended: some instruments can be used directly, others have to be harmonized, and for the core concepts it may be necessary to develop equivalent measures. Some of these very time-consuming tasks can be eliminated by looking for existing 'sets of data sets,' that are based on standardized questionnaires and harmonized response codes. For instance, the *International Committee for Research into Elections and Representative Democracy* brought together the major election studies in a number of democratic countries. This enterprise resulted in an extensive overview of the distinct European studies available for reanalysis (cf. Mochmann, Oedegaard, and Mauer 1998). An example of a more sophisticated approach is the *Integrated Documentation And Retrieval Environment for Statistical Aggregates (IDARESA)*, which aims at meta-analyses and not primarily at secondary analyses (for Web site see Appendix 2). The development phase of this project provides data from several national school-leavers surveys as well as time series in public financing of research and development.

Another way to reduce the costs of *ex post* handling of existing data is to use data from multipurpose surveys. In the last two decades, several projects have been designed which offer groups of researchers the opportunity to study several themes simultaneously by combining their measurement instruments in a single survey or a program of surveys. Table 19.1 provides an overview of important data sets available for reanalysis in comparative survey research.

The availability of data from multipurpose surveys attests to the disappearance of a clear demarcation between primary research and the use of existing data (Hakim 1982, 3). As Hyman's definition of secondary analyses quoted earlier reflects (1972, 1), traditional approaches emphasize differences between the goals and purposes of principal investigators and those of researchers reanalyzing the data. In the case of multipurpose surveys, data are collected and made available to the scientific community and usually no specific purpose or goal for the entire project is formulated. In line with this development, more recent demarcations in this area avoid referring to purposes and goals (cf. Dale, Arber, and Procter 1988, 3–4).

19.5   THE FUTURE OF PUBLISHED DATA

Almost three decades ago, Hyman began his seminal treatment of secondary analysis of survey data with the observation that the vast amount of information stored to illuminate many scientific problems was not widely used: "Why has relatively little scientific wealth been extracted from the figurative mountain of gold?" (Hyman 1972, 2). With the enormous increase in the amount of data stored since Hyman published his book, the use of existing data has certainly grown substantially. However, in spite of this growth and advancements in the professionalization of data archiving, the potential for using existing data is evidently still far from being exhausted.

An understandable answer to Hyman's question could be that the complications of using existing data are commonly underestimated and that 'a mountain of gold' can be a rather risky place. As Jacob remarks, published data ". . . are like the apple in the

TABLE 19.1.   Important Cross-National Survey Programs and Data Bases

Program or Study	Main Area or Themes	Countries	Period
International Social Survey Programme (ISSP)	Theme varies annually; government, inequality, family, environment, religion, etc.	All five continents; majority Western industrial countries (38, 2002)	1985–
World Values Survey (WVS)[a]	Value orientations, religion, environment, general political attitudes	22 countries (mainly Western) 42 countries 66 countries 96 samples	(1) 1981–1983 (2) 1990–1993 (3) 1995–1997 (4) 1999–2001
Eurobarometer (EB)	General attitudes toward European integration and the EU; specific topics in each wave	All EU member states (6 to 15) plus occasional additional	1970–1974 yearly[b] Since 1975 at least twice a year
Central and Eastern Eurobarometer (CEEB)	General attitudes toward economic and political chances and toward European integration	Countries in Central and Eastern Europe (up to 20)	1990–
Panel Comparability Project of Household Panel Studies (PACO)	Socio-economic indicators on income, housing, education, time use, etc.	Several (10)	1968–
Comparative Analysis of Social Mobility in Industrial Nations (CASMIN)	Social class, education, household composition	12 countries	1970–1995
Luxembourg Employment Study (LES)	Employment status	16 countries	1990s–
Luxembourg Income Study (LIS)	Sources of income, household composition	26 countries	1960s–
Multinational Time Use Study (MTUS)	Time-use	20 countries	1960s, 1970s, 1980s, 1990s
Comparative study of electoral systems (CSES)	Political identification, attitudes toward the government, etc.	Currently 32	(1) 1996–2000 (2) 2001–2005
European Community Household Panel (ECHP)	Income, living conditions, housing, health and work, household composition	15 countries	1994–
Third International Mathematics and Science Study (TIMSS)	Teacher, student, school survey; attitudes, beliefs, organization, curriculum, performance, etc.	Initially 12 countries, 38 countries in 1999 survey	1960, 1970, 1981, 1982, 1983, 1986, 1995, 1999

[a]Parts of the WVS are also known as the European Values Survey (EVS).
[b]The European Community Studies of 1970, 1971, and 1973 are usually integrated in Eurobarometer data bases. See, too, Latino-, Asia-, and Afro-Barometer Web sites.

Garden of Eden: tempting but full of danger." Fortunately, he continues by stressing that there is no ground for 'discouragement' and that in the field of using published data: "For almost every problem there is a solution or at worst an acceptable compromise" (Jacob 1984, 9). Many texts on using existing data contain similarly optimistic remarks. Kiecolt and Nathan (1985, 75) note that "[m]ost of the difficulties inherent in secondary analysis, however, may be handled satisfactorily and are far outweighed by the opportunities for research using preexisting data." This positive conclusion can be confirmed here on various grounds. First, an incredible amount of data has now been stored and is waiting to be reanalyzed by the creative and skeptical researcher. In the field of comparative survey research, specific projects to harmonize data and to develop multipurpose surveys imply an expansion of such research opportunities. Second, the fact that an increasing body of data is becoming available means that analyses of social change and dynamic processes are now possible that researchers of an earlier era could only dream of. In several instances, replication strategies and secondary analyses today can be based on trend data covering several decades and a large number of countries. One example of the possibilities in this area is the *Beliefs in Government* project, in which a large number of political scientists reanalyzed available data on public opinion and political behavior in Western Europe since the Second World War (cf. Kaase and Newton 1995). The 'revolution' in 'life studies' described by Elder, Pavalko, and Clipp (1993) is another good example. Third, attempts to deal with methodological pitfalls and complications in using existing data appear to be successful—or at least very promising. Several strategies to harmonize data or to establish functional equivalence of different measures in different contexts have been discussed in the present chapter. Other examples are the treatment of reliability and validity problems in comparative research (cf. Saris and Münnich 1995). Although considerable work remains to be done in this area, it is clear that much can be gained from developing more sophisticated methods. Fourth, modern communication and information technologies make the rapid dissemination of fresh data relatively simple and have already greatly enhanced the opportunities for locating useful data. Data archives are flourishing and their holdings can be easily searched via the World Wide Web; 'databases of databases' make rapid accessibility ever easier (see Appendix 1). Finally, the growing practice among principal investigators to offer direct access to their data encourages the use of existing data, as do the stricter regulations journals and publishers now have on accessibility of data before accepting manuscripts. For instance, the data on social and political participation collected by Verba, Lehman Scholzmann, and Brady (1995) are offered for reanalysis in the introductory pages of the main report. Lijphart (1994) offers the same service for readers interested in his data on electoral systems and party systems. These developments all point out to the increasing importance of using existing data in social research.

The glowing prospects for using existing data, however, might be dimmed by the very growth in use and in access to these data. Precisely because the growth in available data is enormous and most practical obstacles to access have been removed, some researchers might be tempted to attune their research questions to data already available. As Kiecolt and Nathan remarked: "Perhaps the biggest temptation [of secondary analysis] is to formulate questions solely because a particular data set is available" (1985, 75). In this way, the huge supply of directly available high quality data simply might create its own demand. The French economist Jean Baptiste Say's dictum is apparently at work here: general overproduction is impossible in an exchange economy, since supply and demand are indissolubly bound together. Although the relationships between data archives (the main suppliers) and researchers looking for exiting data (their major customers) are not those of a pure exchange economy, it is clear that with the development and institutionalization of archives, the risks of overproduction are multiplied. This is particularly clear in those cases where archives or groups of researchers provide integrated or harmonized data *ex ante* independent of the demands of future users. Since every supply will indeed create its own demand, we can expect to find researchers tailoring their research questions to fit the opportunities offered by these nicely prepared data.

Although the practice of offering *ex ante* solutions can occasionally be fruitful, no serious researcher should let practical considerations, complications, or *ex ante* solutions define the goal of his or her project. Both the opportunities for using the 'golden mountain' of existing data and the prospects of improving *ex post* strategies to deal with the major comparability problems call for a clear distinction between services provided by archives and the formulation of research goals and purposes by researchers. Only if each side of the 'market' accepts its unique responsibilities will the use of published data retain and strengthen its significant role in our attempts to improve our understanding of the social world.

Note
1. Martin Elff harmonized the occupational codes as part of a DFG-funded research project (Grant De 630/2-1) and kindly provided the data for Figures 1 and 2.

APPENDIX 1:

How to Find Published Data

Some of the comparative projects mentioned have their own Web sites with information about data collection and how to access data (see Table 19.1). Archives publish catalogues of their holdings. It is relatively easy to locate data and other information through simple keyword searches. Many Western countries set up national data archives in the 1960s and 1970s, and the role of archives has grown steadily since then.

This development of data archives within national contexts certainly fostered empirical research within specific countries but was less advantageous for the development of cross-cultural and cross-national projects. A number of archives took on the responsibility for archiving data which included data from other countries. The ICPSR (*Inter-University Consortium for Political and Social Research*) in Ann Arbor (Michigan, USA) has played a central role from the beginning. The 'ICPSR Direct' option on their Web page provides member institutes with direct access to ICPSR data holdings. In Europe, the British archive in Essex (Web site: http://www.data-archive.ac.uk/) and its German counterpart in Cologne (Web site: http://www.gesis.org/ZA/index.htm) play similar roles. Each of these three archives serves as an excellent starting point for searching for existing data.

Life became even easier for the inquisitive researcher when CESSDA (*Council of European Social Science Data Archives*) was founded in 1977 as a coordinating institute for different national archives, especially in Europe. CESSDA has promoted the acquisition, archiving, and distribution of electronic data (not only survey data) for social science teaching and research in Europe. The CESSDA Web site (www.nsd.uib.no/cessda) allows easy access to the catalogues of many key data archives (16 institutes in Europe, 14 in North America, and five in the rest of the world). The most important European archives can be searched with a single query for a specific title or topic.

Recently, several projects have been launched to build 'databases of databases' to help users trace public use social science data for secondary analyses and teaching purposes even more easily (cf. odwin.ucsd.edu/idata/). These 'databases of databases' provide direct links to many data archives.

APPENDIX 2:

Information on the surveys may be found at the Web sites below. These Web sites were valid at the time of writing but addresses sometimes change rapidly.

- Central and Eastern Eurobarometer (CEEB)
  http://www.gesis.org/en/data_service/eurobarometer/ceeb/index.htm
- Comparative Analysis of Social Mobility in Industrial Nations (CASMIN)
  http://www.sowi.uni-mannheim.de/lehrstuehle/lesas/casmin/casmin.html
- Comparative Study of Electoral Systems (CSES)
  http://www.umich.edu/~nes/cses/cses.htm
- Eurobarometer (EB)
  http://europa.eu.int/comm/dg10/epo/eb.html
  http://www.gesis.org/en/data_service/eurobarometer/standard_eb/index.htm
- European Community Household Panel (ECHP)
  http://europa.eu.int/comm/eurostat/Public/datashop/print-catalogue/EN?catalogue=Eurostat
- International Social Survey Programme (ISSP)
  http://www.issp.org
- Luxembourg Employment Study (LES)
  http://lisweb.ceps.lu/lestechdoc.htm
- Luxembourg Income Study (LIS)
  http://dpls.dacc.wisc.edu/apdu/lis_country.html
- Multinational Time Use Study (MTUS)
  http://www.iser.essex.ac.uk/mtus/
- Panel Comparability Project of Household Panel Studies (PACO)
  http://www.ceps.lu/paco/pacopres.htm
- Third International Mathematics and Science Study (TIMSS)
  http://nces.ed.gov/timss/
- World Values Survey (WVS)
  http://wvs.isr.umich.edu/index.html
- Integrated Documentation And Retrieval Environment for Statistical Aggregates (IDARESA)
  http://www.socresonline.org.uk/2/1/brannen_lamb.html)

Chapter 20

DOCUMENTING COMPARATIVE SURVEYS FOR SECONDARY ANALYSIS

PETER PH. MOHLER
ROLF UHER

20.1 INTRODUCTION

Survey documentation serves two main purposes, one for those involved in conducting a survey (data producers), the other for those who will work with the data (data users). Without the first, *internal project documentation*, it will be difficult or even impossible to produce the second, *study documentation for secondary data analysis*. The needs of end users are related to various aspects of using edited data sets, whilst those of researchers and staff involved in producing data are related to monitoring and managing the survey process and to meeting disclosure requirements. Various chapters in this volume reflect the importance of providing documentation on different components of a study, such as sampling design and implementation, questionnaire design, translation, field work monitoring and survey outcomes (e.g., contacts attempts, response and nonresponse).

Professional associations have developed detailed codes of best practice and have published manuals for survey quality management outlining disclosure requirements for data to permit evaluation and replication. (For examples in English, see the American Association for Public Opinion Research (AAPOR) *Best Practices for Survey and Public Opinion Research*: http://www.aapor.org; the ESOMAR *Guide to Opinion Polls http://*www.esomar.nl/ guidelines/opolls.htm; and the British Market Research Quality Standard Association (MRQSA) scheme for quality management: http://www.bmra.org.uk)

This chapter focuses on the documentation needed by users or reviewers of data, in particular users not centrally involved in a study. Study documentation needs to be exhaustive enough for others to evaluate or replicate the findings or to be able to use the data for their own research. So-called secondary data analysts have none of the common ground (Clark and Schober 1992) and collective memory of primary data producers. Thus the information required to be able to evaluate and replicate findings is considerable, as indicated shortly.

The chapter begins with several general issues pertinent to documenting data from multicultural surveys, followed by an explanation of two major components in current study documentation, *machine-readable codebooks* and *codeplans*. It concludes with a brief consideration of data harmonization from a general data editor's and data merger's point of view (see, also, Chapter 7).

20.2   DOCUMENTATION REQUIREMENTS

Documentation of cross-national projects is of very varying detail and quality, as other contributions to the volume reflect. Harkness (1999) notes:

> The sobering truth is that information currently available about individual components of cross-national projects is generally well below that required by national standards of best practice . . . Moreover, research publications and study documentation materials suggest that disclosure requirements may frequently be below good practice standards.

The general framework for documentation should include the following basic elements:

- A comprehensive description and explanation of a project's theoretical goals;
- Documentation of the operationalization of theoretical concepts, including how concepts map to latent constructs and constructs relate to indicators (see Chapter 1), as well as documentation of translations and any adaptation of a source questionnaire;
- A detailed description of all the relevant aspects of design, implementation, and outcomes, plus assessment of the quality of the outcomes;
- A "thick" description of the socio-cultural contexts of a cross-national study and the different 'fields,' including all the relevant contextual statistics. Some of these are standard and basic (e.g., population statistics), others are determined by the scope and topic of a study (e.g., divorce rates or GNP). This information should also cover regulations and infrastructure information relevant for understanding respondent answers in a given context (e.g., divorce laws, abortion regulations).

By way of amplification, AAPOR recommendations for disclosure documentation are presented below. Despite the fact that the AAPOR list basically only addresses the first and third points of the list above, it is considerably longer. Similarly detailed lists can be drawn up for the other points above. However, there are practical limits to all documentation efforts. The AAPOR document points out that its list is longer than the minimum list of any single agency.

The AAPOR recommendations are specifications of points relevant in *one* survey context, the American. The list indicates, nevertheless, the type and degree of detail required to provide enough information about a study for anyone to evaluate it, replicate and test the findings, or to conduct their own analyses.

Apart from headings in capitals we have added, the AAPOR details are presented verbatim as currently available on the AAPOR Web site.

PEOPLE and AGENCIES
Sponsors and fielding agencies—who sponsored the survey, and who conducted it";
TOPIC
The purpose of the study, including specific objectives;
MATERIALS
The questionnaire and/or the exact, full wording of all questions asked, including any visual exhibits and the text of any preceding instruction or explanation to the interviewer or respondents that might reasonably be expected to affect the response;
SAMPLE and IMPLEMENTATION
A definition of the universe the population under study which the survey is intended to represent, and a description of the sampling frame used to identify this population (including its source and likely bias); a description of the sample design, including cluster size, number of callbacks, information on eligibility criteria and screening procedures, method of selecting sample elements, mode of data collection, and other pertinent information; a description of the sample selection procedure, giving a clear indication of the methods by which respondents were selected by the researcher, or whether the respondents were entirely self-selected, and other details of how the sample was drawn in sufficient detail to permit fairly exact replication; size of samples and sample disposition the results of sample implementation, including a full accounting of the final outcome of all sample cases: e.g., total number of sample
Elements contacted, those not assigned or reached, refusals, terminations, non-eligibles, and completed interviews or questionnaires;
Documentation and a full description, if applicable, of any response or completion rates cited (for quota designs, the number of refusals), and (whenever available) information on how nonrespondents differ from respondents;
DATA DOCUMENTATION
A description of any special scoring, editing, data adjustment or indexing procedures used;
A discussion of the precision of findings, including, if appropriate, estimates of sampling error with references to other possible sources of error so that a misleading impression of accuracy or precision is not conveyed and a description of any weighting or estimating procedures used;
A description of all percentages on which conclusions are based;
A clear delineation of which results are based on parts of the sample, rather than on the total sample;
INTERVIEW/INTERVIEWERS
Method(s), location(s), and dates of interviews, fieldwork or data collection;
Interviewer characteristics;
Copies of interviewer instructions or manuals, validation results, codebooks, and other important working papers; and any other information that a layperson would need to make a reasonable assessment of the reported findings.
(http://www.aapor.org/ethics/best.html)

Detail of this kind makes it clear that documentation needs to be an integral part of the study design and to be recorded carefully as the study progresses, not drawn together on the basis of diverse documents once field work has been completed. Documentation needs to be driven by disclosure requirements and study goals, not determined on the basis of what happens to be available. This also makes clear why documentation is such an expensive, time-consuming, and possibly unloved activity. While modern technology may encourage us to raise our expectations for provision of documentation, it also offers us technical aids. Proper planning, protocol templates, and data base software can greatly reduce the tedium, if not the effort involved, in supplying quality documentation in user-friendly formats.

20.3   ARCHIVE REQUIREMENTS

Social science data archives have standard requirements for documentation of studies that are related to their role as holders and publishers of data. A detailed example is the *Guide to Social Science Data Preparation and Archiving* provided by the Inter-University Consortium for Political and Social Research (http://www.icpsr.umich.edu/ ACCESS/deposit.html). At the same time, we are not aware of any readily available how-to-do guidelines for researchers on how *they* can tackle documentation for comparative studies. New developments in documentation standardization such as those of the Data Documentation Initiative (see 20.4.1) also start from a national data perspective. Experts and archives can adapt these, however, for multinational projects.

Not all data archived and made available for secondary analysis also meet archive requirements. If data are only acquired by an archive once a study is completed, desirable pieces of information may be lacking. If this information can no longer be located, or was not recorded at all, documentation will necessarily be inferior.

Archives are sometimes themselves involved in ongoing survey research projects; the Norwegian NSD, for example, implements a number of national studies as well as the Norwegian ISSP, while the Cologne Zentralarchiv is the official archive of ISSP data. In such cases, archive requirements can also be specifically tailored to the needs of a project. The ISSP archiving requirements are a good example of this (Uher 2000).

Archives can only devote a limited amount of time and resources to documenting individual studies. Among data sets made available for general distribution those of importance to a large scientific community will usually be given priority. These may include, for example, widely used longitudinal research programs such as the American General Social Survey (GSS), multinational projects such as the ISSP, and national elections studies. For these, archives often produce official codebooks or data base systems. Small and one-time studies, on the other hand, are likely to gain less archival attention.

The degree of detail required for documentation is also partly determined by the study design itself. Thus, for example, a multistage stratified sampling design calls for a fuller description than does a single random design. Documentation for widely used standardized instruments that already have extensive published documentation of their conception, details of prior use, and evaluation of measurement properties can, obviously, refer to this instead of documenting it again.

20.4 MODULAR VIEW OF DOCUMENTATION

Relational databases have much to offer those documenting multinational studies, since they require systematically defined entries and fields. These encourage researchers producing national documentation in countries involved in a multinational project to provide the same information and to use identical keywords. This greatly facilitates the merging of culture-specific entries in a combined database. In addition, reports can be produced using standardized document structures. Changes needed can then be made at any stage without rewriting whole sections.

Figure 20.1 presents a modular view of documentation that is compatible with both linearly organized hard-copy documentation structures and with the flexible hypertext structure of documentation we expect will increasingly replace electronic formats based on linear structures (e.g., PDF files of text or computer printouts). At the top of the figure is the executive summary of the project, covering all the essential components in brief—research team, funding agencies, survey topic(s), theoretical concepts, project goals, project design, plus cultures and populations covered, etc.

Below this follow information slots for specific components, to be filled in for each culture under observation. The slots are organized in three broad vertical tracks, to the left, the center, and to the right. The lines are intended to suggest major information flows between components. In the shaded area, for instance, the source questionnaire is driven by the latent constructs and itself both prompts general translation notes and is the basis for culture-specific translations.

The three tracks represent a) the process of sampling, fielding, and survey outcomes in each culture, b) instrument design and data generation, and c) culture-specific context considerations. Other organizing principles are, of course, possible. A different arrangement could, for example, emphasize the mediating function of the culture-specific components between design and implementation.

Country-specific information (contextual or collateral), such as institutions and regulations, is of particular relevance for cross-national studies because it helps analysts to interpret data collected outside of their own cultural context. It provides the information that strengthens researchers' understanding of contexts with which they are less intimately familiar. Unfortunately, contextual and collateral information is often neglected in comparative survey documentation. In national documentation, it is of relatively minor significance, since researchers 'know' their own cultural context.

The specifics needed for such contextual and collateral information will differ from study to study. However, documentation could be established on a cumulative basis, so that in time researchers who need to know the distribution of languages, religions, laws on divorce and abortion, or whatever, could simply access the database or the relevant links leading to the required information.

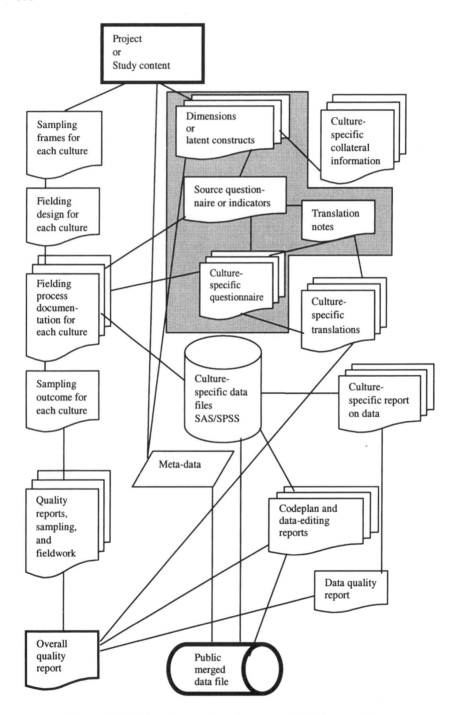

Figure 20.1. Schematic overview of process-oriented documentation.

20.4.1 Relational Data Bases

Relational data base management systems such as the Data Documentation Initiative (DDI) dissolve the dividing line between documentation and data by providing a dynamic form of documentation that allows researchers to investigate specific aspects of interest. The DDI is a social science archive initiative that

> ... is an effort to establish an international criterion and methodology for the content, presentation, transport, and preservation of "metadata" about datasets in the social and behavioral sciences. Metadata (data about data) constitute the information that enables the effective, efficient, and accurate use of those datasets.
>
> In the social sciences, metadata are often called "codebooks," although they have not literally been books for many years. With the achievements of the DDI, codebooks can now be created in a uniform, highly structured format that is easily and precisely searchable on the Web, that lends itself well to simultaneous use of multiple datasets, and that will significantly improve the content and usability of metadata. Further, this specification may have far-reaching implications for improvement of the entire process of data collection, data dissemination, and data analysis.
> (http://www.icpsr.umich.edu/DDI/ORG/index.html)

The DDI has developed what is known as a Document Type Definition (DTD) for 'marking up' social science codebooks. It uses the eXtended Mark up Language (XML), which is a derivative of the Standard General Mark up Language (SGML). While the stated original goal was "to replace the obsolete OSIRIS codebook/ dictionary format" (DDI 2000), the changing technological and information-sharing landscape has led the initiative to extend this goal considerably:

> Now we see codebooks as potentially structuring and supporting the entire data collection, distribution, and analysis process throughout the social and behavioral sciences (including in experiments), and the XML DDI as providing the "glue" that will bring that process together.
>
> We see opportunities for interoperability that cannot be achieved today. We see the DDI as enabling a new mode of doing comparative and other research that uses multiple datasets, a growing trend throughout the social and behavioral sciences. Users will be able to document a complex dataset to statistical packages through its codebook, as they once could with OSIRIS files for simple datasets. SPSS, SAS, and STATA "setups" will no longer be required. The information available to the software from the DDI will be much richer and more complete than could ever be accommodated by these setups. Codebooks will be structured in such a way that importing data into online subsetting and analysis systems will be expedited. Thus we see the DDI as offering not only data producers and data archivists but also data users a new power and flexibility to do their work and to do it effectively and efficiently. (*ibid*)

While the DDI seems set to shape the future for documentation (currently focusing on monocultural documentation), we turn in the next section to immediate needs for comparative studies on the basis of what a standard 'classical' codebook contains and how this format can be improved using relational database management systems and hypertext structures.

20.5   CODEBOOKS

Machine-readable codebooks are the standard device to document studies, whether monocultural or multipopulation. Both their design and later development were closely linked to the emergence of social science data archives during the 1960s and 1970s. From the very beginning, codebooks have provided

> ...information on the structure, contents, and layout of a data file. Typically, a codebook includes: column locations and widths for each variable; definitions of different record types; response codes for each variable; codes used to indicate nonresponse and missing data; exact questions and skip patterns used in a survey; and other indications of the content of each variable. Many codebooks also include frequencies of response. In addition, codebooks may also contain background information such as survey objectives, concept definitions, a description of the survey design, a copy of the survey questionnaire (if applicable), and information on data collection, data processing, and data quality. (http://www.icpsr.umich.edu/NACJD/codebook_descrip.html)

A specific codebook *routine* developed for OSIRIS (Rattenbury 1974) marks a milestone in the history of social science data documentation. This routine enabled archives and researchers to combine a labeled data file (numerical data with variable labels and values for each variable) with textual data (question wordings, response scales, etc.). In addition, since frequency distributions and the codebook could now be compiled directly from the data file (i.e., instead of being typed out by hand), copying errors were reduced and documentation processed more speedily.

20.5.1 Example of a Cross-Cultural 'Classical' Standard Codebook

For monocultural surveys, the codebook provides marginals/frequencies for each variable; for cross-cultural and multinational surveys, marginals are provided for each culture/nation. In other words, in linearly organized codebooks, the reference bringing together the data and the codebook information consists of cross-tabulations for each variable by country or culture. Figure 20.2 reproduces a page from the 1991 OSIRIS codebook of the ISSP study on Religion. We have added information 'balloons.' The page is divided essentially into two parts. The top section contains textual information about a variable in the data set (here V12). The bottom section is a tabulation of the frequencies (marginals) of that variable for all countries. The topmost textual information relating to the variable V12, also included in the data set as the 'variable label' is a highly condensed rendering of the question wording. Next follows a transcript of the source questionnaire wording, also indicating that V12 represents Q7 in the questionnaire. (The exact wordings for each country are provided as separate PDF files.) Question 7 is a Likert-type battery of items.

Thus the next text component (at Q7a) is a transcript of the source questionnaire wording of the first item. This is followed by the list of predefined codes for the response options offered respondents. Codes 1 through 4 are valid responses, codes 8, 9, and 0 indicate missing values. Below this, for the initiated at least, "Germany, Slovenia: 0. Not available" indicates that no German and Slovenian data are available on this item.

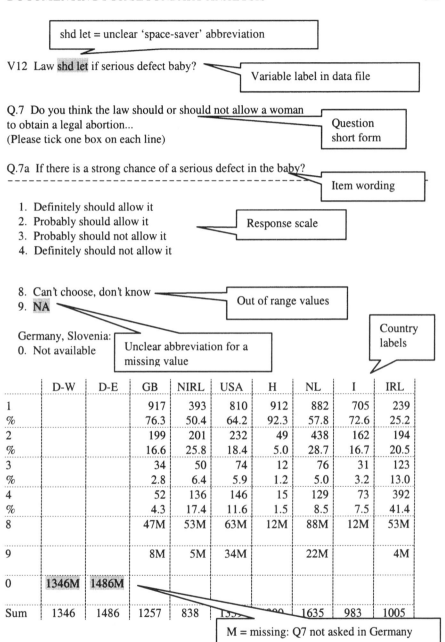

	D-W	D-E	GB	NIRL	USA	H	NL	I	IRL
1			917	393	810	912	882	705	239
%			76.3	50.4	64.2	92.3	57.8	72.6	25.2
2			199	201	232	49	438	162	194
%			16.6	25.8	18.4	5.0	28.7	16.7	20.5
3			34	50	74	12	76	31	123
%			2.8	6.4	5.9	1.2	5.0	3.2	13.0
4			52	136	146	15	129	73	392
%			4.3	17.4	11.6	1.5	8.5	7.5	41.4
8			47M	53M	63M	12M	88M	12M	53M
9			8M	5M	34M		22M		4M
0	1346M	1486M							
Sum	1346	1486	1257	838	1359	990	1635	983	1005

M = missing: Q7 not asked in Germany

D-W = West Germany, D-E = East Germany, GB = Great Britain, NIRL = Northern Ireland, USA = United States, H = Hungary, NL = The Netherlands, I = Italy, IRL = Ireland

Figure 20. 2. Excerpt from a standard multinational study codebook—ISSP 1991 Zentralarchiv Cologne.

In the second half of the page, the first column of the tabulation contains the value labels (1–9 and 0). The rows aligned to each value label contain counts of the responses in each country, while the rows aligned to the percent signs provide the respective percentages. Country acronyms are contained in the topmost row ("D-W," for instance, stands for West Germany, "NIRL" for Northern Ireland). Figures followed by an "M" at values 8, 9, and 0 are counts of the respective missing values. The bottommost row contains the sum of cases per country.

Given the potential of hypertext links and relational data bases, standard codebook technology has a number of important drawbacks and limitations for complex project documentation. The information in the codebook can only be accessed and treated sequentially and cannot be directly linked to other information or to the data set. Figure 20.2, for example, provided marginal frequencies for each country. The diagram for the same question on abortion in Figure 20.3 below, on the other hand, indicates the kind and variety of informational components that could be accessed in a relational database with hypertext buttons. Researchers would then be able to move back and forth from a question to the numerical data base and on to a variety of hyperlinked information components, such as question formulations for each country and/or language, definitions of the underlying dimensions, past performance in one or more countries, or publications using the variable in question.

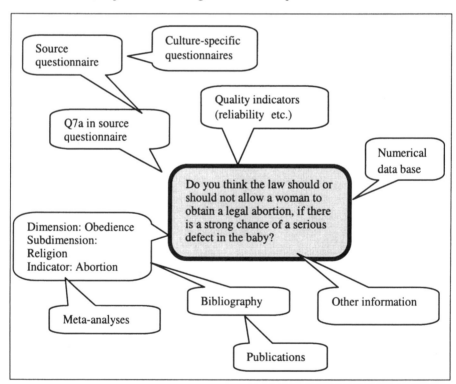

Figure 20.3. Information map for an abortion item.

20.6 EDITING A MULTICULTURAL DATA SET

Whatever the conception, scope, and format of documentation, three basic steps are involved in producing a codebook or data documentation tool. We present these here within the framework of a comparative project:

(a) Creating a codeplan that links the data sets with the source questionnaire and country-specific questionnaires;
(b) Editing and screening the data at national and multinational level;
(c) Harmonizing culture-specific variables at multinational level.

20.6.1 Codeplan

A codeplan has sometimes been seen as the 'beginnings' of a codebook—the first core information that forms the basis of what later becomes the codebook. Since 'codebooks' are today complex documentation tools, it is useful to distinguish between the two.

A codeplan defines the rules which assign specific numeric codes to response categories and open-format responses in the questionnaire or computer-assisted application.

Coders key in the codes assigned to each response option selected by a respondent, (optical scanners 'read' them and computer-assisted applications transfer key strokes). Any information provided by respondents in an interview or on a questionnaire that does not fall within the domain determined by the codeplan is, as a rule, excluded. Comments, glosses, and explanations that respondents provide in self-completion formats and in interaction with interviewers are, for example, rarely coded.

Assigning codes to response categories at the questionnaire design stage simplifies and improves data entry routines. Data set-ups can then be made available to participating countries, a strategy that reduces omissions and miscoding at the national level for the combined (merged) multinational data set.

In the codebook page depicted in Figure 20.2, respondents' answers to "Q7" (question 7 in the questionnaire) are assigned the variable name "V12" in the data set. A second set of rules defines the codes assigned to each response option available to respondents. This is indicated by the numbers 1 to 8 alongside the response options, including the off-scale response category, "can't choose/ don't know."

Q.7a If there is a strong chance of a serious defect in the baby?
----

1. Definitely should allow it
2. Probably should allow it
3. Probably should not allow it
4. Definitely should not allow it
8. Can't choose, don't know
9. NA

The ninth code, "NA," is included to accommodate "No coded answer is available." Codes such as "can't choose" or "NA" indicate missing values that can result from errors in instrument administration, from item nonresponse, or as the result of questions skipped because they are not applicable to the respondent. Questions on employment, for example, would be 'systematically missing' for a respondent who has never been employed.

Codes are assigned sequentially in the codeplan but researchers decide which response categories are assigned high or low codes. Well-designed codeplans are relatively straightforward to understand and use, provided the conventions are known and followed. Nonetheless, mistakes in codeplan design and incorrect assignments on the part of coders or data entry applications (e.g., optical scanner outputs) are not uncommon. Assigning codes is not difficult once general principles are recognized and applied. We give only two illustrations of these here.

(1)   As a rule, more codes are assigned than are offered as response categories (e.g., the '7' for item refusal, '8' for don't know, '9' for NA, and '0' for a systematic skip). Thus response scales with more than six valid options mean that the extra codes will be double digit ('97' to '99').

(2)   It is advisable to avoid using '0' as a valid code since a genuinely valid value of zero ('0') can occur in response options or as part of an open-format entry (representing NIL). In a question on earned income, for example, if '0' is used for valid codes as well as for missing values, the code would have two potential meanings: on the one hand, 'income is zero' and on the other, 'systematically filtered, person not employed, thus no income.'

In the cross-national/cross-cultural context, the lead or meta-codeplan for the merged data set needs to be harmonized with or matched against the individual national codeplans. If, for example, the meta-codeplan assigns value labels in descending order and a national codeplan assigns them in ascending order and fails to document this, problems will obviously arise. The country-specific descending order needs to be reversed in the merged data set.

Table 20.1 presents the actual codes assigned to British and German response options to a Likert-type battery in 1997 (ISSP). The British codeplan, which functioned as the lead codeplan in the study, assigned codes in ascending order (1 to 4); the German codeplan assigned codes in descending order (4 to 1). If this had not been documented, the wrong data could have been merged and the distributions in Germany would have been highly skewed. Since the data were merged in Germany, the archive staff merging the data would also have been able to read the codelplan/questionnaire and identify a mistake like this quickly. If the questionnaire and response categories had been in Chinese, however, the checking process would have been considerably more complicated.

This question, incidentally, involves a response scale which translates poorly and is known to have measurement problems in German (Harkness 2001).

TABLE 20.1. British (Lead) Codeplan versus GERMAN (National) Codeplan

SOURCE QUESTIONNAIRE ITEM/QUESTION	GERMANY: ITEM/QUESTION	VARIABLE NAME AND VARIABLE LABELS IN CODEBOOK	COMMENTS FOR DATA MERGING EDITOR
Q.8 Do you personally think it is wrong or not wrong for a woman to have an abortion . . .	Q.7. Halten Sie es persönlich für schlimm oder nicht schlimm, wenn eine Frau einen Schwangerschaftsabbruch vornehmen läßt . . .	VARNAME= V17	
Q.8a If there is a strong chance of a serious defect in the baby?	Q.7a. Wenn das Baby mit hoher Wahrscheinlichkeit eine ernsthafte Schädigung haben wird	All value labels identical to questionnaire.	

SOURCE RESPONSE SCALE AND ANSWER OPTIONS	C O D E	GERMANY: RESPONSE SCALE AND ANSWER OPTIONS	GERMANY. COMMENTS ON NUMERIC VALUES	REVERSED RESPONSE CATEGORY ORDER IN GERMANY
Always wrong	1	Nie schlimm	Code 4 in German data	Verify recode in merged data
Almost always wrong	2	Nur manchmal schlimm	Code 3 in German data	Verify recode in merged data
Wrong only sometimes	3	Fast immer schlimm	Code 2 in German data	Verify recode in merged data
Not wrong at all	4	Immer schlimm	Code 1 in German data	Verify recode in merged data
REFUSED	6		Out of range value	
NOT ASKED IN COUNTRY	7		Out of range value	
RESPONDENT CHOSE DK CATEGORY	8		Out of range value	
NO ENTRY (NA)	9		Out of range value	
FILTERED / SKIP	0		Out of range value	

Adapted from the ISSP 1997 codeplan.

20.7   EDITING AND DATA VERIFICATION

A raw data file is edited before being made available for analysis. The purpose of editing is first to provide information on the quality of the survey data (recorded in part in the codebook) second, to 'clean up' loose ends in the data, and third, to identify any weaknesses in the survey vehicle.

Data editing is a time-consuming and costly business, but one of central importance in establishing the quality of data. In cross-national/cross-cultural projects, data editing can be undertaken at the national level or by one central body also responsible for merging national data sets. The preferred option is to undertake the main checks at national level with further checks during the supranational merging process.

The rationale for this is simple. The original questionnaires, contact protocols, interviewer instructions and other original documentation are all available at national level. Even if these were collected at a central agency, interpreting the various languages could pose a considerable linguistic problem. Incidentally, it is important to ensure that documents available at national level also remain available until they are no longer needed at supra-national level. Countries differ in their views on the importance of storing documents versus data.

If data editing is indeed performed twice, the degree of editing to be undertaken (how 'clean' is 'clean') requires to be agreed beforehand. Views differ, for example, on how to respond to query edits (edits that point to records which are potentially in error). Heavy editing of data which records a respondent as having 'no earned income' but also 'working full time' would set the variables at 'missing' because of a logical inconsistency. People can, however, work full time as volunteers.

Searching for errors is very time consuming. In a comparative data set with about 30 cultures and about 150 variables, more than 25,000 table cells have to be checked and proof-read against the original national data files and against the context and collateral documentation. It is also not uncommon for countries to differ in standards of 'cleanliness' and methods of 'cleaning'. In the 1986 ISSP merged data set, for example, a number of cases indicated that respondents in the United States stated both that their mother was dead and that they met their mother weekly. The German data had very few such cases, and it is now not possible to determine whether the German data were edited 'clean' or simply did not have anomalies that for some reason appeared (and were left) in the American data set.

Depending on the documentation accompanying a data set, a variety of checks can be made, ranging from monitoring field contact protocols to complex relationship verifications. Checks of various kinds are applied to identify missing, invalid, or inconsistent entities. Some are *logical consistency* checks or formal checks for 'wild codes', that is, searching for codes in the data file that are not defined in the codeplan. For instance, if the codeplan stipulates a four-point response scale with the codes 1 through 4 for valid response codes and 6 through 9 for missing values and a respondent's answer is coded '5' in the data set, this would be a wild code.

Q.3	There are many ways people or organizations can protest against a government action they strongly oppose. Please show which you think should be allowed and which should not be allowed by ticking a box on each line.
Q.3c	Organising protest marches and demonstrations
1.	Definitely allowed
2.	Probably allowed
3.	Probably not allowed
4.	Definitely not allowed

Figure 20.4. British English codebook entry—ISSP 1985.

Such kinds of discrepancy should appear in early stages of data editing. If they turn up when data are being merged, two options are open—either a national editor is asked to check the questionnaire in question or the code is set to a 'missing' value.

Checks are also made for systematic errors in skip patterns, since coders sometimes systematically but wrongly change codes or an optical scanner sets wrong skip patterns. Careful reading of marginal distributions helps detect such errors. If there are too many skips or dubious distributions, merging editors have to check with those holding the questionnaires or computer applications to clarify the situation. Errors can also be found by applying commonsense, pragmatic reasoning—codes assigning impossible ages or unlikely numbers of children are examples of the kind of errors that can be noticed in these checks. Occasionally, data anomalies point to translation errors. The first ISSP survey in 1985 provides an illustration of a blatant mistranslation. Figure 20.4 shows the British English (source) question formulation, as given in the codebook. Table 20.2 shows the distribution of answers between different countries for this question; it is evident that the German and Austrian distributions for code 1 (bold in table) differ substantially from those of other countries.

Examination of the original questionnaires showed that Germany and Austria (who shared a translation) had translated the English 'organizing protest marches and demonstrations' as 'organizing protest marches *which block traffic*' (Protestmärsche organisieren, *die den Verkehr behindern*). This clearly explains the difference in answer distributions. Such plausibility checks are only possible late in the processing stage when the culture-specific data sets are merged. Plausibility checks, we note, do not normally turn up translation problems.

TABLE 20.2. Marginals from the ISSP 1985 Codebook

	Germany	Austria	Australia	Great Britain	USA	Italy
1. Definitely allowed	**10%**	**11%**	37%	38%	37%	37%
2. Probably allowed	21%	22%	32%	32%	29%	31%
3. Probably not allowed	39%	40%	17%	12%	19%	19%
4. Definitely not allowed	31%	28%	14%	18%	14%	12%

20.7.1 Harmonization and Data editing

Harmonization is the name given to procedures which transform country-specific formats of an indicator into a derived comparative variable. Many substantive questions—such as attitude or opinion questions and some behavior questions—are assumed to be 'equivalent' and not to need harmonization. Country-specific elements in questions, such as references to a school-leaving age, are also treated as 'equivalent,' wording differences are simply noted in the codebook. In contrast to these, behavioral or factual data about respondents and their social context, such as data relating to 'degree of education', the 'political parties' in a country, left-right scales, or aspects of health or environmental behavior, can be asked in country-specific fashions. For the merged data set, editors are required to construct a 'derived variable' in which recoded national codes are aligned and harmonized, thus permitting this variable to function as a single comparative variable. Harmonization of background variables (see Chapter 7) can begin at national level—when codes preparing for a re-coding of national information into an international scheme are assigned. Harmonization is then completed late in editing when the national data sets are merged and documented. Harmonization procedures can be complex (see Chapter 7) but even comparatively simple harmonization procedures need careful consideration. Trade union membership, for example, can be used for some countries as an indicator of political commitment. The membership question has often a dichotomous format—respondents are asked to indicate whether they are a member or are not a member. Working on the premise that trade union membership will be affected by age cohort and employment status, a number of German ISSP surveys asked respondents about current and past trade union membership with three possible codes: current member, former member, never member. In merging this data with dichotomous data from other countries, the archive harmonized the three codes to the two-code codeplan by setting respondents who were no longer members to the code for nonmembers. However, if membership is indeed an indicator for political commitment and membership (not political commitment) has varying life course relevance, a case could also be made for assigning former trade union members to the code for 'member' rather than 'nonmember.'

20.8 CONCLUSION

Well-documented data is an integral part of study quality and a key factor in helping to move research forward in many of the areas discussed in this volume. Modern survey documentation comprises survey process documentation and complete documentation of the merged data file. Documentation strategies and procedures require to be detailed in a study design. Tools such as the DDI Data Type Definitions and emerging data base management systems can greatly facilitate survey documentation. Documenting multicultural, multipopulation studies involves often complex coordination across survey implementations and country-specific documentation. New projects are under way to support the special needs for comparative survey research (e.g., LIMBER: http://www.limber.rl.ac.uk/; FASTER: http://www.faster-data.org/; see also a critical description at SIDOS http://www.sidos.ch/data/projects/e_d_p_nesstar.html). Since the numbers of researchers making their data available outside the archive framework (and on small data-editing budgets) is likely to increase, a detailed listing of disclosure requirements such as presented at the outset here can only be the first step. Simple and sustainable how-and-what-to-do guidelines on what to document and how to incorporate this into a study design will be invaluable for a host of hitherto uncatered-for researchers.

Chapter 21

THE USE OF META-ANALYSIS IN CROSS-NATIONAL STUDIES

EDITH D. DE LEEUW
JOOP J. HOX

21.1 INTRODUCTION

Meta-analysis offers excellent tools for researchers in the field of cross-national and cross-cultural studies. It provides researchers with methods to combine the outcomes of different studies and analyze the results to investigate potential differences between countries or cultures. Meta-analysis can help those conducting cross-national and cross-cultural research and users of international databases decide if results are comparable across countries and thus unravel general and country-specific information.

Meta-analysis has its origin and some of its earliest applications in the educational sciences. The term 'meta-analysis' was introduced by Glass (1976) in his presidential address to the American Educational Research Association (AERA) as ". . . the analysis of the results of statistical analysis for the purpose of drawing general conclusions." One of the first meta-analyses published was on the effect of class size on educational achievement (Glass and Smith 1979). Meta-analysis was adopted quickly in psychology and bio-medical research, and soon meta-analyses were published on topics as varied as the effectiveness of psychotherapy (Smith, Glass, and Miller 1980) and the effect of aspirin on the occurrence of heart attacks (Barnett et al. 1988). Interest in meta-analysis increased dramatically in the eighties and nineties. The topics broadened from summary studies estimating an overall effect (e.g., does class size affect achievement) to studies focused on differential effects for different cultural or ethnic subgroups. Examples are studies on cultural differences in child competitiveness (Strube 1981), gender and cognitive performance (Signorella and Jamieson 1986), and race effects in performance evaluation (Ford, Kraiger, and Schechtman 1986). Interest in using meta-analysis for theoretical explanation rather than summary description is now increasing (Cook et al. 1994).

Despite its name, meta-analysis is not a single type of method or analysis but a set of methods: a methodology for the systematic combination of information from several different sources. In other words, meta-analysis is a systematic approach for integrating the outcomes of a set of studies, summarizing what is common and analyzing what is different. It started as a formal method for systematic literature review. The medical sciences used meta-analysis to combine the results of so-called 'multi-site' experiments or clinical trials, thereby establishing an important second use of the techniques. At present, there are two important applications of meta-analysis: (1) systematic literature review, and (2) systematic analysis of data collected at different sites. Both approaches are valuable in cross-national and cross-cultural studies. Below, we describe these two approaches then discuss the application of meta-analysis in cross-cultural and cross-national research.

21.2   TWO APPROACHES TO META-ANALYSIS

21.2.1 Meta-analysis for Literature Review Summaries

Meta-analysis started as a method for reviewing research literature. The traditional narrative literature reviews no longer met the needs of researchers (Glass, Smith, and McGaw, 1981; Rosenthal 1984). Narrative reviews were considered subjective and an inefficient way to extract useful information from the literature (Light and Pillemer 1984). Quantitative procedures therefore were developed to integrate the outcomes of individual studies. This approach prescribes formal methods not only for the statistical summary of results, but also for the earlier stages, such as collecting the relevant literature and coding the results of the individual studies. These formal methods need to be described clearly in the meta-analysis report, in order to allow the analysis to be replicated and evaluated (Light and Pillemer 1984; Cooper and Hedges 1994).

Problem Formulation
The first step in a scientific meta-analysis is the *problem formulation*. One should start with a clear description of the research problem, including the universe to which one wants to generalize. For example, in a medical application, such as the Barnett et al. (1988) aspirin study, the typical problem formulation would focus on the effectiveness of the intervention: does aspirin help to prevent heart attacks? In other applications, the focus may be on the role of moderator variables. For example, when a researcher is interested in the role of incentives in raising survey response rates, she wants not only to know if incentives have worked in the past (descriptive meta-analysis) but also to be able to generalize to future and probably somewhat different studies. Thus, while the basic question is still whether incentives will increase response rates in later surveys, additional research questions might include the following: "Does the effect vary by mode of data collection?", "What is the best strategy: prepaid or promised incentive?", "What type of incentive works better: money or gift?" Here the interest is not only in the combined outcome but also in study-level explanatory variables that moderate the effect of incentives on response rates (see Singer et al. 1999 for a good example).

In comparative meta-analyses, the most important aspects of the problem formulation are explicit research questions on differences between countries and/or cultures within countries; examples are Daly (1996), Schimmack (1996), and Van de

Vijver (1997). Daly's research question is whether companies act in manners consistent with behaviorist or cognitive organizational theories. She compares American and international studies to find out if there are intercultural differences. Her findings are that companies tend to behave in accordance with cognitive theories, with no discernable intercultural differences. Schimmack investigates whether the structure of the correlations between the recognition of facial expressions of emotions is equivalent across different countries. He concludes that there may be real cross-cultural differences with respect to the recognition of sadness and fear. Finally, Van de Vijver presents a comparative meta-analysis of cross-cultural comparisons of cognitive test scores. Again, the focus of the meta-analysis is on explaining cross-cultural differences. His comparative meta-analysis illustrates that a meta-analysis does not need to be restricted to simple research questions. He investigates several models that may explain cross-cultural differences in cognitive test performance. He finds that differences in cognitive performance are positively related to the degree of affluence of cultural groups and that these differences increase with age and education. The performance differences are larger on common Western tasks and smaller on locally developed non-Western tasks. There were no differences in abstract thinking. Finally, only intranational studies show a relation with the complexity of the task. In general, he finds that intranational comparisons are related to different explanatory variables than are cross-national comparisons.

Retrieval of Relevant Literature

The second step is the data collection phase: *retrieval of literature*. In this phase, the researcher conducts a literature search for relevant studies. Based on the research question of interest, the researcher has to formulate a search strategy and define key concepts and key words. This problem is analogous to the problem of sampling respondents in a survey. The researcher must begin by defining the universe of interest. In meta-analysis, this involves decisions about the time period to be searched (e.g., are older publications still relevant?), whether *all* studies are to be included or only studies that meet certain methodological requirements (e.g., do we include nonrandomized experiments or trials?), and whether unpublished and 'gray' literature is included. After defining the population of studies, a search strategy is defined, based on the research question of interest, and key concepts and key words are defined. Reference databases are a good starting point. There are specialized databases on CD-ROM for each field of research (e.g., Ageline for gerontology, DRUGINFO for interdisciplinary research on substance abuse, ERIC for the educational sciences, MEDLINE for healthcare, PsycINFO for psychology, Sociological Abstracts for sociology, SRM for methodology in the broadest sense).

A generic problem with computerized databases is that they are compiled by information scientists who cannot know the problem formulation of the actual meta-analysis, so the databases are not tailored to the needs of a given meta-analysis. Typically, these databases focus almost exclusively on literature published in English, and mostly in American research journals. Special efforts therefore are needed to avoid the resulting Anglo-Saxon selection bias, perhaps especially for comparative research. Thus, it is wise to search several sources. Studying the indices of specialized journals and the references of earlier review papers provides an additional check on the effectiveness of the search. The increased availability of Internet-based resources makes Internet searches a valuable resource, especially for non-U.S. based studies. Good additional strategies are appeals in newsletters and on pertinent e-mail discussion lists, formal appeals to scholars in the field, and in informal conversations at conferences. Since not all countries share the strong Anglo-Saxon pressure to 'publish-or-perish' in international journals, it is important to search for unpublished reports, conference papers, and recent studies (the gray literature). Fink (1998) provides suggestions for such 'multiple approach' strategies.

Coding of the Studies

Systematic coding of the studies: The goal of this step is to transform the corpus of studies into a data matrix. Each row then represents a study or a case in statistical terms and each column a variable, such as study characteristics and outcome variables. A detailed coding schedule is needed, which also must be described in the methods section of the meta-analytic report; this is often added as an Appendix.

A good starting point for designing a coding scheme is to inspect schemes used in other meta-analyses on related topics. For example, de Leeuw (1992) took a coding scheme from Sudman and Bradburn (1974) as the inspiration and starting point for her meta-analysis of the influence of data collection method on data quality. The general study characteristics, background variables, and the outcome variables of interest are coded. Usually, part or all of the coding is carried out by two independent coders, and a measure of intercoder reliability (e.g., Cohen's kappa) is reported.

Study characteristics and background variables. In meta-analysis, background variables that relate to the research report are coded, such as year of publication, country where the study was done, and journal or type of publication (e.g., dissertation, conference paper). In addition, meta-analysis codes characteristics that describe the study and its design. For instance, in de Leeuw's (1992) meta-analysis on data collection modes, the study characteristics included type of sample, sample size, topic of survey, saliency, and equivalence of questionnaire in different modes.

A special background or study characteristic is 'quality of study'; this is often a general evaluation based on coded study characteristics, such as experimental design (e.g., true random experiment or not, control for attrition, use of placebo, a double blind treatment in medical research, etc.; cf. Wortman 1994). A good example is the meta-analysis by Wortman and Bryant (1985). They criticized earlier meta-analyses on the effect of desegregation on educational attainment of minority groups in the United States that suggested that desegregation had beneficial effects. These early meta-analyses included all the studies found, including studies with relatively weak research designs. In response to the ensuing debate, Wortman and Bryant reanalyzed the studies *but* included methodological criteria in the analysis. Studies that fell below a predefined quality level were excluded. Wortman and Bryant concluded that methodologically strong studies tended to find a smaller effect of desegregation. This smaller effect in the stronger studies was nonetheless large enough to be of substantive importance and to have valid implications for educational policy.

Outcome variables. Attention in meta-analysis focused first on statistical significance and the combination of significance levels (*p*-values). The main outcome variable coded in this case was the *p*-value of the statistical test in the original publications. The statistical combination of *p*-values (Becker 1994) focuses on the research question "Is there a statistically significant effect?" Soon attention in meta-analysis shifted to the effect size (see Cohen 1969, 1988) and methods for combining effect sizes. This approach focuses on the research question "How large is the effect?"

There are several formulae for effect size, and the specific effect size chosen depends on the research question and the design of the original studies. For example, for experimental studies the standardized mean difference between the experimental and control group (the difference between the group means divided by their common standard deviation) is a common estimator of effect size:

$$d = \left(\bar{x}_E - \bar{x}_C\right)/s .$$ (21.1)

In correlational research, the usual effect size is the correlation coefficient. For studies that rely on categorical outcome variables, effect size estimates use proportions—for example, the relative risk:

$$RR = P_E/P_C ,$$ (21.2)

is defined as the ratio of the proportion of a specified outcome in the experimental and control group (cf. Cornell and Mulrow 1999).

The use of effect size estimators in meta-analysis is based on the assumption that the studies all estimate the same parameter and that this parameter of interest can be defined using a common metric (e.g., standardized mean difference, correlation, relative risk). The goal of the meta-analysis is then to combine all effect sizes into one overall 'superoutcome' and at the same time give a description of its sampling variability. In the end, the meta-analyst reports an estimate of the overall effect size and a confidence interval, considering all available information.

A good example of combining outcomes is de Leeuw's (1992) meta-analysis of the differences between mail, telephone, and face-to-face surveys. She found 67 articles and papers that compared at least two of these data collection modes on data quality. Three articles reanalyzed earlier studies, one study contained severe design flaws, and ten articles did not provide enough information for coding. In the end, 52 studies were available. The quality criteria coded were response validity, item nonresponse, number of statements in reply to open questions, social desirability, and similarity of response distributions. Not all the studies compared all three modes, and not all presented results for all quality criteria. This is typical for a meta-analysis involving several comparisons and multiple outcome variables. De Leeuw (1992) presents the results in the form of pairwise comparisons of the modes, using correlation coefficients as the common effect size measure. For example, for social desirability, comparing face-to-face and telephone surveys, she finds a mean $r = -0.01$ (95% confidence interval: -0.03 to $+0.01$), based on 14 studies, and comparing mail and face-to-face surveys, she finds a mean $r = +0.09$ (95% confidence interval: $+0.07$ to $+0.11$), based on 13 studies. Finally, comparing mail and telephone surveys, she finds a mean $r = +0.06$ (95% confidence interval: $+0.03$ to $+0.09$), based on five studies. De Leeuw concludes that concerning social desirability bias, mail surveys perform somewhat better than face-to-face and telephone surveys, which do not differ from one another in this respect.

Statistical Analysis of the Coded Data

The fourth step in meta-analysis is *statistical analysis of the coded data.* Basic statistical techniques are used to summarize the results. More sophisticated techniques are used to investigate potential heterogeneity in the data. The latter are very important in cross-national studies, as they address the question of whether the outcomes of the studies are comparable across countries.

The basic statistical approaches are (1) summarize p-values (summarize significance levels of individual studies) and (2) summarize outcomes. There are many methods to *combine p-values* (Becker 1994). One of the most robust is Stouffer's method. In this, each p-value is first transformed to a z-score via a standard normal transformation, then these z-scores are summed up across all studies and the sum is divided by the square root of the number of p-values. Finally, this overall z-score is transformed back to a p-value. If the combined p-value is smaller than a chosen significance level, the null hypothesis of no effect is rejected, and one may conclude that there is an effect in the studies investigated.

For each effect size estimator, there is a statistical procedure to *combine the effect sizes* into one overall effect size and calculate the corresponding standard error (Hedges and Olkin 1985). We illustrate these procedures to combine effect size with an example of the combination of an effect size d (e.g., the standardized difference between females and males in reasoning) over studies. The statistical procedure has two important assumptions. First, one assumes that all studies estimate the same effect (e.g., all studies study a homogeneous domain such as reasoning, and *no other* type of intelligence test). Second, one assumes that the outcomes are homogeneous, that is, that the variation in study outcomes is exclusively attributed to sampling variance and that there is *no systematic variance* associated with other sources. Each study coded provides an estimate of the effect size d_i and an accompanying standard error se_i. These are statistically combined into one overall outcome $d = \hat{\delta}$ with corresponding standard error, which are used to decide whether there is a statistically significant effect over studies and to estimate a confidence interval.

To combine the effect sizes, a weighted integration method is used, which weights each study with the inverse of its sampling variance given by:

$$w_i = 1/se_i^2 .$$ (21.3)

Thus, the weighted integration method estimates the combined effect size as

$$\bar{d} = \sum w_i d_i \Big/ \sum w_i .$$ (21.4)

The basic meta-analytical statistics above follow the *fixed effect model*, which is only valid under the assumption of *homogeneity*. Homogeneity means that all variance between studies is only sampling variance. If there are real differences between the studies, other sources contribute to the variance, and the fixed effect model is no longer valid. This means that the *random effects* model should be adopted, which includes this extra source of variation in the appropriate statistical formulas. This is obviously of central relevance for cross-cultural studies, and we discuss it in more detail in section 21.3. First, we describe another very useful application of meta-analysis: the combination of data from multi-site studies.

21.2.2 Meta-analysis for the Combination of Data: Multi-Site Studies

Especially in the biomedical sciences, meta-analysis is used not only for combining existing results but also in the design of prospective studies. In medical research, the number of cases available at each site (hospital) is often too small to permit the essential analyses. Therefore, data from different sites are combined using meta-analytic procedures. Essential in such *multi-site studies* is that all the experiments are replications, investigating the same research question and using the same methods. To achieve comparability, an explicit protocol is used that describes all the important aspects of the study's design and data collection procedures. The data are collected in medical practices and hospitals located in different places all over the world. Since populations may be different at different sites and organizational and cultural differences between the sites may necessitate small deviations in the protocol, background variables that describe the site and the procedures involved may be used to resolve uncertainties when such deviations occur.

Richard Peto pioneered prospective uses of meta-analysis for clinical trials. He persuaded 5,000 British doctors to participate in a multi-site clinical trial on the effectiveness of aspirin in preventing heart attacks. The Cochran Collaboration is another well-known international project, supported by 15 medical centers worldwide (Cornell and Mulrow 1999, see also http://hiru.mcmaster.ca/cochrane).

In international studies, it is essential that collateral background information on countries involved is collected and coded. For example, different countries may differ in health regulations or dietary customs, and protocols on how to treat 'dropout' may differ between sites. In statistical terminology, there are potential sources for heterogeneity, a situation also common in cross-cultural and cross-national studies. In meta-analysis, this is solved by coding available background variables and modeling the heterogeneity. In prospective meta-analysis, the need for background variables to explain differences between countries should be anticipated by including the collection of important background data in the study design.

21.3   STATISTICAL METHODS TO ANALYZE HETEROGENEITY

The goal in meta-analysis is to summarize the results of many studies, preferably in *one* clear 'summary' outcome. To justify this, *homogeneity* must be assumed: all studies must estimate the same fixed parameter, and all variance is assumed to be only sampling variance. If the study outcomes are heterogeneous, a simple fixed effect model to summarize the outcomes is no longer valid. For instance, if study outcomes differ for different countries, the country of origin is an additional source of variation. This additional, systematic, variation should be included in the statistical model; a *random effects* model should be adopted. In view of their relevance for comparative research, heterogeneity and the random effects model are discussed in more detail.

We illustrate the statistical procedures for meta-analysis and tests for homogeneity by analyzing a small and manageable data set that illustrates the main points to be made about meta-analysis in general. We follow classical meta-analysis methods as implemented in the program META (Schwarzer 1989). The data set consists of six studies assessing the effect of taking aspirin after a heart attack comparing an experimental group and a control group which was administered a placebo (Draper et al. 1992). The data are presented in Table 21.1. Note that study six has a strongly significant effect in the opposite direction, which results in a very high one-sided p-value. This could be an indication of heterogeneity, and that can be tested using a formal chi-square test for homogeneity.

Table 21.1.   Mortality Rates for Aspirin and Placebo Control Group

	Aspirin (E)		Placebo (C)		Comparison		
Study	N_E	Mortality	N_C	Mortality	d	s.e.(d)	p
1	615	.0797	624	.1074	−.1666	.0569	.0017
2	758	.0580	771	.0830	−.1866	.0513	.0001
3	317	.0852	309	.1036	−.1096	.0800	.0853
4	832	.1226	850	.1482	−.1179	.0488	.0079
5	810	.1049	406	.1281	−.1187	.0609	.0256
6	2267	.1085	2257	.0970	+.0643	.0297	.9847

We use the standardized effect size d, which for the comparison of two proportions is calculated as

$$d = Z_{p_E} - Z_{p_C},$$ (21.5)

where Z_p is the inverse of the standard normal distribution corresponding to the proportion p (Hedges and Olkin 1985). The sampling variance of d is $(n_E + n_C) / (n_E n_C) + d^2/2(n_E + n_C)$, where n_E and n_C are the sample sizes of the experimental and control groups. The p-values in Table 21.1 are the left-sided p-values for $Z = d/\text{s.e.}(d)$.

21.3.1 Classical Meta-analysis

Classical meta-analysis contains a variety of complementary approaches. One simple approach combines the p-values of the studies into one overall p-value for the collection of studies. For the aspirin example, the Stouffer method gives a combined Z of 4.17, with $p < .001$.

The combined p-value gives us proof that an effect exists but no information on the size of the experimental effect. The next step combines the effect sizes of the six studies into one overall effect size and establishes the significance or a confidence interval for this combined effect. Using a fixed effect model, the weighted integration method estimates the combined effect size as −0.06, with a 95% confidence interval extending from −0.10 to −0.02. However, the deviant outcome of Study 6 strongly suggests that the effects are heterogeneous, that is, that the effects differ across studies. In this event, the random effects model is more appropriate for combining the studies. The usual homogeneity test is a chi-square test on the residuals, which for our example leads to $\chi^2 = 30.2$, $df = 5$, $p < .001$. This is significant, so we conclude that the outcomes are strongly heterogeneous.

We used a formal homogeneity test to decide whether a fixed or a random effects model is appropriate. Hunter and Schmidt (1990) propose an additional criterion: the size of the between-studies variance. They suggest that if this is more than 25% of the total variance, it is important. In our example, the proportion of systematic between-studies variance v_θ is estimated as 0.60, much larger therefore than the lower limit of 0.25 that Hunter and Schmidt propose.

Since the between-studies variance is large and significant, random effects meta-analysis must be used. First, the between-studies variance v_θ is estimated. This is added to all the individual variances, so the estimated variance for each study becomes $v_i^* = v_\theta^* + v_i$. This leads to different weights and different estimates.

Classical meta-analysis uses a simple estimate for the between-studies variance (Lipsey and Wilson 2001, 119). This leads to an overall effect estimate of $\delta = -0.10$, with a 95% confidence interval from −0.17 to −0.03. These results differ from the results of the fixed effect model, because the random effects model takes into account the systematic variation between the studies when the studies' outcomes are averaged. The interpretation of the random effects estimates is also different. The overall effect size of −0.10 is the average of the distribution of effects in the population of studies. The average outcome is negative, that is, the aspirin group has significantly fewer heart attacks than the placebo group.

When the outcomes are heterogeneous, the classical approach is to divide the studies into clusters that have different average effect sizes but are internally homogeneous. In our example, this cluster analysis produces two clusters. The first consists of Studies 1 through 5, and the second consists of Study 6. In the first cluster, the variances are homogeneous, which means across these five studies there is no systematic between-study variance. The lack of available background information prevents further analysis of the aspirin data, a problem typical for meta-analysis (Lipsey and Wilson 2001).

In many cases, it would be impractical for primary researchers to publish their raw data. There is a real need for a publication practice that encourages researchers to publish sufficient statistics and their corresponding standard errors for various subgroups. The American Psychological Association already supports such publication practices and recommends that researchers retain their raw data for possible use by others (APA 1994). Developments in data archiving and using the Internet also will help.

21.3.2 Multilevel Meta-analysis

Multilevel regression analysis can be used to estimate the random effects model for meta-analysis, including available explanatory variables (Raudenbush 1994; Hox 2002). In the multilevel approach to meta-analysis, we recognize the two-level structure that is implicit in Table 21.1. We have six studies, with a total of 10,816 patients. If we had access to the raw data of all studies, we could set up a standard multilevel regression analysis with patients at the lowest level, studies at the highest level, and explanatory variables that describe patient or study characteristics. Since we do not have access to the raw data, a special multilevel model is used instead (Kalaian and Raudenbush 1996; Hox 2002).

Multilevel regression analysis is a regression model with a complicated error structure. For an introduction to multilevel modeling, see Hox (1995, 2002). Since we have no access to the raw data, we set up a model for the sufficient statistics: the d's and their standard errors. In such a model, the lowest level sampling variance is not estimated, because it is known from the sampling errors provided in Table 21.1. The term 'variance known' model (Bryk and Raudenbush 1992) aptly describes this characteristic feature of multilevel meta-analysis.

A multilevel meta-analysis of the six aspirin studies produces virtually the same results as the classical meta-analysis reported above. The starting model is a model without explanatory variables. The intercept, which is the overall outcome, is estimated as −0.10, with a 95% confidence interval from −0.18 to −0.02. The null hypothesis of homogeneous outcomes is rejected. The chi-square test on the significance of the between-studies variance yields a chi-square of 34.34 ($df = 5$, $p < .001$). The proportion of systematic variance is estimated as 0.73, a bit higher than in the classical meta-analysis.

The power of multilevel meta-analysis becomes clear when we model the differences in the study outcomes. We have no real background variables that describe the studies. Given the deviant outcome of Study 6, we decide to model the heterogeneity by including a dummy variable that represents Study 6 as an explanatory variable in the model. The regression coefficient for this variable is 0.21 ($p < .01$), which means that in Study 6, the difference between the experimental and control group is 0.21 larger than in the other studies but in an unexpected direction. The intercept, which in this model is the average outcome of the other five studies, is –0.14, with a standard error of 0.025 ($p < .001$). After including the dummy variable as an explanatory variable in the model, the residual between-studies variance is no longer significant ($\chi^2 = 1.50$, $df = 4$, $p = .83$). Thus, the heterogeneity is completely explained by the deviant outcome of Study 6.

In contrast to the cluster analysis reported above, we now have a formal hypothesis test for the hypothesis that Study 6 is an outlier and for the assertion that after controlling for this variable, there is no between-study variance left. If we had had more background variables that code for study characteristics, we could have added them to the model in the same manner.

Table 21.2 presents the results from the analyses of the aspirin data. The fixed effect model yields the smallest estimate of the combined effect $\overline{d}$. The classical random effects model and the multilevel meta-analysis null-model are equivalent, but the estimation procedures differ, since multilevel analysis uses Maximum Likelihood estimates. The differences are small but noticeable. In the multilevel model that includes a dummy for the unusual Study 6, the variance between the studies is estimated as zero. In this model, the between-studies variance is an estimate of the residual variance, and the chi-square test is a test for the significance of this residual variance. Thus, after taking into account the distinction between Study 6 and the others, there is no between-study variance left.

Table 21.2. Results from Classical and Multilevel Meta-analysis on Aspirin Data

Model:	Fixed Effect			Random Effects			Multilevel Null Model			Multilevel with Study 6 Dummy		
	$\overline{d}$	(s.e.)	p	$\overline{d}$	(s.e.)	p	coef.	(s.e.)	p	coef.	(s.e.)	p
Intercept	–.06	(.019)	.00	–.10	(.035)	.00	–.10	(.043)	.02	–.14	(.035)	.00
Study 6		N/A			N/A			–		+.21	(.039)	.00
Study var.		N/A		.005			.008			.000		
χ^2 (df) p	30.2	(5)	.00	30.2	(5)	.00	34.3	(5)	.00	1.5	(4)	.83

21.4   META-ANALYSIS IN COMPARATIVE RESEARCH

The following section describes two different meta-analyses of cross-national data sets that illustrate important issues in cross-national comparisons. The first example is a cross-national literature review and the second a cross-national multi-site study.

21.4.1 Multicultural and Multinational Approaches in Literature Reviews

The goal of meta-analysis for literature reviews is generalization over studies to summarize outcomes. In cross-cultural and cross-national research, we routinely expect heterogeneous results. The modern approach to heterogeneous results is to investigate if explanatory variables influence this general outcome. This is especially important in cross-cultural and cross-national comparisons. In these comparisons, it is important to control for any differences in the data collection methods and instrumentation used in the different countries. In addition, we need explanatory variables to explain real conceptual differences between different countries. Thus, testing and modeling heterogeneity become the central issue. This makes multilevel meta-analysis a powerful tool in cross-cultural and cross-national meta-analyses, because testing and modeling heterogeneity is an integral part of multilevel modeling.

A study by de Leeuw and de Heer (2001) on international nonresponse trends illustrates this. The study combines the results from reports on nonresponse in 16 different countries. Summary data were solicited from the official statistical offices on the response rates of a number of surveys over as many years as were available. Information about the sampling design, survey design, fieldwork strategy, interviewer corps, and survey climate also were collected. Three outcome variables—proportion response, proportion refusals, and proportion noncontacts—were used to investigate the nonresponse process in some detail. The study addresses three research questions: (1) Does nonresponse differ between countries? (2) Does nonresponse increase over time? (3) Is the increase different between countries?

De Leeuw and de Heer did not have raw data at their disposal. Thus multilevel meta-analysis was used to analyze similarities and differences across the 16 countries. The authors find that countries differ significantly in response rate, noncontact rate, and refusal rate, and that type of survey influences response rate and refusal rate but does not influence noncontact rate. There were strong trends over time, but these were not the same for all countries, as Figure 21.1 from de Leeuw and de Heer (2001) clearly shows. De Leeuw and de Heer conclude that countries differ in response rate and that response rates have indeed been declining over the years. However, the trends differ across countries, and the differences in response trends are caused by differences between countries in the rate at which refusals are increasing. Some design and fieldwork factors appear to have an effect.

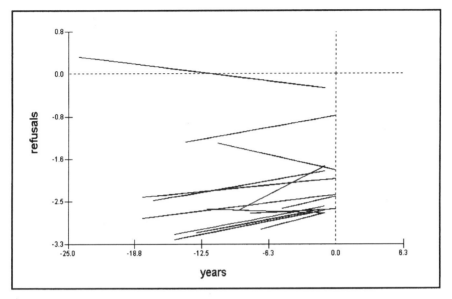

Figure 21.1   Refusals Across Years for Different Countries (1998 = 0; logit scale).

De Leeuw and de Heer's study illustrates some problems typical for meta-analysis. Ideally, to compare nonresponse trends internationally, the data set should contain long and detailed time series, covering a range of survey types and organizations, with all countries providing data for all surveys. In fact, in their data set, countries differ in the surveys they report on, in the interval between two subsequent surveys, and in the length of the time series. To cope with these problems, de Leeuw and de Heer carry out several analyses, using different subsets of the data. They use multilevel regression analysis, which allows inclusion of multiple control and explanatory variables. By comparing different models, they show that their findings point to real differences between the countries.

21.4.2 Meta-analysis to Combine Multi-Site Results from Cross-national Studies

In biomedical research, multi-site studies are an accepted tool to collect data in different places to achieve a sufficient sample size. The goal of such studies is to combine data into one outcome, which assumes homogeneity. In cross-national studies, the interest is often different; heterogeneity is routinely expected, and if found, one would want to explain the heterogeneity. Good examples are the large studies initiated by the International Association for the Evaluation of Educational Achievement (IEA) to compare educational achievement across countries. Here, the interest is squarely on differences between countries and their potential causes.

Two types of variables are important in a comparative study: methodological variables and conceptual or theoretical variables. Methodological variables are used to control statistically for differences in methods between the countries, such as differences in response or data collection methods. The primary aim of introducing methodological variables is to filter out 'method variance' or systematic error. The conceptual variables are used to investigate real differences between the countries, such as differences in educational systems or hours devoted to teaching.

Table 21.3.   Reading Achievement Data from 27 Countries

Country	Reading	s.e.	Age	Grade	Economic	Health	Literacy
Finland	569	3.40	9.7	3	.76	.52	1.26
U.S.A.	547	2.80	10.0	4	1.25	.03	.31
Sweden	539	2.80	9.8	3	1.56	.83	1.17
France	531	4.00	10.1	4	.55	.46	.09
Italy	529	4.30	9.9	4	.02	.18	−.37
N. Zealand	528	3.30	10.0	5	−.43	.31	.53
Norway	524	2.60	9.8	3	1.34	.83	1.26
Iceland	518	3.26	9.8	3	.69	1.04	1.29
Hong Kong	517	3.90	10.0	4	−.62	.75	−.74
Singapore	515	1.00	9.3	3	−.50	−.28	−.76
Switzerland	511	2.70	9.7	3	2.15	.61	1.10
Ireland	509	3.60	9.3	4	−.57	.37	.03
Belgium/Fr	507	3.20	9.8	4	.39	.31	.18
Greece	504	3.70	9.3	4	−1.07	.40	−.74
Spain	504	2.50	10.0	4	−.80	.75	−.65
Germany/W	503	3.00	9.4	3	.76	.31	.59
Canada/BC	500	3.00	8.9	3	1.00	.40	.29
Germany/E	499	4.30	9.5	3	−.19	−.21	1.37
Hungary	499	3.10	9.3	3	−1.14	−1.54	.26
Slovenia	498	2.60	9.7	3	−.97	−.43	−.84
The Netherlands	485	3.60	9.2	3	.43	.83	.48
Cyprus	481	2.30	9.8	4	−.90	.14	−.95
Portugal	478	3.60	10.4	4	−1.16	−.49	−1.65
Denmark	475	3.50	9.8	3	.88	.09	.63
Trinidad	451	3.40	9.6	4	−.81	−1.09	−.35
Indonesia	394	3.00	10.8	4	−1.52	−3.77	−2.70
Venezuela	383	3.40	10.1	4	−1.09	−1.32	−1.08

Table 21.3 presents the results of the IEA study into reading achievement in 27 countries, assembled from information given by Elley (1992, Table 2.1). We have the mean reading achievement of nine-year-old school children, the associated standard error (adjusted for complex survey design), the mean age when tested, the grade when tested, and three indicators of the countries' economic, health, and literacy status. For convenience, the data are sorted by average reading score of each country.

Although the IEA went to great effort to insure that data collection methods were comparable across countries (Elley 1992), Table 21.3 makes clear that children were tested at slightly different ages and in different school grades. These two variables are clear examples of methodological variables. Thus, before comparing countries on reading achievement, we first must adjust the reading scores for differences in mean age and school grade.

Table 21.4 shows the results of three multilevel meta-analysis models on these data. For each model, the table presents the regression coefficient(s) of the explanatory variables in the model, the corresponding standard error (in parentheses), and the resulting p-value. In addition, for each model, the between-country variance is given, as is the value of the chi-square for its significance test.

The first model in Table 21.4 is a model without explanatory variables. Here, the intercept coefficient represents the average reading score, which is 500. It shows a large variance (1,594) between countries which is significant using the chi-square test. The reading scale was standardized to a mean of 500 and a standard deviation of 100, across all pupils and countries. Thus the total variance is 10,000, and the between-country variance of 1,594 is about 16.0% of the total variance. The second model includes the methodological variables mean age and school grade. Neither has a significant effect, and the residual variance is still large (1,567) and significant. The residual variance has dropped by 30, which means that about 1.9% of the original between-country variance has been explained. Clearly, the small differences in age and grade do *not* explain the differences between countries.

Table 21.4. Meta-analysis Models for Reading Data

Model:	Intercept only			Methodological vars.			+ Conceptual vars.		
Fixed part Predictor	coef.	(s.e.)	p	coef.	(s.e.)	p	coef.	(s.e.)	p
Intercept	500	(7.7)	0.00	792	(198.9)	0.00	452	(170.8)	0.02
Mean age				−29	(22.7)	0.21	2.8	(19.4)	0.88
Grade				−2.5	(15.6)	0.88	4.6	(13.6)	0.73
Economic							1.0	(10.3)	0.97
Health							22.0	(8.4)	0.02
Literacy							12.0	(10.7)	0.27
Random part									
Variance		1,597			1,567			816	
χ^2 (df) p	4,274	(26)	0.00	3,604	(24)	0.00	2,701	(21)	0.00

The third and final model adds conceptual variables that describe theoretically important differences between the countries: in our example, economic status, average health, and overall literacy level. Only the health indicator shows a significant effect. In the third model, the residual variance is much smaller (816) but still significant. Compared to the empty first model, 48.9% of the variance is now explained. The methodological variables therefore explain a (nonsignificant) 1.9% of the variance, while the conceptual variables describing differences between countries explain a (significant) 47% of the variance.

This analysis is intended as an example only. It illustrates the importance of controlling for methodological differences in the study design in international multi-site studies. Even if cross-national studies are intended to be similar in design, in practice some differences are almost inevitable. Consequently, such differences must be controlled for at the analysis stage.

21.5   CONCLUSION

Meta-analysis is of itself comparative, and it becomes international in many cases simply by including all the relevant publications. In cross-cultural or cross-national comparative meta-analyses, cross-comparisons are the central theoretical issue. This is reflected in both the design and the analysis of such meta-analyses.

Implicit in all meta-analytic comparisons is the assumption that the data are comparable. Thus, the data collection procedures in the individual studies must be similar, and the measures (which often imply translating materials) must be equalized. Bechger, van Schooten, de Glopper, and Hox (1998) introduce the general issues involving such comparisons. In a comparative meta-analysis, the ensuing complications should be anticipated. This means that in the design and coding phase of the meta-analysis, indispensable control variables must be explicitly defined and operationalized. In planned multi-site comparisons, the study design should include measures that improve the comparability of the studies. Elley's (1992, 95–111) careful description of the measures taken in the IEA reading study to increase comparability provides a good example.

In the analysis of comparative meta-analytic data, the first step should be to estimate the between-country variability and its significance. The second step then investigates whether any differences found are attributable to methodological differences in the procedures. The third step investigates explanatory variables at the country (or culture) level. This last step is the most interesting one in cross-cultural or cross-national research. The variance between countries must be decomposed into sampling variance, variance due to methodological differences, and systematic and substantively interesting variance. This requires advanced statistical modeling techniques. A classical approach is weighted regression analysis of study outcomes (Hedges and Olkin 1985; Lipsey and Wilson 2001). A powerful alternative is multilevel meta-analysis. This uses iterative Maximum Likelihood estimation, which is an improvement on weighted regression analysis, and is available in standard multilevel software.

A problem that occurs frequently in cross-national comparisons is the multi-collinearity of the predictor variables. For instance, one of the problems in the comparison of countries is the large impact of Gross National Product (GNP). Many country-level explanatory variables, such as educational level, health status, or average income, are related to GNP. Even average temperature correlates with GNP. Since these correlations are typically high, it is not possible to include all these variables in the analysis. This creates a problem of choice, since these variables are to some degree interchangeable. The problem is magnified because in cross-national comparisons, the sample size for study-level variables is the number of countries. This is typically not very high, which poses other restrictions on the number of country-level variables that can be included. The issues involved here are no different from the issues in disentangling the contribution of correlated explanatory variables in ordinary multiple regression analysis. They are more difficult to solve because the small samples frequently limit the analysis. Tabachnick and Fidell (2001) discuss possible solutions in the context of multivariate regression analysis, such as step-down analyses.

In the last thirty years, many articles and books have been published on how to do meta-analysis. The classical statistical treatment of meta-analysis is the monograph by Hedges and Olkin (1985). A good general introduction, including a nontechnical discussion of classical analysis methods and how to implement these in SPSS, is the recent book by Lipsey and Wilson (2001). The handbook edited by Cooper and Hedges (1994) is an excellent reader that gives a thorough overview.

The basic statistics for meta-analyses can be programmed easily or calculated by hand. One dedicated program for meta-analysis is by Ralf Schwarzer (Schwarzer 1989). It is user-supported software, meaning that any user may copy and distribute it as long as no charge is made. Lipsey and Wilson (2001) describe procedures for meta-analysis using SPSS, including a method for weighted least squares regression analysis. However, for modeling strongly heterogeneous data, multilevel meta-analysis is superior, because it estimates the between-studies variance using maximum likelihood methods instead of a simple plug-in estimate. Multilevel meta-analysis can be carried out with any multilevel software that supports imposing a constraint on the lowest level variance component. This is possible in the generally available dedicated multilevel software HLM (Raudenbush et al. 2000) and MLwiN (Rasbash 2000) and in the mixed model procedure in the general package SAS (Littell et al. 1996). For details, see Hox (2002) or Hox and de Leeuw (2001).

GLOSSARY

TIMOTHY P. JOHNSON

Glossary entries reflect key concepts and methodologies discussed in this volume that are important for understanding and conducting cross-cultural survey research. Definitions are also provided for cross-referenced items identified in *italics*.

Adaptation

Adapted questions are derived from existing questions. Some component is deliberately changed to make a question more suitable. Adaptations are distinct from unavoidable changes in translation but may become necessary because a question needs to be translated. Adaptations may be substantive, relate to question design, or consist of slight wording modifications. Regardless of the form of change, the aim of adaptation is to render questions culturally or linguistically appropriate.
See *adoption.*

Administration Bias

A form of *method bias* in which systematic differences in survey measurements across two or more cultures are a consequence of differences in the effects of survey procedures on individual responses. For example, cultural differences in the effects of interviewer characteristics or mode of data collection would result in administration bias.
See also *bias.*
Contrast with *concept bias*, *instrument bias*, *item bias* and *sample bias*.

Adoption

Adopting a question or instrument, consists, essentially, of translating items or using the original items verbatim if translation is not necessary.
See also *translation, adaptation.*

Ask-the-Same-Question Approach (ASQ)

Questions from a common *source questionnaire* that are implemented across cultural groups and languages on the basis of *close translation*.
See also *lexical meaning.*

Back Translation

A form of translation assessment which operates on the premise that 'what goes in will come out.' The translated questionnaire is translated back into the source questionnaire language. Then the *source questionnaire* and back translated version in the source language are compared for difference or similarity. The back translated text is taken as an indicator of the quality of the target language translation, which is not, itself, assessed. If a back translated question is 'like' the original source question, the translated question is considered to be good.
See also *source questionnaire*.
Contrast with *committee assessment, comprehension assessment, double administration assessment*, and *split ballot assessment*.

Bias

A systematic effect that distorts a survey estimate from its true value differentially across two or more cultures. For example, a measure is said to be culturally biased if it has different meanings in different cultural groups.
See also *administration bias, concept bias, instrument bias, item bias, method bias*, and *sample bias*.

Bias Analysis

See item *bias analysis*.

Close Translation

In survey research, an approach to translating that emphasizes close adherence to the semantic or *lexical meaning* of the wording of *source questionnaires*. The level of closeness varies. A close translation may also aim to follow syntactic arrangements and other formal features of a *source questionnaire*. The terms 'close', 'literal' and 'semantic' translation have multiple uses and meanings in other disciplines (e.g., the translation sciences). Also referred to as literal translation.
See also *ask-the-same-question approach*.
Contrast with *decentering* and *on-the-fly translation*.

Collectivist Cultures

Cultures where collective well-being is emphasized over personal interests and individual independence. Group obligations and interests take priority over the personal goals and identity of the individual.
Contrast with *individualist cultures* and *interdependent self-construal*.

Committee Translation and Assessment

In committee approaches to translation and assessment, a good part of the work, possibly including first translations, is done working in a group.
See also TRAPD.
Contrast with *back translation, comprehension assessment, double administration assessment.*

Comparability

The degree to which survey data represent similar phenomena across two or more cultures.
Contrast with *equivalence.*

Comprehension Assessment

A method of questionnaire *translation assessment* in which the adequacy of a translation is evaluated by the ability of representative members of the target population to correctly interpret the translated document, in some cases, demonstrated by their ability to follow instructions, etc.
Contrast with *back translation, committee assessment, double administration assessment.*

Concept Bias

A survey question contains concept bias when it does not systematically represent the same underlying latent concept across two or more cultures.
See also *bias* and *construct underrepresentation.*
Contrast with *administration bias, instrument bias, item bias, method bias,* and *sample bias.*

Construct Equivalence

Construct equivalence assures that a defined construct exists in two or more cultures and can be measured using similar or different survey questions.
Also commonly referred to as 'functional equivalence' and 'structural equivalence.'
See also *equivalence.*
Contrast with *measurement equivalence,* and *scalar equivalence.*

Construct Underrepresentation

When a measurement instrument fails to include all the domains and dimensions relevant to represent a given construct in a given culture.
See also *concept bias, culturally anchored methodologies* and *research imperialism.*

Convergence Approach

A method of addressing *concept bias* by which survey measures are developed independently within two or more cultures and each measure is subsequently administered within each culture of interest.

Courtesy Bias

A form of *administration bias* common in cultural groups that emphasize harmonious and polite interactions, and the avoidance of negative or offending topics. It is associated with the production of socially desirable and/or acquiescent responses as respondents attempt to determine what an interviewer is looking for and adjusting answers to assist the interviewer in finding it. In some Latin American countries, courtesy bias is referred to as '*simpatia*.'

Cross-Cultural Research

Investigations conducted across two or more cultures.
Contrast with *cross-national research* and *mono-cultural research*.

Cross-National Research

Investigations conducted across two or more national political boundaries.
Contrast with *cross-cultural research* and *mono-cultural research*.

Cultural Complexity

The degree of organization or diversification within a cultural group, commonly assessed by the number and types of social roles that a society is organized into, its degree of occupational stratification, its physical size or level of affluence.

Cultural Distance

The degree to which norms, beliefs, values, and life experiences are shared or not shared between two cultural groups.

Culturally Anchored Methodologies

Methods and procedures which are conceptually bound to the cultural context for which they were developed.
See also *mono-cultural research*.

Culture-Sensitive Context Effects

Differential reactions to variations in question wording, format or order across two or more cultures.

Decentering

In classical decentering models, different cultures are asked the 'same' questions, but the questions are developed simultaneously in each language; there is no sequential or hierarchical arrangement of source questionnaire followed by target language questionnaire. The decentering process removes culture-specific elements.

Contrast with *close translation, committee assessment* and *on-the-fly translation, source document.*

Dif

See *item bias.*

Difference-Oriented Studies

See *level-oriented studies.*

Differential Item Functioning

See *item bias.*

Double Administration (Double Ballot) Assessment

A method of questionnaire *translation assessment* in which bilinguals complete both *source questionnaire* and *target questionnaire* versions of an instrument and results are then compared to determine their *equivalence.* Contrast with *back translation, committee assessment, comprehension assessment,* and *split ballot assessment.*

Emic

Concepts or constructs that are culture-specific. Also referred to as 'idiographic.' Contrast with *etic.*

Equivalence

The degree to which survey measures or questions are able to assess identical phenomena across two or more cultures.

See also *construct equivalence, measurement equivalence,* and *scalar equivalence.*

Etic

Concepts or constructs that are universal or equally understood across multiple cultures. Also referred to as 'nomothetic' and 'pan-cultural.'
Contrast with *emic* and *pseudo-etic.*

Full Score Equivalence

See *scalar equivalence*.

Functional Equivalence

Multiple definitions of functional equivalence exist. In cross-cultural psychology, it is a synonym for *construct equivalence*. Functional equivalence is not used in a single, specific manner in this volume.

Harmonization

A method for equating conceptually similar but operationally different variables that are collected as part of separate surveys for purposes of *cross-cultural* or *cross-national research*. Also referred to as '*ex post* harmonization.' Harmonization is typically applied to certain background variables.

Idiographic

See *emic*.

In-Groups

Those social groups to which an individual belongs that are most important to his/her social identity and which can command immediate allegiance and cooperation without expectations of reciprocal treatment.
Contrast with *out-groups*.

Independent Self-Construal

A personal identity based upon beliefs of self-determination and independence of action. The fulfillment of personal goals are emphasized over group objectives.
Contrast with *individualist cultures* and *interdependent self-construal*.

Individualist Cultures

Cultures where personal interests and individual independence are emphasized over collective well-being. Personal identity and goals are independent of and take priority over group obligations and interests.
Contrast with *collectivist cultures* and *independent self-construal*.

Instrument Bias

A form of *method bias* in which systematic differences in survey responses across cultural groups are a consequence of differences in the effects of survey instrumentation on individual responses. For example, cultural differences in familiarity with the survey or response devices would result in instrument bias.
See also *bias*.
Contrast with *administration bias, concept bias, item bias and sample bias*.

Interdependent Self-Construal

A personal identity based upon beliefs of group membership and dependence. The fulfillment of group goals is emphasized over personal objectives.

Contrast with *collectivist cultures* and *independent self-construal*.

International Research

Research projects carried out in two or more countries but without any intention of conducting cross-national comparisons of data.

Contrast with *cross-cultural research* and *cross-national research*.

Item Bias

A survey question contains item bias when it systematically produces different responses across cultures as a consequence of some feature of the item itself, such as poor translation or ambiguous wording. Also referred to as 'differential item functioning' or as 'dif.'

See also *bias*.

Contrast with *administration bias*, *concept bias*, *instrument bias*, *method bias*, and *sample bias*.

Item Bias Analysis

A family of statistical techniques used to detect survey questions that have *item bias*. Examples of techniques used to identify item bias include analysis of variance (ANOVA), the Mantel-Haenszel statistic, and item response theory (IRT). Also referred to as 'bias analysis.'

See also *bias*.

Level-Oriented Studies

Statistical analyses concerned with determining if cross-cultural or cross-national similarities or differences exist in responses to survey measures. Examples of procedures used to conduct level-oriented studies include descriptive statistics, analysis of variance (ANOVA), *t*-tests, and multiple linear regression. Also referred to as 'difference-oriented studies.'

Contrast with *structure-oriented studies*.

Lexical Meaning

The semantic meaning associated with words as presented, for example, in dictionaries, and in potential contrast to their meaning in a single, specific context.

See also *close translation*.

Contrast with *pragmatic meaning*.

Literal Translation

See *close translation.*

Measurement Equivalence

A survey question measured on an interval-level scale has measurement equivalence if, in all cultures, its response options have the same intervals. Also referred to as 'measurement unit equivalence.'
See also *equivalence*.
Contrast with *construct equivalence* and *scalar equivalence*.

Measurement Unit Equivalence

See *measurement equivalence.*

Method Bias

A measure contains method bias when practical factors associated with the collection of survey data, rather than the underlying construct of interest, produce systematic differences across two or more cultures. Three forms of method bias are distinguished here: *administration bias*, *instrument bias*, and *sample bias*.
See also *bias*.
Contrast with *concept bias* and *item bias*.

Monocultural Research

Investigations conducted exclusively within a single cultural group.
Compare to *cross-cultural research* and *cross-national research*.

Nomothetic

See *etic.*

On-the-Fly Translation

Oral translation of questionnaires during an interview. This unmonitored approach to translation is common when the number of persons to be interviewed in a specific language are held to be too few to justify producing a translated questionnaire. The procedure leaves open what is asked and what is answered. The quality of translation is therefore unknown. It can also vary between translators and, for one translator, from respondent to respondent.
Contrast with *close translation*, *committee approach* and *decentering*.

One-Size-Fits-All

A study design approach that targets comparability by keeping design components and implementations as similar as possible across cultures.
See also *ask-the-same-question approach, culturally anchored methodologies* and *research imperialism*.

Out-Groups

Those social groups to which an individual does not belong and which do not command their immediate allegiance and cooperation.
Contrast with *in-groups*.

Pancultural

See *etic*.

Pragmatic Meaning

Pragmatic meaning is meaning in context, that is, what words mean in use in a given context for participants; e.g., "I love this" understood to mean the opposite.
Contrast with *lexical meaning*.

Pseudo-Etic

Emic concepts or constructs that are incorrectly believed to be *etic*.

Research Imperialism

The imposition of research methodologies developed in one culture or nation to another without concern as to whether or not they can be appropriately applied in that context.
See also *culturally anchored methodologies*.

Sample Bias

A form of *method bias* in which systematic differences in survey measurements across cultural groups are a consequence of differences in the composition of the samples being compared.
See also *bias*.
Contrast with *administration bias*, *concept bias*, *instrument bias* and *item bias*.

Scalar Equivalence

A survey question measured on an interval-level scale has scalar equivalence if in all cultures its response options have the same intervals and the same origin.
Also referred to as 'full score equivalence.'
See also *equivalence*.
Contrast with *construct equivalence* and *measurement equivalence*.

Source Language

The language out of which a translation is made.

Source Questionnaire

The final version of a questionnaire for a project which is used as the text from which all translations are made. Also referred to as 'source language questionnaire.'
See also *target questionnaire, back translation* and *translation assessment.*
Contrast with *decentering.*

Split Ballot Assessment

A collection of questionnaire *translation assessment* methods in which groups of bilinguals complete either a *source questionnaire* or a *target questionnaire* (or portions of each) and results are compared to determine their *equivalence.*
Contrast with *back translation, committee assessment, comprehension assessment,* and *double administration.*

Structural Equivalence

See *construct equivalence.*

Structure-Oriented Studies

Statistical analyses concerned with determining if survey measures have *construct equivalence* across two or more cultures. Examples of procedures used to conduct structure-oriented studies include exploratory factor analysis, structural equation modeling, latent class analysis, cluster analysis and multidimensional scaling.
Contrast with *level-oriented studies.*

Survey Climate

The general acceptance of participation in surveys as a valued or appropriate social activity among persons in a given society, nation, language or cultural group. Also referred to as 'survey-taking climate.'

Survey Literacy

The degree to which a surveyed population is familiar with the concept of social surveys, with participating in surveys, and with their roles as respondents.

Target Questionnaire

A target questionnaire is a questionnaire translated from a *source questionnaire* into whichever language is required for fielding.
See also *source questionnaire* and *target language.*
Contrast with *decentering.*

Target Language

The language into which a translation is made.
See *target questionnaire*.

Translation Assessment

Procedures used in survey research (of various standing and orientation) to assess the quality of a translation. Text-based assessment evaluates the translation; statistical assessment assesses the performance of items in translation—which may point to problems in translations. Assessment methods currently in use include *back translation, committee approaches, comprehension assessment, double administration assessment, split ballot assessment*, and *TRAPD*.

TRAPD

Acronym for five basic procedures that are involved in producing a final version of a questionnaire: Translation, Review (assessing the translation), Adjudication (final decision phase on a translation), Pretesting (of a version decided upon), and Documentation of the translation and decision-making process. Several loops through the whole or parts of the process are possible.

REFERENCES

AAPOR, 2000. *Standard Definitions: Final Dispositions of Case Codes and Outcome Rates for Surveys*. Ann Arbor: AAPOR. http://www.aapor.org/ethics/stddef.html.

Abe J., Zane N.W.S, 1990. Psychological Maladjustment Among Asian and White American College Students: Controlling for Confounds. *Journal of Counseling Psychology*, **37**, 437–444.

Abelson R.P., 1981. Psychological Status of the script Concept. *American Psychologist*, **36**, 715–729.

Abramson P.R., Claggett W., 1986. Race-Related Differences in Self-Reported and Validated Turnout in 1984. *Journal of Politics*, **48**, 412–422.

Abramson P.R., Inglehart R., 1995. *Value Change in Global Perspective*. Ann Arbor: University of Michigan Press.

Achen C.H., Shively W.P., 1995. *Cross-Level Inference*. Chicago: University of Chicago Press.

Acquadro C., Jambon B., Ellis D., Marquis P., 1996. Language and Translation Issues. In: *Quality Life and Pharmacoeconomics in Clinical Trials*. Spilker B. (ed.). 2nd edition. Philadelphia: Lippincott-Raven.

Adorno T., Frenkel-Brunswik E., Levison D., Sanford R., 1950. *The Authoritarian Personality*. New York: Harper & Row.

Agar M., 1994. *Language Shock: Understanding the Culture of Conversation*. New York: William Morrow.

Al-Issa I. (ed.), 1995. *Handbook of Culture and Mental Illness: An International Perspective*. Madison, WI: International Universities Press.

Aldenderfer M.S., Blashfield R.K., 1984. *Cluster Analysis*. Beverly Hills, CA: Sage.

Allen J., 1998. Personality Assessment with American Indians and Alaska Natives: Instrument Considerations and Service Delivery Style. *Journal of Personality Assessment*, **70**, 17–42.

Almond G.A., Verba S., 1963. *The Civic Culture. Political Attitudes and Democracy in Five Nations*. Princeton, NJ: Princeton University Press.

Altbach P.G., Arnove R.F., Kelly G.P., 1991. *Comparative Education*. New York: Macmillan.

Altemeyer B., 1988. *Enemies of Freedom. Understanding Right-Wing Authoritarianism*. San Francisco: Jossey-Bass.

Alwin D.F., 1992. Information Transmission in the Survey Interview: Number of Response Categories and the Reliability of Attitude Measurement. In: *Sociological Methodology*. Marsden P.V. (ed.). Cambridge: Blackwell.

Alwin D.F., 1997. Feeling Thermometers versus 7-Point Scales: Which are Better? *Sociological Methods and Research*, **25**, 318–340.

Alwin D.F., Krosnick J.A., 1991. The Reliability of Survey Attitude Measurement: The Influence of Question and Respondent Attributes. *Sociological Methods and Research*, **20**, 139–181.

Andersen R., Kasper J., Frankel M.R., 1979. *Total Survey-Error*. San Francisco: Jossey-Bass.

Anderson B.A., Silver B.D., Abramson P.R., 1988. The Effect of Race of the Interviewer on Race-Related Attitudes of Black Respondents in SRC/CPS National Election Studies. *Public Opinion Quarterly*, **52**, 289–324.

Andrews F.M., 1984. Construct Validity and Error Components of Survey Measures: A Structural Modeling Approach. *Public Opinion Quarterly*, **48**, 409–422.

Angoff W.H., 1982. Use of Difficulty and Discrimination Indices for Detecting Item Bias. In: *Handbook of Methods for Detecting Item Bias*. Berk R.A. (ed.). Baltimore: Johns Hopkins University Press.

APA, 1994. *Publication Manual of the American Psychological Association*. 4th edition. Washington, DC: APA.

Apter D.E., 1971. Comparative Studies: A Review with Some Projections. In: *Comparative Methodology on Sociology — Essays on Trends and Applications*. Vallier I. (ed.). Berkeley, CA: UCA-Press.

Aquilino W.S., 1992. Telephone versus Face-to-Face Interviewing for Household Drug Use Surveys. *International Journal of the Addictions, 27*, 71–91.

Aquilino W.S., 1994. Interviewer Mode Effects in Surveys of Drug and Alcohol Use. *Public Opinion Quarterly, 58*, 210–240.

Aquilino W.S., Wright D.L., 1996. Substance Use Estimates from RDD and Area Probability Samples — Impact of Differential Screening Methods and Unit Nonresponse. *Public Opinion Quarterly, 60*, 563–573.

Armer A., 1973. Methodological Problems and Possibilities in Comparative Research. In: *Comparative Social Research: Methodological Problems and Strategies*. Armer M., Grimshaw A.D. (eds.). New York: Wiley.

Armer A., Grimshaw A.D., 1973. Preface. In: *Comparative Social Research: Methodological Problems and Strategies*. Armer M., Grimshaw A.D. (eds.). New York: Wiley.

Artiola i Fortuny L., Mullaney H., 1997. Neuropsychology with Spanish-Speakers: Language Use and Proficiency Issues for Test Development. *Journal of Clinical and Experimental Neuropsychology, 19*, 615–623.

Arvey R.D., Strickland W., Drauden G., Martin C., 1990. Motivational Components of Test Taking. *Personnel Psychology, 43*, 695–716.

Austin J., 1962. *How to Do Things with Words*. Oxford: Clarendon Press.

Ayidiya S.A., McClendon M.J., 1990. Response Effects in Mail Surveys. *Public Opinion Quarterly, 54*, 229–247.

Bachman J.G., O'Malley P., 1984. Black-White Differences in Self-Esteem: Are They Affected by Response Styles? *American Journal of Sociology, 90*, 624–639.

Backstrom C.H., Hursh-César G., 1981. *Survey Research*. New York: Wiley.

Barnes S.H., Kaase M. (eds.), 1979. *Political Action: Mass Participation in Five Western Democracies*. Beverly Hills, CA: Sage.

Barnett H., Bousser M.-G., Boysen G., Breddin K., Britton M., Cairns J., Canner P., Collins R., Cortellaro M., Daniels A., Deykin D., Elwood P., Elwin E.-E., Eschwege E., Farrell B., Fields W.S., Gent M., Gary R., Guiraud-Chaumeil B., Hennekens C.H., Klimt C., Lewis H.D., Loria Y., Lowenthal A., MacMahon C., Miller J., Peto R., Qizilbash N., Reuther R., Richards S., Rosen A., Sandercock P., Sorensen P.S., Sweetnam D., Taylor W., Thibult N., Van Gijn J., Vogel G., Warlow C., Yusuf S., 1988. Secondary Prevention of Vascular Disease by Prolonged Antiplatelet Therapy. *British Medical Journal, 296*, 320–331.

Baron R.M., Kenny D.A., 1986. The Moderator-Mediator Variable distinction in Social Psychological Research: Conceptual, Strategic, and Statistical Considerations. *Journal of Personality and Social Psychology, 51*, 1173–1182.

Barrett P.T., Petrides K.V., Eysenck S.B.G., Eysenck H.J., 1998. The Eysenck Personality Questionnaire: An Examination of the Factorial Similarity of P, E, N, and L across 34 Countries. *Personality and Individual Differences, 25*, 805–819.

Bartram J.G., Yelding D., 1973. The Development of an Empirical Method of Selecting Phrases Used in Verbal Rating Scales: A Report on a Recent Experiment. *Journal of the Market Research Society,* **15**, 151–156.

Bates B.A., 1998. Standard Demographic Classification. In: *The ESOMAR Handbook of Market and Opinion Research.* McDonald C., Vangelder P. (eds.). Amsterdam: ESOMAR.

Bechger T.M., Van Schooten E., De Glopper C., Hox J.J., 1998. The Validity of International Studies of Reading Literacy: The Case of the IEA Reading Literacy Study. *Studies in Educational Evaluation,* **24**, 2, 99–125.

Beck A.T., Ward C.H., Mendelson M., Mock J., Erbaugh J., 1961. An Inventory for Measuring Depression. *Archives for General Psychiatry,* **4**, 561–571.

Becker B.J., 1994. Combining Significance Levels. In: *The Handbook of Research Synthesis.* Cooper H., Hedges L.V. (eds.). New York: Russell Sage Foundation.

Belson W., 1981. *The Design and Understanding of Survey Questions.* London: Gower.

Berry J.W., 1980. Introduction to Methodology. In: *Handbook of Cross-Cultural Psychology.* Triandis H.C., Berry J.W. (eds.). Boston: Allyn & Bacon.

Berry J.W., 1989. Imposed Etics–Emics–Derived Etics: The Operationalization of a Compelling Idea. *International Journal of Psychology,* **24**, 721–735.

Berry J.W. (ed.), 1997. *Handbook of Cross-Cultural Psychology.* 2nd edition. Boston: Allyn & Bacon.

Berry J.W., 1999. Emics and Etics: A Symbiotic Conception. *Culture and Psychology,* **5**, 165–171.

Biemer P.P., Groves R.M., Lyberg L.E., Mathiowetz N.A., Sudman S., 1991. *Measurement Errors in Surveys.* New York: Wiley.

Bien W., Marbach J., 1991. Haushalt - Verwandschaft - Beziehungen. Familienleben als Netzwerk. In: *Die Familie in Westdeutschland. Stabilität und Wandel familiärer Lebenformen.* Bertram H. (ed.). Opladen: Leske + Budrich.

Bijnen E.J., Van der Net T.Z., Poortinga Y.H., 1986. On Cross-Cultural Comparative Studies with the Eysenck Personality Questionnaire. *Journal of Cross-Cultural Psychology,* **17**, 3–16.

Billiet J., 1995. Church Involvement, Individualism and Ethnocentrism among Catholics: New Evidence of a Moderating Effect. *Journal of the Scientific Study of Religion,* **34**, 224–233.

Billiet J., Eisinga R., Scheepers P., 1996. Ethnocentrism in the Low Countries: A Comparative Perspective. *New Community,* **22**, 401–416.

Billiet J., Loosveldt G., Waterplas L., 1986. *Het survey-interview onderzocht: Effecten van het ontwerp en gebruik van vragenlijsten op de kwaliteit van de antwoorden.* Leuven: Sociologisch Onderzoeksinstituut KU Leuven.

Billiet J., McClendon M.J., 2000. Modeling Acquiescence in Measurement Models for Two Balanced Sets of Items. Structural Equation Modeling. *An Interdisciplinary Journal,* **7**, 608–629.

Bishop G.F., Tuchfarber A.J., Oldendick R.W., 1986. Opinions on Fictitious Issues: The Pressure to Answer Survey Questions. *Public Opinion Quarterly,* **50**, 240–250.

Bodenhausen G.V., 1992. Information-Processing Functions of Generic Knowledge Structures and Their Role in Context Effects in Social Judgements. In: *Context Effects in Social and Psychological Research.* Schwarz N., Sudman S. (eds.). New York: Springer.

Bohrenstedt G.W., Knoke D., 1988. *Statistics for Social Data Analysis.* 2nd edition. Itasca, IL: F.E. Peacock.

Bollen K.A., 1989. *Structural Equations with Latent Variables.* New York: Wiley.

Bollen K.A., Entwistle B., Alderson A.S., 1993. Macrocomparative Research Methods. *Annual Review of Sociology,* **19**, 321–351.

Bollen K.A., Long J.S., 1987. Tests for Structural Equation Models. *Sociological Methods and Research,* **2**, 123–131.

Bollen K.A., Long J.S. (eds.), 1993. *Testing Structural Equation Models.* Newbury Park, CA: Sage.

Bolton R.N., Bronkhorst T.M., 1996. Questionnaire Pretesting: Computer-Assisted Coding of Concurrent Protocols. In: *Answering Questions: Methodology for Determining Cognitive and Communicative Processes in Survey Research.* Schwarz N., Sudman S. (eds.). San Francisco: Jossey-Bass.

Bond M.H., Fu P.P., Pasa S.F., 2001. A Declaration of Independence for Editing a New International Journal of Cross Cultural Management? *International Journal of Cross Cultural Management,* **1**, 24–30.

Bond R., Smith P.B., 1996. Culture and Conformity: A Meta-Analysis of Studies Using Asch's (1952b, 1956) Line Judgement Task. *Psychological Bulletin,* **119**, 111–137.

Borg I., 1998. A Facet-Theoretical Approach to Item Equivalency. In: *ZUMA-Nachrichten Spezial No.3. Cross-Cultural Survey Equivalence.* Harkness J.A. (ed.). Mannheim: ZUMA.

Borg I., Groenen P., 1997. *Modern Multidimensional Scaling: Theory and Applications.* New York: Springer.

Bradburn N.M., 1983. Response Effects. In: *Handbook of Survey Research.* Rossi P.H., Wright J.D. (eds.). New York: Academic Press.

Bradburn N.M., Danis C., 1984. Potential Contributions of Cognitive Research to Questionnaire Design. In: *Cognitive Aspects of Survey Methodology: Building a Bridge Between Disciplines.* Jabine T., Straf M., Tanur J., Tourangeau R. (eds.). Washington, DC: National Academy Press.

Bradburn N.M., Frankel L.R., Baker R.P., Pergamit M.R., 1991. *Information Technology in Survey Research.* Chicago: NORC.

Bradburn N.M., Sudman S., 1979a. *Improving Interview Method and Questionnaire Design.* San Francisco: Jossey-Bass.

Bradburn N.M., Sudman S., 1979b. Reinterpreting the Marlowe-Crown Scale. In: *Improving Interview Method and Questionnaire Design.* Bradburn N.M., Sudman S. (eds.). San Francisco: Jossey-Bass.

Braun M., 1993. *Potential Problems of Functional Equivalence in ISSP 88 (Family and Changing Gender Roles).* Paper presented to the Annual ISSP Research Conference, Chicago, May, 1993.

Braun M., 1998. Gender Roles. In: *Comparative Politics. The Problem of Equivalence.* Van Deth J.W. (ed.). London, New York: Routledge, 111–134.

Braun M., forthcoming. *Funktionale Äquivalenz in interkulturell vergleichenden Umfragen. Mythos und Realität.*

Braun M., 1994. The International Social Survey Programme (ISSP). In: *Social Statistics and Social Reporting in and for Europe.* Flora P., Kraus F., Noll H.-H., Rothenbacher F. (eds.). Bonn: Informationszentrum Sozialwissenschaften.

Braun M., Müller W., 1997. Measurement of Education in Comparative Research. In: *Comparative Social Research.* Mjøset L., Engelstad F., Brochmann G., Kalleberg R., Leira A. (eds.). Greenwich: JAI Press.

Braun M., Scott J., Alwin D.F., 1994. Economic Necessity or Self-Actualization? Attitudes Towards Women's Labour-Force Participation in East and West Germany. *European Sociological Review*, **10**, 29–47.

Brauns H., Steinmann S., 1999. Educational Reform in France, West-Germany and the United Kingdom: Updating the CASMIN Educational Classification. *ZUMA-Nachrichten*, **44**, 7–44.

Brehm J., 1993. *The Phantom Respondents. Opinion Surveys and Political Representation.* Ann Arbor: University of Michigan Press.

Brewer M.B., Gardner W.L., 1996. Who Is This 'We'? Levels of Collective Identity and Self Representations. *Journal of Personality and Social Psychology*, **71**, 83–93.

Brewer M.B., Miller N., 1996. *Intergroup Relations.* Buckingham, England: Open University Press.

Brick J.M., Kalton G., 1996. Handling Missing Data in Survey Research. *Statistical Methods in Medical Research*, **5**, 215–238.

Brislin R.W., 1970. Back-Translation for Cross-Cultural Research. *Journal of Cross-Cultural Psychology*, **1**, 185–216.

Brislin R.W., 1976. Introduction. In: *Translation: Applications and Research.* Brislin R.W. (ed.). New York: Gardner.

Brislin R.W., 1980. Translation and Content Analysis of Oral and Written Material. In: *Handbook of Cross-Cultural Psychology, Vol. 2.* Triandis H.C., Berry J.W. (eds.). Boston: Allyn & Bacon.

Brislin R.W., 1986. The Wording and Translation of Research Instruments. In: *Field Methods in Cross-Cultural Research.* Lonner W.J., Berry J.W. (eds.). Newbury Park, CA: Sage.

Bronner A.E., 1988. Surveying Ethnic Minorities. In: *Sociometric Research.* Saris W.E., Gallhofer I.N. (eds.). London: Macmillan.

Bros L., De Leeuw E.D., Hox J.J., Kurvers G., 1995. Nonrespondents in a Mail Survey: Who Are They? In: *International Perspectives on Nonresponse.* Laakson S. (ed.). Helsinki: Statistics Finland (RR219).

Brown P., Levinson S.C., 1999. Politeness: Some Universals in Language Usage. In: *The Discourse Reader.* Jaworski A., Coupland N. (eds.). London: Routledge.

Bruner J.S., 1957. Going Beyond the Information Given. In: *Contemporary Approaches to Cognition.* Gruber H., Terrell G. (eds.). Cambridge, MA: Harvard University Press.

Bryant F.B., Wortman P.M., 1978. Secondary Analysis. The Case for Data Archives. *American Psychologist*, **33**, 381–387.

Bryk A.S., Raudenbush S.W., 1992. *Hierarchical Linear Models: Applications and Data Analysis.* Newbury Park, CA: Sage.

Bulmer M., 1983. Sampling. In: *Social Research in Developing Countries. Surveys and Censuses in the Third World.* Bulmer M., Warwick D.P. (eds.). New York: Wiley.

Bulmer M., Bales K., Sklar K. (eds.), 1991. *The Social Survey in Historical Perspective, 1880–1940.* Cambridge: Cambridge University Press.

Bulmer M., Warwick D.P. (eds.), 1983. Social Research in Developing Countries. London: UCL Press.

Byrne B.M., 1994. *Structural Equation Modelling with EQS and EQS/Windows: Basic Concepts, Application, and Programming.* Thousand Oaks, CA: Sage.

Byrne B.M., 1998. *Structural Equations Modelling with LISREL, PRELIS, and SIMPLIS: Basic Concepts, Applications, and Programming.* New York: Springer.

Byrne B.M., 2001. *Structural Equation Modelling with AMOS: Basic Concepts, Applications, and Programming.* Thousand Oaks, CA: Sage.

Byrne B.M., Shavelson R.J., Muthén B., 1989. Testing for the Equivalence of Factor Covariance and Mean Structures: The Issue of Partial Measurement Invariance. *Psychological Bulletin,* **105**, 456–466.

Camilli G., Shepard L.N., 1994. *Methods for Identifying Biased Test Items.* Thousand Oaks: Sage.

Campbell A., Converse P.E., Rodgers W.L., 1976. *The Quality of American Life: Perceptions, Evaluations, and Satisfactions.* New York: Russell Sage Foundation.

Campbell D.T., 1986. Science`s Social System of Validity-Enhancing Collective Believe Change and the Problems of the Social Sciences. In: *Metatheory in Social Science.* Fiske D.W., Shweder R.A. (eds.). Chicago: University of Chicago Press.

Campbell D.T., Fiske D.W., 1959. Convergent and Discriminant Validation by the Multimethod-Multitrait Matrix. *Psychological Bulletin,* **56**, 833–853.

Cannell C.F., Miller P.V., Oksenberg L., 1981. Research on Interviewing Techniques. In: *Sociological Methodology 1970.* Leinhardt S. (ed.). San Francisco: Jossey-Bass.

Carmines E.G., Zellerm R.A., 1979. *Reliability and Validity Assessment.* Newbury Park, CA: Sage.

Carr L.G., Krause N., 1978. Social Status, Psychiatric Symptomatology, and Response Bias. *Journal of Health and Social Behavior,* **19**, 86–91.

Carstensen L.L., Cone J.D., 1983. Social Desirability and the Measurement of Psychological Well-Being in Elderly Persons. *Journal of Personality,* **38**, 713–715.

Chan W., Ho R.M., Leung K., Cha D.K.S., Yung Y.-F., 1999. An Alternative Method for Evaluating Congruence Coefficients with Procrustes Rotation: A Bootstrap Procedure. *Psychological Methods,* **4**, 378–402.

Chen C., Lee S., Stevenson H.W., 1995. Response Style and Cross-Cultural Comparisons of Rating Scales Among East Asian and North American Students. *Psychonomic Science,* **6**, 170–175.

Chen S., Chaiken S., 1999. The Heuristic-Systematic Model in Its Broader Context. In: *Dual-Process Theories in Social Psychology.* Chaiken S., Trope Y. (eds.). New York: Guilford Press.

Cheung F.M., Leung K., Fan R.M., Song W.Z., Zhang J.X., Zhang J.P., 1996. Development of the Chinese Personality Assessment Inventory. *Journal of Cross-Cultural Psychology,* **27**, 181–199.

Cheung G.W., Rensvold R.B., 2000. Assessing Extreme and Acquiescence Response Sets in Cross-Cultural Research Using Structural Equations Modeling. *Journal of Cross-Cultural Psychology,* **31**, 187–212.

Chidambaram V.C., Cleland J.G., Verma V., 1980. *Some Aspects of World Fertility Survey Data Quality: A Preliminary Assessment.* Voorburg, The Netherlands: International Survey Statistical Institute. World Fertility Survey Comparative Studies 16, May.

Chun K.-T., Campbell J.B., Yoo J.H., 1974. Extreme Response Style in Cross-cultural Research: A Reminder. *Journal of Cross-Cultural Psychology,* **5**, 465–480.

Church A.H., 1993. Estimating the Effect of Incentives on Mail Survey Response Rates: A Meta-Analysis. *Public Opinion Quarterly,* **57**, 62–79.

Church A.T., 1987. Personality Research in a Non-Western Setting: The Philippines. *Psychology Bulletin,* **102**, 272–292.

Church A.T., Katigbak M.S., Reyes J.A.S., Jensen S.R., 1998. Language and Organization of Filipino Emotion Concepts: Comparing Emotion Concepts and Dimensions Across Cultures. *Cognition and Emotion,* **12**, 63–92.

Clark H.H., Schober M.F., 1992. Asking Questions and Influencing Answers. In: *Questions about Questions: Inquiries into the Cognitive Bases of Surveys*. Tanur J. (ed.). New York: Springer.

Clark T.N., 1998. Is There Really a New Political Culture: Evidence from Major Historical Developments of Recent Decades. In: *The New Political Culture*. Clark T.N., Hoffmann-Martinot V. (eds.). Boulder: Westview Press.

Clark T.N., Inglehart R., 1998. The New Political Culture: Changing Dynamics of Support for the Welfare State and Other Policies in Postindustrial Societies. In: *The New Political Culture*. Clark T.N., Hoffmann-Martinot V. (eds.). Boulder, CO: Westview Press.

Clarke H.D., Kronberg A., McIntyre C., Bauer-Kaase P., Kaase M., 1999. The Effect of Economic Priorities on the Measurement of Value Change. *American Political Science Review*, **93**, 637–647.

Cleary T.A., Hilton T.L., 1968. An Investigation of Item Bias. *Educational and Psychological Measurement*, **28**, 61–75.

Cliff N., 1959. Adverbs as Multipliers. *Psychological Review*, **66**, 27–44.

Clogg C.C., 1982. Using Association Models in Sociological Research: Some Examples. *American Journal of Sociology*, **88**, 114–134.

Clogg C.C., 1984. Some Statistical Models for Analyzing Why Surveys Disagree. In: *Surveying Subjektive Phenomena*. Turner C.F., Martin E. (eds.). New York: Russell Sage Foundation.

Cohen J., 1969 (2nd edition 1988). *Statistical Power Analysis for the Behavioral Sciences*. New York: Academic Press.

Cohen J., Cohen P., 1975. *Applied Multiple Regression/correlation Analysis for the Behavioral Sciences*. Hillsdale, NJ: Erlbaum.

Commandeur J.J.F., 1991. *Matching Configurations*. Leiden: DSWO Press Leiden.

Converse J.M., Presser S., 1986. *Survey Questions: Handcrafting the Standardized Questionnaire*. Beverly Hills, CA: Sage.

Cook T.D., Cooper H., Cordray D.S., Hartmann H., Hedges L.V., Light R.J., Louis T.A., Mosteller F., 1994. *Meta-Analysis for Explanation* . New York: Russell Sage Foundation.

Cooper H., Hedges L.V., 1994. Research Synthesis as a Scientific Enterprise. In: *The Handbook of Research Synthesis*. Cooper H., Hedges L.V. (eds.). New York: Russell Sage Foundation.

Cornell J., Mulrow C., 1999. Meta-Analysis. In: *Research Methodology in the Social, Behavioral, and Life Science*. Ader H.J., Mellenbergh G.J. (eds.). London: Sage.

Cortese M., Smyth P., 1979. A Note on the Translation to Spanish of Measure of Acculturation. *Hispanic Journal of Behavioral Sciences*, **1**, 65–68.

Cotter P.R., Cohen J., Coulter P.B., 1982. Race-of-Interviewer Effects in Telephone Interviews. *Public Opinion Quarterly*, **46**, 278–284.

Couper M.P., Groves R.M., 1992. The Role of Interviewer in Survey Participation. *Survey Methodology*, **18**, 263–277.

Couper M.P., Singer E., Kulka R.A., 1998. Participation in the 1990 Decennial Census: Politics, Privacy, Pressures. *American Politics Quarterly*, **26**, 59–80.

Cox T.F., Cox M.A.A., 1994. *Multidimensional Scaling*. London: Chapman & Hall.

Crandall V.C., Crandall V.J., 1965. A Children's Social Desirability Questionnaire. *Journal of Consulting Psychology*, **29**, 27–36.

Crespi L.P., 1981. *Semantic Guidelines to Better Survey Reportage*. Office of Research, International Communication Agency, Memorandum, August 11.

Crino M.D., Svoboda M., Rubenfield S., White M.C., 1983. Data on the Marlowe-Crowne and Edwards Social Desirability Scales. *Psychological Reports,* **53**, 963–968.

Cronbach L.J., Meehl P.E., 1955. Construct Validity in Psychological Tests. *Psychological Bulletin,* **52**, 281–302.

Crowne D.P., Marlowe D., 1960. A New Scale of Social Desirability Independent of Psychopathology. *Journal of Consulting Psychology,* **24**, 349–354.

Crowne D.P., Marlowe D., 1964. *The Approval Motive.* New York: Wiley.

D'Andrade R., 1987. A Folk Model of Mind. In: *Cultural Models in Language and Thought.* Holland D., Quinn N. (eds.). Cambridge: Cambridge University Press.

Daamen D.D.L., De Bie S.E., 1992. Serial Context Effects in Survey Items. In: *Context Effects in Social and Psychological Research.* Schwarz N., Sudman S. (eds.). New York: Springer.

Dale A., Arber S., Procter M., 1988. *Doing Secondary Analysis.* London: Unwin Hyman.

Daly S.P., 1996. Toward a Comprehensive Model of Marketing Channels Behavior and Performance: A Meta-Analytical Test of the Behavioral and Cognitive Psychology Paradigms. *International Section A: Humanities and Social Sciences,* **57**, 25–76.

Darity W.A., Turner C.B., 1972. Family Planning, Race Consciousness and the Fear of Genocide. *American Journal of Public Health,* **62**, 1454–1459.

Data Documentation Initiative (DDI), 2000. *Final Report to National Science Foundation "Electronic Preservation of Data Documentation: Complementary SGML and Image Capture".* ISR: Ann Arbor, MI. www.icpsr.umich.edu/DDI/codebook/papers/papers.html.

Davis D.W., 1997. Nonrandom Measurement Error and Race of Interviewer Effects Among African Americans. *Public Opinion Quarterly,* **61**, 183–207.

Davison M., 1983. *Multidimensional Scaling.* New York: Wiley.

Day S.X., Rounds J., 1998. Universality of Vocational Interest Structure Among Racial and Ethnic Minorities. *American Psychologist,* **53**, 728–736.

De Grott A., Koot H.M., Verhulst F.C., 1994. Cross-Cultural Generalizability of the Child Behavior Checklist Cross-Informant Syndromes. *Psychological Assessment,* **6**, 225–230.

De Heer W., 1999a. International Response Trends: Results of an International Study. *Journal of Official Statistics,* **15**, 129–142.

De Heer W., 1999b. *Survey Practices in European Countries.* Statistics Netherlands: Report to the Eurolit Expertgroup.

De Heer W., 2000. Survey Practices in European Countries. In: *Measuring Adult Literacy.* London: Office of National Statistics.

De Leeuw E.D., 1992. *Data Quality in Mail, Telephone, and Face-to-Face Surveys.* Amsterdam: TT-Publikaties.

De Leeuw E.D., 2001. Reducing Missing Data in Surveys: An Overview of Methods. *Quality & Quantity,* **35**, 147–160.

De Leeuw E.D., Collins M., 1997. Data Collection and Survey Quality: An Overview. In: *Survey Measurement and Process Quality.* Lyberg L.E., Biemer P.P., Collins M., De Leeuw E.D., Dippo C., Schwarz N., Trewin D. (eds.). New York: Wiley.

De Leeuw E.D., De Heer W., 2001. Trends in Household Survey Nonresponse: A Longitudinal and International Comparison. In: *Survey Nonresponse.* Groves R.M., Dillman D.A., Eltinge J.L., Little R.J.A. (eds.). New York: Wiley.

De Leeuw E.D., Hox J.J., 1988. The Effects of Response-Stimulating Factors on Response Rates and Data Quality in Mail Surveys: A Test of Dillman's Total Design Method. *Journal of Official Statistics,* **4**, 241–249.

De Leeuw E.D., Nicholls II W.L., 1996. Technological Innovations in Data Collection: Acceptance, Data Quality and Costs. *Sociological Research Online*, **1**, www.socresonline.org.uk/socresonline/1/4/leeuw.html.

De Leeuw E.D., Van der Zouwen J., 1988. Data Quality in Telephone and Face-to-Face Surveys: A Comparative Analysis. In: *Telephone Survey Methodology*. Groves R.M., Biemer P.P., Lyberg L.E., Massey J.T., Nicholls II W.L., Wakesberg J. (eds.). New York: Wiley.

De Leeuw E.D., Van Leeuwen R., 1999. *I am not Selling Anything: Experiments in Telephone Introductions*. Paper presented at the International Conference on Survey Nonresponse, Portland, October, 1999.

De Leeuw E.D., Mellenbergh G.J., Hox J.J., 1996. The Influence of Data Collection Method on Structural Models. *Sociological Methods and Research*, **24**, 443–472.

De Vera M.V., 1985. *Establishing Cultural Relevance and Measurement Equivalence Using Emic and Etic Items*. Unpublished dissertation. Urbana: University of Illinois.

De Vries R.E., 1998. Can the Library and Data Archive Meet in the Active Support of Research in Social Sciences? The Case of Ilses. *IASSIST Quarterly*, **22**, 10–11.

De Witte H., Verbeeck G., 1998. Een rechts radicalisme met twee snelheden. Hoe het verschil verklaren tussen het succes van uiterst partijen in Vlaanderen en Franstalig België. In: *Racisme: Een element in het conflict tussen Vlamingen en Franstaligen*. Morelli A. (ed.). Berchem: EPO.

Delbanco S., Lundy J., Hoff T., Parke M., Smith M.D., 1997. Public Knowledge and Perceptions about Unplanned Pregnancies in Three Countries. *Family Planning Perspectives*, **29**, 70–75.

DeMaio T.J., 1984. Social Desirability and Survey Measurement: A Review. In: *Surveying Subjective Phenomena*. Turner C.F. (ed.). New York: Russell Sage Foundation.

Devereux T., Hoddinott J. (eds.), 1992. *Fieldwork in Developing Countries*. New York: Harvester Wheatsheaf.

Deville Y.-C., 1991. A Theory of Quota Surveys. *Survey Methodology*, **17**, 163–181.

Dijksterhuis G.B., Van Buuren S., 1989. *PROCRUSTES-PC v2.0 User Manual*. Utrecht: OP&P Software.

Dijkstra W., Van der Zouwen J., 1982. *Response Behaviour in the Survey-Interview*. London: Academic Press.

Dillman D.A., 1978a. *Mail and Telephone Surveys: The Total Design Method*. New York: Wiley.

Dillman D.A., 1978b. The Design and Administration of Mail Surveys. *Annual Review of Sociology*, **17**, 225–249.

Dillman D.A., 1991. *The Design and Administration of Mail Surveys. The Tailored Design Method*. New York: Wiley.

Dillman D.A., 2000. *Mail and Internet Surveys: The Tailored Design Method*. New York: Wiley.

Dillman D.A., Sangster R.L., Tarnai J., Rockwood T., 1996. Understanding Differences in People's Answers to Telephone and Mail Surveys. In: *New Directions for Evaluation Series, 70 (Advances in Survey Research)*. Braverman M.T., Slater J.K. (eds.). San Francisco: Jossey-Bass.

Dogan M., Pélassy D., 1984. *How to Compare Nations: Strategies in Comparative Politics*. New Jersey: Chatham House Publishers.

Dohrenwend B., 1966. Social Status and Psychological Disorder: An Issue of Substance and an Issue of Method. *American Sociological Review*, **31**, 14–34.

Doob L.W., 1968. Tropical Weather and Attitude Surveys. *Public Opinion Quarterly*, **32**, 423–430.

Dorans N.J., Kulick E., 1986. Demonstrating the Utility of the Standardization Approach to Assessing Unexpected Differential Item Performance on the Scholastic Aptitude Test. *Journal of Educational Measurement*, **23**, 355–368.

Draper D., Gaver D.P., Goel P.K.Jr., Greenhouse J.B., Hedges L.V., Morris C.N., Tucker J.R., Waternaux C.M., 1992. *Combining Information. Statistical Issues and Opportunities for Research*. Washington, DC: National Academy Press.

Duncan O.D., 1961. "A Socio-Economic Index for all Occupations" and "Properties and Characteristics of the Socioeconomic Index". In: *Occupations and Social Status*. Reiss A.J. (ed.). Glencoe, IL: Free Press.

Dyal J.A., 1984. Cross-Cultural Research with the Locus of Control Construct. In: *Research with the Locus of Control Construct (Vol.3)*. Lefcourt H.M. (ed.). New York: Academic Press.

Earley C., 1989. Social Loafing and Collectivism: A Comparison of the United States and the People's Republic of China. *Administrative Science Quarterly*, **34**, 565–581.

Edwards A.L., 1953. The Relationship Between the Judged Desirability of a Trait and the Probability That the Trait Will be Endorsed. *Journal of Applied Psychology*, **37**, 90–93.

Edwards A.L., 1957. *The Social Desirability Variable in Personality Assessment and Research*. New York: Druiden Press.

Edwards D., Riordan S., 1994. Learned Resourcefulness in Black and White South African University Students. *Journal of Social Psychology*, **134**, 665–675.

Ekman G., 1955. Dimensions of Emotion. *Acta Psychologica*, **11**, 279–288.

Elder G.H., Pavalko E.K., Clipp E.C., 1993. *Working with Archival Data. Studying Lives*. Newbury Park, CA: Sage.

Elley W.B., 1992. *How in the World Do Students Read?* Den Haag, The Netherlands: IEA.

Ellis B.B., Becker P., Kimmel H.D., 1993. An Item Response Theory Evaluation of an English Version of the Trier Personality Inventory (TPI). *Journal of Cross-Cultural Psychology*, **24**, 133–148.

Embretson S.E., 1983. Construct Validity: Construct Representation versus Nomothetic Span. *Psychological Bulletin*, **93**, 179–197.

Entwistle B., Mason W.M., 1985. Multilevel Effects of Socioeconomic Development and Family Planning in Children Ever Born. *American Journal of Sociology*, **91**, 616–649.

Erikson R., Goldthorpe J.H., 1992. *The Constant Flux: A Study of Class Mobility in Industrial Societies*. Oxford: Clarendon Press.

Erikson R., Goldthorpe J.H., Portocarero L., 1979. Intergenerational Class Mobility in Three Western European Societies: England, France and Sweden. *British Journal of Sociology*, **30**, 415–451.

Espe H., 1985. A Cross-Cultural Investigation of the Graphic Differential. *Journal of Psycholinguistic Research*, **14**, 97–111.

Esser H., 1986. Können Befragte lügen? Zum Konzept des "wahren Wertes" im Rahmen der handlungstheoretischen Erklärung von Situationseinflüssen bei der Befragung. *Kölner Zeitschrift für Soziologie und Sozialpsychologie*, **38**, 314–336.

Esser H., 1990. "Habits", "Frames" und "Rational Choice": Die Reichweite von Theorien der rationalen Wahl (am Beispiel der Erklärung des Befragtenverhaltens). *Zeitschrift für Soziologie*, **4**, 231–247.

Esser H., 1993. Response Set: Habit, Frame or Rational Choice. In: *New Directions in Attitude Measurement*. Krebs D., Schmidt P. (eds.). Berlin: de Gruyter.

European Science Foundation, 1999. *The European Social Survey (ESS) — A Research Instrument for Social Sciences in Europe*. Strasbourg, France: ESF.

Eurostat, 1996. *European Community Household Panel (ECHP): Survey Methodology and Implementation Vol. 1 — Survey Questionnaires*. Luxembourg: Eurostat.

Eurostat, 1998. *Labour Force Survey. Methods and Definitions*. Luxembourg: Eurostat.

Eurostat, 1999. *European Community Household Panel (ECHP): Longitudinal Users` Database Manual. Waves 1, 2 and 3*. Luxembourg: Eurostat.

Eurostat, 2000. *Harmonisation of Recommended Core Units, Variables and Classifications*. Luxembourg: Eurostat.

Everitt B., 2001. *Cluster Analysis*. 2nd edition. London: Heinemann Educational Books.

Eysenck H.J., Eysenck S.B.G., 1964. *Manual of the Eysenck Personality Inventory*. London: University Press.

Eysenck H.J., Eysenck S.B.G., 1975. *Manual of the Eysenck Personality Questionnaire*. London: Hodder and Stoughton.

Eysenck H.J., Eysenck S.B.G., 1983. Recent Advances in the Cross-Cultural Study of Personality. In: *Advances on Personality Assessment*. Butcher J.N., Spielberger C.D. (eds.). Hillsdale, NJ: Erlbaum.

Fabricius W.V., Noyes C.R., Bigler K.D., Alexander J.M., 1994. The Organization of Mental Verbs and Folk Theories. *Journal of Memory & Language*, 33, 376–395.

Featherman D.L., 1993. What Does Society Need from Higher Education? *Items*, 47, 38–43.

Fendrich M., Johnson T.P., Sudman S., Wislar J.S., Spiehler V., 1999. Validity of Drug Use Reporting in a High-Risk Community Sample: A Comparison of Cocaine and Heroin Survey Reports with Hair Tests. *American Journal of Epidemiology*, 149, 955–962.

Fennema M., 1997. Some Conceptual Issues and Problems in the Comparison of Anti-Immigrant Parties in Western Europe. *Party Politics*, 3, 473–492.

Fennema M., Tillie J., 1994. *The Extreme Right as Perverse Cargo Cult. Ethnic Nationalsim, Social Efficacy and the Extreme Right in The Netherlands*. Paper prepared for the workshop 'electoral research', Politicologenetmaal, Soesterberg, May, 1994.

Fink A., 1998. *Conducting Research Literature Reviews*. Newbury Park, CA: Sage.

Finkel S.E., Guterbock T.M., Borg M.J., 1991. Race-of-Interviewer Effects in a Presidential Poll: Virginia 1989. *Public Opinion Quarterly*, 55, 313–330.

Fisher G., 1967. The Performance of Male Prisoners on the Marlowe-Crowne Social Desirability Scale: II. Differences as a Function of Race and Crime. *Journal of Clinical Psychology*, 23, 473–475.

Fiske A.P., Kitayama S., Markus H.R., Nisbett R.E., 1998. The Cultural Matrix of Social Psychology. In: *The Handbook of Social Psychology*. Gilbert D.T., Fiske S.T., Lindzey G. (eds.). 4th edition. Boston: McGraw-Hill.

Fiske S.T., Taylor S.E., 1991. Social Cognition. 2nd edition. New York: McGraw-Hill.

Foddy W., 1993. *Constructing Questions for Interviews and Questionnaires: Theory and Practice in Social Research*. Cambridge: Cambridge University Press.

Fontaine J., 1999. *Culturele vertekening in Schwartz`waardeninstrument: Een exemplarisch onderzoek naar culturele vertekening in sociaal- psychologische en persoonlijkheidsvragenlijsten [Cultural Bias in Schwartz`s Value Instrument: An Exemplary Study into Bias in Social-Psychological and Personality Questionnaires]*. University of Leuven, Belgium.

Fontaine J., Poortinga Y.H., Setiadi B., Markam S.S., 1998, unpublished work. *Invariance of Cognitive Emotion Structures and Location Differences of Specific Emotion Terms: 'Shame' and 'Guilt' in Indonesian and Dutch Emotion Structures.*

Fontaine J., Poortinga Y.H., Setiadi B., Suprapti S.M., 1996. The Cognitive Structure of Emotions in Indonesia and The Netherlands: A Preliminary Report. In: *Key Issues in Cross-Cultural Psychology: Selected Papers from the Twelfth International Congress of the International Association for Cross-Cultural Psychology.* Grad H., Blanca A., Georgas J. (eds.). Lisse, The Netherlands: Swets & Zeitlinger.

Fontaine J., Poortinga Y.H., Setiadi B., Suprapti S.M. 2002. Cognitive Structure of Emotion Terms in Indonesia and The Netherlands. *Cognition and Emotion, 16,* 61–86.

Ford J.K., Kraiger K., Schechtman S.L., 1986. Study of Race Effects in Objective Indices and Subjective Evaluations of Performance: A Meta-Analysis of Performance Criteria. *Psychological Bulletin, 99,* 330–337.

Formann A.K., Piswanger K., 1979. *Wiener Matrizen-Test. Ein Rasch-skalierter sprachfreier Intelligenztest [The Viennese Matrices Test. A Rasch-Calibrated Non-Verbal Intelligence Test].* Weinheim, Germany: Beltz Test.

Forsyth B., Lessler J.T., 1991. Cognitive Laboratory Methods: A Taxonomy. In: *Measurement Errors in Surveys.* Biemer P.P., Groves R.M., Lyberg L.E., Mathiowetz N.A., Sudman S. (eds.). New York: Wiley.

Foster K., Bushnell D., 1994. *Non-Response Bias on Government Survey in Great Britain.* Paper presented at the Fifth International Workshop on Household Survey Non-Response, Ottawa, Canada, September 1994.

Fowler F.J.Jr., 1991. Reducing Interviewer-Related Error Through Interviewer Training, Supervision, and Other Means. In: *Measurement Errors in Surveys.* Biemer P.P., Groves R.M., Lyberg L.E., Mathiowetz N.A., Sudman S. (eds.). New York: Wiley.

Fowler F.J.Jr., 1992. How Unclear Terms Affect Survey Data. *Public Opinion Quarterly, 56,* 218–231.

Fowler F.J.Jr., 1993. *Survey Research Methods.* 2nd edition. Newbury Park, CA: Sage.

Fowler F.J.Jr., 1995. *Improving Survey Qustions: Design and Evaluation.* Thousand Oaks, CA: Sage.

Fowler F.J.Jr., Cannell C.F., 1996. Using Behavioral Coding to Identify Cognitive Problems with Survey Questions. In: *Answering Questions: Methodology for Determining Cognitive and Communicative Processes in Survey Research.* Schwarz N., Sudman S. (eds.). San Francisco: Jossey-Bass.

Fowler F.J.Jr., Roman A., Di Z.X., 1998. Mode Effects in a Survey of Medicare Prostate Surgery Patients. *Public Opinion Quarterly, 62,* 29–46.

Fox J., 1997. *Applied Regression Analysis, Linear Models and Related Methods.* Thousand Oaks, CA: Sage.

Foy P., Martin M.O., Kelly D.L., 1996. Sampling. In: *Third International Mathematics and Science Study: Quality Assurance in Data Collection.* Martin M.O., Mullis I.V.S. (eds.). Chestnut Hill, MA: Boston College.

Frankfort-Nachmias C., Leon-Guerrero A., 1999. *Social Statistics for a Diverse Society.* 2nd edition. Thousand Oaks, CA: Sage.

Frazer L., Meredith L., 2000. *Questionnaire Design & Administration.* Brisbane, Australia: Wiley.

Frey F.W., 1963. Survey Peasant Attitudes in Turkey. *Public Opinion Quarterly, 27,* 335–355.

Frey F.W., 1970. Cross-Cultural Survey Research. In: *The Methodology of Comparative Research.* Holt R.T., Turner J.E. (eds.). New York: Free Press.

Fu G., Lee K., Cameron C.A., Xu F., 2001. Chinese and Canadian Adults 'Categorization and Evaluation of Lie- and Truth-Telling about Prosocial and Antisocial Behaviors. *Journal of Cross-Cultural Psychology*, **32**, 720–727.

Gamson W., 1968. *Power and Discontent*. Homewood, IL: Dorsey Press.

Gandek B., Ware J.E.Jr., Aaronson N.K., Alonso J., Apolone G., Bjorner J.B., Brazier J., Bullinger M., Fukuhara S., Kaasa St., Leplege A., Sullivan M., 1998. Test of Data Quality, Scaling Assumptions, and Reliability of the SF-36 in Eleven Countries: Results from the IQOLA Project. *Journal of Clinical Epidemiology*, **51**, 1149–1158.

Ganzeboom H.B.G., De Graaf P.M., Treiman D.J., 1992. A Standard International Socio-Economic Index of Occupational Status. *Social Science Research*, **21**, 1–56.

Ganzeboom H.B.G., Treiman D.J., 1996. Internationally Comparable Measures of Occupational Status for the 1988 International Standard Classification of Occupations. *Social Science Research*, **25**, 201–239.

Gardner W.L., Gabriel S., Lee A., 1999. "I" Value Freedom but "We" Value Relationship: Self-Construal Priming Mirrors Differences in Judgment. *Psychological Science*, **10**, 321–326.

Garland R., 1991. The Mind-Point of a Rating Scale: Is it Desirable? *Marketing Bulletin*, **2**, 66–70.

Gass S.M., Varonis E.M., 1991. Miscommunication in Nonnative Speaker Discourse. In: *Miscommunication and Problematic Talk*. Coupland N., Giles H., Wiemann J.M. (eds.). Newbury Park, CA: Sage, 121–145.

Gauthier A.H., 2000. *The Promises of Comparative Research*. Paper prepared for the European Panel Analysis Group.

Gehm T.L., Scherer K.R., 1988. Factors Determining the Dimensions of Subjective Emotional Space. In: *Facets of Emotion: Recent Research*. Scherer K.R. (ed.). Hillsdale, NJ: Erlbaum.

Gehring U.W., Weins C., 1998. *Grundkurs Statistik für Politologen*. Opladen: Westdeutscher Verlag.

Geis A., Hoffmeyer-Zlotnik J.H.P., 2000. Stand der Berufsvercodung. *ZUMA-Nachrichten*, **47**, 103–128.

Gendre F., Gough H.G., 1982. Note cernant les pourcentages de responses a l'Adjective Check List de H. Gough pour des echantillons Francais et Americains. [Note Concerning the Percentages of Responses to the Adjective Check List of H. Gough for French and American Samples]. *Revue de Psychologie Appliquee*, **32**, 81–90.

Gerber E.R., 1999. The View from Antropology: Ethnography and the Cognitive Interview. In: *Cognition and Survey Research*. Sirken M.G., Herrmann D.J., Schechter S., Schwarz N., Tanur J.M., Tourangeau R. (eds.). New York: Wiley.

Ghorpade J., Hattrup K., Lackritz J.R., 1999. The Use of Personality Measures in Cross-Cultural Research: A Test of Three Personality Scales Across Two Countries. *Journal of Applied Psychology*, **84**, 670–679.

Gile D., 1995. *Basic Concepts and Models for Interpreter and Translator Training*. Amsterdam: John Benjamins.

Glass G.V., 1976. Primary, Secondary, and Meta-Analysis of Research. *Educational Researcher*, **5**, 3–8.

Glass G.V., Smith M.L., 1979. Meta-Analysis of Research on Class Size and Achievement. *Educational Evaluation and Policy Analysis*, **1**, 2–16.

Glass G.V., Smith M.L., McGaw B., 1981. *Meta-Analysis in Social Research*. Beverly Hills, CA: Sage.

Glover J., 1996. Epistemological and Methodological Considerations in Secondary Analysis. In: *Cross-National Research Methods in the Social Sciences*. Hantrais L., Mangen S. (eds.). London, New York: Pinter.

Godambe V.P., 1955. A Unified Theory of Sampling from Finite Populations. *Journal of the Royal Statistical Society*, **17**, 269–278.

Goldstein H., 1987. *Multilevel Models in Educational and Social Research*. London: Griffin.

Goldstein H., 2000. A Commentary on the Scaling and Data Analysis. In: *Measuring Adult Literacy. The International Adult Literacy Survey (IALS) in the European Context*. Carey S. (ed.). London: Office for National Statistics.

Goodman L.A., 1974a. Exploratory Latent Structure Analysis Using Both Identifiable and Unidentifiable Models. *Biometrika*, **61**, 215–231.

Goodman L.A., 1974b. The Analysis of Systems of Qualitative Variables Where Some of the Variables Are Unobserved. Part I — A Modified Latent Structure Approach. *American Journal of Sociology*, **79**, 1179–1259.

Goodwin R., Lee I., 1994. Taboo Topics Among Chinese and English Friends. A Cross-Cultural Comparison. *Journal of Cross-Cultural Psychology*, **25**, 325–338.

Gough H.G., Heilbrun A.B.Jr., 1980. *The Adjective Check List Manual*. Palo Alto, CA: Consulting Psychologists Press.

Gouldner A., 1960. The Norm of Reciprocity. *American Sociological Review*, **25**, 161–179.

Gove W.R., Geerken M.R., 1977. Response Bias in Surveys of Mental Health: An Empirical Investigation. *American Journal of Sociology*, **82**, 1289–1317.

Gove W.R., McCorkel J., Fain T., Hughes M.D., 1976. Response Bias in Community Surveys of Mental Health: Systematic Bias or Random Noise? *Social Science & Medicine*, **10**, 497–502.

Goyder J., 1987. *The Silent Minority*. Cambridge: Blackwell.

Green M.C., Krosnick J.A., 2001. Comparing Telephone and Face-to-Face Interviewing in Terms of Data Quality: The 1982 National Election Studies Method Comparison Project. In: *Seventh Conference on Health Survey Research Methods*. Cynamon M.L., Kulka R.A. (eds.). Maryland: Hyattsville.

Greenfield P.M., 1997. You Can't Take It With You: Why Ability Assessments Don't Cross Cultures. *American Psychologists*, **52**, 1115–1124.

Greenleaf E.A., 1992. Measuring Extreme Response Style. *Public Opinion Quarterly*, **56**, 328–351.

Grice H.P., 1975. Logic and Conversation. In: *Syntax and Semantics; 3. Speech Acts*. Cole P., Morgan J.L. (eds.). New York: Academic Press.

Grill J.J., Bartel N.R., 1977. Language Bias in Tests: ITPA Grammatic Closure. *Journal of Learning Disabilities*, **10**, 229–235.

Grimshaw A.D., 1973. Comparative Sociology: In What Ways Different from Other Sociologies. In: *Comparative Social Research: Methodological Problems and Strategies*. Armer M., Grimshaw A.D. (eds.). New York: Wiley.

Groves R.M., 1989. *Survey Errors and Survey Costs*. New York: Wiley.

Groves R.M., 1991. Measurement Error Across Disciplines. In: *Measurement Errors in Surveys*. Biemer P.P., Groves R.M., Lyberg L.E., Mathiowetz N.A. (eds.). New York: Wiley.

Groves R.M., 1996. How We Know They Think Is Really What They Think. In: *Answering Questions — Methodology for Determining Cognitive and Communicative Processes in Survey Research*. Schwarz N., Sudman S. (eds.). San Francisco: Jossey-Bass.

Groves R.M., 1999. Survey Error Models and Cognitive Theories of Response Behavior. In: *Cognition and Survey Research*. Sirken M.G., Herrmann D.J., Schechter S., Schwarz N., Tourangeau R. (eds.). New York: Wiley.

Groves R.M., Couper M.P., 1998. *Nonresponse in Household Interview Surveys*. New York: Wiley.

Groves R.M., Kahn R.L., 1979. *Surveys by Telephone: A National Comparison with Personal Interviews*. New York: Academic Press.

Groves R.M., McGonagle K.A., 2000, unpublished paper. *A Theory-Guided Training Protocol Regarding Survey Participation*. Ann Arbor: University of Michigan.

Groves R.M., Cialdini R.B., Couper M.P., 1992. Understanding the Decision to Partcipate in a Survey. *Public Opinion Quarterly*, **56**, 475–495.

Groves R.M., Raghunathan T.E., Couper M.P., 1995. *Evaluating Statistical Adjustments for Unit Nonresponse in a Survey of the Elderly*. Paper presented at the Sixth International Workshop on Household Survey Nonresponse, Helsinki, October, 1995.

Grube J.W., 1997. Monitoring Youth Behavior in Response to Structural Changes. Alternative Approaches for Measuring Adolescent Drinking. *Evaluation Review*, **21**, 231–265.

Grunert S.C., Muller T.E., 1996. Measuring Values in International Settings: Are Respondents Thinking 'Real' Life or "Ideal Life". *Journal of International Consumer Marketing*, **8**, 169–185.

Gudykunst W.B., Kim Y.Y. (eds.), 1992. *Communicating with Strangers: An Approach to Intercultural Communication*. 2nd edition. New York: McGraw-Hill.

Guillemin F., Bombardier C., Beaton D., 1993. Cross-Cultural Adaptation of Health-Related Quality of Life Measures: Literature Review and Proposed Guidelines. *Journal of Clinical Epidemiology*, **46**, 1417–1432.

Gutknecht C., Rölle L., 1996. *Translating by Factors*. Albany: State University of New York Press.

Häder S., Gabler S., 1997. Deviations from the Population and Optimal Weights. In: *ZUMA-Nachrichten Spezial No.2. Eurobarometer — Measurements for Opinions in Europe*. Saris W.E., Kaase M. (eds.). Mannheim: ZUMA.

Haberman S.J., 1979. *Analysis of Qualitative Data Vol. 2 New Developments*. New York: Academic Press.

Haberstroh S., Oyserman D., Schwarz N., Kühnen U., Ji L.J., 2002. Is the Independent Self More Sensitive to Question Context than the Independent Self? Self-Construal and the Observation of Conversational Norms. *Journal of Experimental Social Psychology*, **38**, 323–329.

Hagenaars J.A., 1990. *Categorical Longitudinal Data: Loglinear Panel, Trend and Cohort Analysis*. London: Sage.

Hagenaars J.A., 1993. *Loglinear Models with Latent Variables*. New York: Wiley.

Hakel M.D., 1968. How Often Is Often? *American Psychologists*, **23**, 533–534.

Hakim C., 1982. *Secondary Analysis in Social Research*. London: Allen & Unwin.

Hall E.T., 1976. *Beyond Culture*. New York: Doubleday.

Hambleton R.K., 1993. Translating Achievement Tests for Use in Cross-National Studies. *European Journal of Psychology Assessment (Bulletin of the International Test Commission)*, **9**, 57–68.

Hambleton R.K., 1994. Guidelines for Adapting Educational and Psychological Tests: A Progress Report. *European Journal of Psychology Assessment (Bulletin of the International Test Commission)*, **10**, 229–244.

Hambleton R.K., forthcoming. Issues, Designs, and Technical Guidelines for Adapting Tests in Multiple Languages and Cultures. In: *Adapting Educational and Psychological Tests in Cross-Cultural Assessment*. Hambleton R.K., Merenda P.F., Spielberger C.D. (eds.). Hillsdale, NJ: Erlbaum.

Hambleton R.K., Swaminathan H., 1985. *Item Response Theory: Principles and Applications.* Dordrecht: Kluwer.

Hambleton R.K., Merenda P.F., Spielberger C.D. (eds.), forthcoming. *Adapting Educational Tests and Psychological Tests for Cross-Cultural Assessment.* Hillsdale, NJ: Erlbaum.

Hambleton R.K., Patsula L., 1998. Adapting Tests for Use in Multiple Languages and Cultures. *An International and Interdisciplinary Journal of Quality-of-Life Measurement,* **45**, 153–171.

Hansen M.H., Hurwitz W.N., Madow W.G., 1953. *Sample Survey Methods and Theory. Vol. I, Methods and Applications. Vol. II, Theory.* New York: Wiley.

Harkness J.A., 1996. *Thinking Aloud about Survey Translation.* Paper presented at the International Sociological Association Conference on Social Science Methodology, Colchester, England, August 1996.

Harkness J.A. (ed.), 1998. *ZUMA-Nachrichten Spezial No. 3. Cross-Cultural Survey Equivalence.* Mannheim: ZUMA.

Harkness J.A., 1999. In Pursuit of Quality: Issues for Cross-National Survey Research. *International Journal of Social Research Methodology,* **2**, 125–140.

Harkness J.A., 2001. German Respondents Comments on the 1998 ISSP Module: What Do They Say? And What, if Anything, Do They Tell Us? In: *Observatoire des Religions en Suisse*, Working Paper, Lausanne University.

Harkness J.A., Schoua-Glusberg A., 1998. Questionnaires in Translation. In: *ZUMA-Nachrichten Spezial No. 3. Cross-Cultural Survey Equivalence.* Harkness J.A. (ed.). Mannheim: ZUMA.

Harkness J.A., Langfeldt B., Scholz E., 2000. *The International Social Study Programme. Study Monitoring Report 1996–1998.* ZUMA: Mannheim.

Harkness J.A., Langfeldt B., Scholz E., 2001. *ISSP Study Monitoring 1996–1998. Reports to the ISSP General Assembly on Monitoring Work Undertaken for the ISSP by ZUMA, Germany.* Mannhein: ZUMA.

Harkness J.A., Mohler P.Ph., Thomas R., 1997. *General Report on Study Programme for Quantitative Research (SPQR).* Mannheim: ZUMA.
www.gesis.org/en/cooperation/research/compass/spqr.htm

Harkness J.A., Hönig H., Pennell B.-E., Kussmaul P., Schoua-Glusberg A., forthcoming. Guidelines and Materials for Translating Questionnaires (working title). *European Social Survey Report to the European Commission.* Mannheim: ZUMA.

Harman H.H., 1976. *Modern Factor Analysis.* 3rd edition. Chicago: University of Chicago Press.

Harris-Kojetin B., Tucker C., 1999. Exploring the Relation of Economic and Political Conditions with Refusal Rates to the Government Survey. *Journal of Official Statistics,* **15**, 167–184.

Hayashi E., 1992. Belief Systems: The Way of Thinking, and Sentiments of Five Nations. *Behaviormetrica,* **19**, 127–170.

Hayashi C., Scheuch E.K., 1996. *Quantitative Social Research in Germany and Japan.* Opladen: Leske + Budrich.

Hayashi C., Suzuki T., Sasaki M., 1992. *Data Analysis for Comparative Social Research: International Perspectives.* Amsterdam: North-Holland.

Heath A., Martin J., 1997. Why Are There so Few Formal Measuring Instruments in Social and Political Research? In: *Survey Measurement and Process Quality.* Lyberg L.E., Biemer P.P., De Leeuw E.D., Dippo C., Schwarz N., Trewin D. (eds.). New York: Wiley.

Hedges L.V., Olkin I., 1985. *Statistical Methods for Meta-Analysis.* Boston: Academic Press.

Heinemann W., Viehweger D., 1991. *Textlinguistik: Eine Einführung. Reihe Germanistische Linguistik.* Tübingen: Niemeyer.

Heise D.R., Bohrnstedt G.W., 1970. Validity, Invalidity, and Reliability. In: *Sociological Methodology.* Borgatta E.F., Bohrnstedt G.W. (eds.). San Francisco: Jossey-Bass.

Hermalin A.I., Entwistle B., Myers L.G., 1985. Some Lessons from the Attempt to Retrieve Early KAP and Fertility Surveys. *Population Index, 51,* 194–208.

Herrmann D.J., Raybeck D., 1981. Similarities and Differences in Meaning in Six Cultures. *Journal of Cross-Cultural Psychology, 12,* 194–206.

Herrnson P.S., 1995. Replication, Verification, Secondary Analysis, and Data Collection in Political Science. *Political Science & Politics, 28,* 452–455.

Higbee K.R., Roberts R.E., 1994. Reliability and Validity of a Brief Measure of Loneliness with Anglo-American and Mexican American Adolescents. *Hispanic Journal of Behavioral Sciences, 16,* 459–474.

Hinkle D.E., Wiersma W., Jurs St.G., 1998. *Applied Statistics for the behavioral Sciences.* 4th edition. Boston: Houghton Mifflin.

Hippler H.-J., Schwarz N., 1986. Not Forbidding Isn't Allowing: The Cognitive Basis of the Forbid-Allow Asymmetry. *Public Opinion Quarterly, 50,* 87–96.

Ho D.Y.F., 1996. Filial Piety and Its Psychological Consequences. In: *Handbook of Chinese Psychology.* Bond M.H. (ed.). Hong Kong: Oxford University Press.

Hoffmeyer-Zlotnik J.H.P., Warner U., 1998. Die Messung von Einkommen im nationalen und internationalen Vergleich. *ZUMA-Nachrichten, 42,* 30–65.

Hofstede G., 1980. *Culture's Consequences. International Differences in Work-Related Values.* Beverly Hills, CA: Sage.

Hofstede G., 1998. *Masculinity and Femininity. The Taboo Dimension of National Cultures.* Thousand Oaks, CA: Sage.

Hofstede G., 2001. *Culture's Consequences.* 2nd edition. Thousand Oaks, CA: Sage.

Hohn M.L. (ed.), 1989. *Cross-National Research in Sociology.* Newbury Park, CA: Sage.

Holland J.L., 1957. A Theory of Vocational Choice. *Journal of Counseling Psychology, 6,* 35–45.

Holland J.L., 1997. *Making Vocational Choices: A Theory of Vocational Personalities and Work Environments.* 3rd edition. Odessa, FL: Psychological Assessment Resources.

Holland P.W., Thayer D.T., 1988. Differential Item Performance and the Mantel-Haenszel Procedure. In: *Test Validity.* Wainer H., Braun H.I. (eds.). Hillsdale, NJ: Erlbaum.

Holland P.W., Wainer H. (eds.), 1993. *Differential Item Functioning.* Hillsdale, NJ: Erlbaum.

Holleman B.C., 1999. The Nature of the Forbid/Allow Asymmetry. Two Correlational Studies. *Sociological Methods and Research, 28,* 209–244.

Holt R.T., Turner J.E., 1973. Series Preface. In: *Comparative Social Research: Methodological Problems and Strategies.* Armer M., Grimshaw A.D. (eds.). New York: Wiley.

Holz-Mänttäri J., 1984. Sichtbarmachung und Beurteilung translatorischer Leistungen bei der Ausbildung von Berufstranslatoren [The Elucidation and Evaluation of Translation Performances in Translator Training]. In: *Die Theorie des Übersetzens und ihr Aufschlußwert für die Übersetzungs- und Dolmetschdidaktik.* Wilss W., Thome G. (eds.). Tübingen: Narr.

Hougland J.G., Johnson T.P., Wolf J.G., 1992. A Fairly Common Ambiguity: Comparing Rating and Approval Measures of Public Opinion. *Sociological Focus,* **25**, 257–271.

Hox J.J., 1995. *Applied Multilevel Analysis.* Amsterdam: TT-Publikaties.

Hox J.J., forthcoming. *Multilevel Analysis of Regression and Structural Equation Models.* Mahwah, NJ: Erlbaum.

Hox J.J., De Leeuw E.D., 1994. A Comparison of Nonresponse in Mail, Telephone, and Face-to-Face Surveys, Applying Multilevel Modeling to Meta-Analysis. *Quality & Quantity,* **28**, 329–344.

Hox J.J., De Leeuw E.D., 1999. The Handling of Incomplete Multivariate Data, Special Issue of Kwantitatieve Methoden. *Journal of the Netherlands Society for Statistics and Operational Research,* **62**, 37–39.

Hox J.J., De Leeuw E.D., 2001. The Influence of Interviewers` Attitude and Behaviour on Household Survey Nonresponse: An International Comparison. In: *Survey Nonresponse.* Groves R.M., Dillman D.A., Eltinge J.L., Little R.J.A. (eds.). New York: Wiley.

Hox J.J., De Leeuw E.D., Kreft I.G.G., 1991. The Effect of Interviewer and Respondent Characteristics on the Quality of Survey Data: A Multilevel Model. In: *Measurement Errors in Surveys.* Biemer P.P., Groves R.M., Lyberg L.E., Mathiowetz N.A. (eds.). New York: Wiley.

Hox J.J., De Leeuw E.D., Vorst H., 1995. Survey Participation as Reasoned Action: A Behavioral Paradigm for Survey Nonresponse. *Bulletin de Méthodologie Sociologique,* **47**, 52–67.

Hser Y.I., 1997. Self-Reported Drug Use: Results of Selected Empirical Investigations of Validity. In: *The Validity of Self-Reported Drug Use: Improving the Accuracy of Survey Estimates. NIDA Research Monograph 167.* Harrison L., Hughes A. (eds.). Rockville, MD: National Institute on Drug Abuse.

Huang C.D., Church A.T., Katigbak M.S., 1995. Identifying Cultural Differences in Items and Traits: Differential Item Functioning in the NEO Personality Inventory. *Journal of Cross-Cultural Psychology,* **28**, 192–218.

Hubert M.C., Bajos N., Sandfort Th.F.M., 1998. *Sexual Behaviour and HIV/AIDS in Europe.* London: UCL Press.

Hue P.T., Sager E.B., 1975. Interviewer`s Nationality and Outcome of the Survey. *Perceptual and Motor Skills,* **40**, 907–913.

Hui C.H., Triandis H.C., 1985. The Instability of Response Sets. *Public Opinion Quarterly,* **49**, 253–260.

Hui C.H., Triandis H.C., 1989. Effects of Culture and Response Format on Extreme Response Style. *Journal of Cross-Cultural Psychology,* **20**, 296–309.

Hulin C.L., 1987. A Psychometric Theory of Evaluations of Item and Scale Translation: Fidelity across Languages. *Journal of Cross-Cultural Psychology,* **18**, 115–142.

Humphrey D.C., 1973. Dissection and Discrimination: The Social Origins of Cadavers in America, 1760–1915. *Bulletin of the New York Academy of Medicine,* **49**, 819–827.

Hunter J., Schmidt F., 1990. *Methods of Meta-Analysis: Correcting Error and Bias in Research Findings.* Beverly Hills, CA: Sage.

Hyman H.H., 1972. *Secondary Analysis of Sample Surveys: Principles, Procedures, and Potentialities*. New York: Wiley.

Inglehart R., 1977. *The Silent Revolution. Changing Values and Political Styles Among Western Publics*. Princeton, NJ: Princeton University Press.

Inglehart R., 1993. *World Values Survey 1990–1991. WVS Program*. Madrid, J.D. Systems, S.L. ASEP S.A.

Inglehart R., 1997. *Modernization and Postmodernization. Cultural, Economic and Political Change in 43 Societies*. Princeton, NJ: Princeton University Press.

Inglehart R., Carballo M., 1997. Does Latin America Exist (And Is There a Confucian Culture)?: A Global Analysis of Cross-Cultural Differences. *Psychonomic Science*, **30**, 34–46.

Inkeles A., Masamichi S., 1995. *Comparing Nations and Cultures: Readings in a Cross-Disciplinary Perspective*. Englewood Cliffs, NJ: Prentice-Hall.

INRA, 1994a. *Codebook Eurobarometer 41.0: Explanatory Note*. Mannheim: ZEUS.

INRA, 1994b. Eurobarometer Experiment: Survey Documentation. In: *ZEUS Eurobarometer Experiment: Integrated Data-Set of EB 41.0 Variables Panel Survey Variables and Variables of the FORSA European Survey*. Mannheim: ZEUS.

International Labour Office, 1990. *International Standard Classification of Occupations: ISCO-88*. Geneva: International Labour Office.

Iwawaki S., Fukuhara M., Hidano T., 1966. Probability of Endorsement of Items in the Yatabe-Guilford Personality Inventory: Replication. *Psychological Reports*, **19**, 249–250.

Jacob H., 1984. *Using Published Data. Errors and Remedies*. Beverly Hills, CA: Sage.

Jacobson E., Kumata H., Gullahorn J.E., 1960. Cross-Cultural Contributions to Attitude Research. *Public Opinion Quarterly*, **24**, 205–223.

Jahoda G., 1995. In Pursuit of the Emic–Etic Distinction: Can We Ever Capture It? In: *The Culture and Psychology Reader*. Goldberger N.R., Veroff J.B. (eds.). New York: New York University Press.

Javeline D., 1999. Response Effects in Polite Cultures: A Test of Acquiesence in Kazakhstan. *Public Opinion Quarterly*, **63**, 1–28.

Jeanrie C., Bertrand R., 1999. Translating Psychological Tests: Keeping Validity in Mind. *European Journal of Psychological Assessment*. **15**, 277-283.

Jenkins C.R., Dillman D.A., 1997. Towards a Theory of Self-Administered Questionnaire Design. In: *Survey Measurement and Process Quality*. Lyberg L.E., Biemer P.P., Collins M., De Leeuw E., Dippo C., Schwarz N., Trewin D. (eds.). New York: Wiley.

Jennings M.K., 1990. *Continuities in Political Action: A Longitudinal Study of Political Orientation in Three Western Democracies*. Berlin: de Gruyter.

Jensen A.R., 1980. *Bias in Mental Testing*. New York: Free Press.

Ji L.J., Schwarz N., Nisbett R.E., 2000. Culture, Autobiographical Memory, and Behavioral Frequency Reports: Measurement Issues in Cross-Cultural Studies. *Personality and Social Psychology Bulletin*, **26**, 586–594.

Jöreskog K.G., Sörbom D., 1989. *Lisrel VII: Users Reference Guide*. Mooresville: Scientific Software.

Jöreskog K.G., Sörbom D., 1993. *LISREL® 8 User's Reference Guide*. Chicago: Scientific Software International.

Jobe J.B., Mingay D.J., 1989. Cognitive Research Improves Questionnaires. *American Journal of Public Health*, **79**, 1053–1055.

Jobe J.B., Mingay D.J., 1991. Cognition and Survey Measurement: History and Overview. *Applied Cognitive Psychology,* **5**, 175–192.

Johnson T.P., O'Rourke D., Chavez N., Sudman S., Warnecke R., Lacey L., Horm J., 1997. Social Cognition and Responses to Survey Questions Among Culturally Diverse Populations. In: *Survey Measurement and Process Quality.* Lyberg L.E., Biemer P.P., Collins M., De Leeuw E.D., Dippo C., Schwarz N., Trewin D. (eds.). New York: Wiley, 87–114.

Johnson T.P., 1998a. Approaches to Equivalence in Cross-Cultural and Cross-National Survey Research. In: *ZUMA-Nachrichten Spezial No. 3. Cross-Cultural Survey Equivalence.* Harkness J.A. (ed.). Mannheim: ZUMA.

Johnson T.P., 1998b. *Empirical Evidence of an Association Between Individualism/ Collectivism and the Trait Social Desirability.* Paper presented at the Third ZUMA Symposium on Cross-Cultural Survey Methodology, Leinsweiler, Germany, December 1998.

Johnson T.P., Harkness J.A., Mohler P.Ph., Van de Vijver F.J.R., Oscan Y.Z., 2001. *Respondent Cultural Orientations and Survey Participation: The Effects of Individualism and Collectivism.* 2000 Proceedings of the Section on Survey Research Methods. Alexandria: American Statistical Association, 941–946.

Jones E.L., 1983. The Courtesy Bias in South-East Asian Surveys. In: *Social Research in Developing Countries.* Bulmer M., Warwick D.P. (eds.). London: UCL Press.

Jones E.L., Forrest J.D., 1992. Underreporting of Abortion in Surveys of U.S. Women: 1976 to 1988. *Demography,* **29**, 113–126.

Jones J.H., 1981. *Bad Blood.* New York: Macmillian.

Jones L.V., Thurstone L.L., 1955. The Psychophysics of Semantics: An Experimental Investigation. *Jounrnal of Applied Psychology,* **39**, 31–36.

Jowell R., 1998. How Comparative is Comparative Research? *American Behavioral Scientist,* **42**, 168–177.

Kaase M. (ed.), 1999. *Quality Criteria for Survey Research.* Memorandum der Deutschen Forschungsgemeinschaft. Berlin: Akademie-Verlag.

Kaase M., Newton K., Scarborough E., 1995. *Beliefs in Government. Vol. 5.* Oxford: Oxford University Press.

Kalaian H.A., Raudenbush S.W., 1996. A Multivariate Mixed Linear Model for Meta-Analysis. *Psychological Bulletin,* **1**, 227–235.

Kalfs N., 1993. *Hour by Hour: Effects of the Data Collection Mode in Time Use Research.* Amsterdam: Nimmo.

Kamano S., 1999. Comparing Individual Attitudes in Several Countries. *Social Science Research,* **28**, 1–35.

Katan D., 1999. *Translating Cultures.* Manchester: St. Jerome Publishing.

Katosh J.P., Traugott M.W., 1981. The Consequences of Validated and Self-Reported Voting Measures. *Public Opinion Quarterly,* **45**, 519–535.

Keesing R.M., 1987. Models, "Folk" and "Cultural": Paradigms regained? In: *Cultural Models in Language and Thought.* Holland D., Quinn N. (eds.). Cambridge: Cambridge University Press.

Keillor B., Owens D., Pettijohn C., 2001. A Cross-Cultural/Cross-National Study of Influencing Factors and Socially Desirable Response Biases. *International Journal of Market Research,* **43**, 63–84.

Kelderman H., Macready G.B., 1990. The Use of Loglinear Models for Assessing Differential Item Functioning Across Manifest and Latent Examinee Groups. *Journal of Educational Measurement*, **27**, 307–327.

Khavari K.A., Mabry E.A., 1985. Personality and Attitude Correlates of Psychosedative Drug Use. *Drug and Alcohol Dependence*, **16**, 159–168.

Kiecolt K.J., Nathan L.E., 1985. *Secondary Analysis of Survey Data*. Beverly Hills, CA: Sage.

Kiers H.A.L., 1990. *SCA: A Program for Simultaneous Components Analysis*. Groningen: IEC ProGamma.

Kim J.-O., Mueller C.W., 1978. *Factor Analysis: Statistical Methods and Practical Issues*. Newbury Park, CA: Sage.

King G., 1995. Replication, Replication. *Political Science & Politics*, **28**, 444–452.

Kish L., 1965. *Survey Sampling*. New York: Wiley.

Kish L., 1994. Multipopulation Survey Designs: Five Types with Seven Shared Aspects. *International Statistical Review*, **62**, 167–186.

Kish L., 1995. Nomands: Enumerating and Sampling. In: *Questions and Answers from the Survey Statistician 1978–1994*.International Association of Survey Statisticians.

Kish L., 1997. Foreword. In: *DHS Analytical Reports No. 3. An Analysis of Sample Designs and Sampling Errors of the Demographic and Health Surveys*. Calverton, MD: Macro International Inc.

Költringer R., 1995. Measurement Quality in Austrian Personal Interview Surveys. In: *The Multitrait-Multimethod Approach to Evaluate Measurement Instruments*. Saris W.E., Münnich A. (eds.). Budapest: Eötvös University Press.

König W., Lüttinger P., Müller W. 1987. *Comparative Analysis of the Development and Structure of Educational Systems: Methodological Foundations and the Construction of a Comparative Educational Scale*. CASMIN Working Paper No. 12. Mannheim: Institut für Sozialwissenschaften.

Klassen D., Hornstra R.K., Anderson P.B., 1975. Influence of Social Desirability on Symptom and Mood Reporting in a Community Survey. *Journal of Consulting and Clinical Psychology*, **43**, 448–452.

Knopp J.C., 1979. Assessing Equivalence of Indicators in Cross-National Survey Research: Some Practical Guidelines. *International Review of Sport Sociology*, **14**, 137–156.

Koch A., 1993. *Die Nutzung demographischer Informationen in den Veröffentlichungen mit ALLBUS-Daten*. ZUMA-Arbeitsbericht 93/09. Mannheim: ZUMA.

Körmendi E., Noordhoek J., 1989. *Data Quality and Telephone Interviews. A Comparative Study of Face to Face and Telephone Data Collecting Methods*. Copenhagen: Danmarks Statistik.

Kohn M.L., 1987. Cross-National Research as an Analytical Strategy. *American Sociological Review*, **52**, 713–731.

Kohn M.L. (ed.), 1989a. *Cross-National Research in Sociology*. Newbury Park, CA: Sage.

Kohn M.L., 1989b. Cross-National Research as an Analytic Strategy. In: *Cross-National Research in Sociology*. Kohn M.L. (ed.). Newbury Park, CA: Sage.

Kozma A., Stones M.J., 1987. Social Desirability in Measures of Subjective Well-Being: A Systematic Evaluation. *Journal of Gerontology*, **42**, 56–59.

Krebs D., Schuessler K.F., 1987. *Soziale Empfindungen: ein interkultureller Skalenvergleich bei Deutschen und Amerikanern*. Frankfurt am Main: Campus-Verlag.

Krosnick J.A., 1991. Response Strategies for Coping with Cognitive Demands of Attitude Measures in Surveys. *Applied Cognitive Psychology*, **5**, 201–219.

Krosnick J.A., 1999. Survey Resaerch. *Annual Review of Psychology,* **50**, 537–567.

Krosnick J.A., Alwin D.F., 1987. An Evaluation of a Cognitive Theory of Response-Order Effects in Survey Measurement. *Public Opinion Quarterly,* **51**, 201–219.

Krosnick J.A., Fabrigar L.R., 1997. Designing Rating Scales for Effective Measurement in Surveys. In: *Survey Measurement and Process Quality.* Lyberg L.E., Biemer P.P., Collins M., De Leeuw E.D., Dippo C., Schwarz N., Trewin D. (eds.). New York: Wiley.

Kruskal J.B., Wish M., 1978. *Multidimensional Scaling.* Beverly Hills, CA: Sage.

Kruskal W., 1991. Introduction. In: *Measurement Errors in Surveys.* Biemer P.P. (ed.). New York: Wiley.

Krysan M., Schuman H., Scott L.J., Beatty P., 1994. Response Rates and Response Content in Mail Versus Face-to-Face Surveys. *Public Opinion Quarterly,* **58**, 381–399.

Kumata H., Schramm W., 1956. A Pilot Study of Cross-Cultural Meaning. *Public Opinion Quarterly,* **20**, 229–238.

Kuechler M., 1998. The Survey Method: An Indispensable Toll for Social Science Research. *American Behavioral Scientist,* **42**, 178–200.

Kussmaul P., 1995. *Training the Translator.* Amsterdam: John Benjamins.

Laaksonen S., 1995. *International Comparability of Business Surveys: Some Key Factors Based on European Experiences.* Bulletin of the International Statistical Institute. Proceeding of the 50th Session, Tome LVI, Beijing, China, 1995, 861–871.

Landsberger H.A., Saavedra A., 1967. Response Sets in Developing Countries. *Public Opinion Quarterly,* **31**, 214–229.

Lass J., Saris W.E., Kaase M., 1997. Sizes of the Different Effects: Coverage, Mode and Nonresponse. In: *ZUMA-Nachrichten Spezial No.2. Eurobarometer. Measurement Instruments for Opinions in Europe.* Saris W.E., Kaase M. (eds.). Mannheim: ZUMA, 73–86.

Laumann E.O., Gagnon J.H., Robert T., Michaelis S., 1994. *The Social Organization of Sexuality: Sexual Practices in the United States.* Chicago: University of Chicago Press.

Lavrakas P.J., 1993. *Telephone Survey Methods. Sampling, Selection, and Supervision.* 2nd edition. Newbury Park, CA: Sage.

Lazersfeld R.F., 1950a. Interpretation of Statistical Relations as Research Operation. In: *Measurement and Prediction.* Stouffer S. (ed.). Princeton, NJ: Princeton University Press.

Lazersfeld R.F., 1950b. The Logical and Mathematical Foundation of Latent Structure Analysis. In: *Measurement and Prediction.* Stouffer S. (ed.). Princeton, NJ: Princeton University Press.

Lee C., Green R.T., 1991. Cross-Cultural Examination of the Fishbein Behavioral Intentions Model. *Journal of Business Studies,* **2**, 289–305.

Lee E.S., Forthofer R.N., Lorimer R.J., 1986. Analysis of Complex Sample Survey Data; Problems and Strategies. *Sociological Methods and Research,* **15**, 69–100.

Lee K., Xu F., Fu F., Cameron C.A., Chen S, 2001. Taiwan and Mainland Chinese and Canadian Children's Categorization and Evaluation of Lie- and Truth-Telling: A Modesty Effect. *British Journal of Developmental Psychology,* **19**, 525–542.

Lessler J.T., Kalsbeek W.D., 1992. *Nonsampling Error in Surveys.* New York: Wiley.

Levinson S.C., 1983. *Pragmatics.* Cambridge: Cambridge University Press.

Levy P.S., Lemeshow S., 1991. *Sampling of Populations.* 2nd edition. New York: Wiley.

Lichtenstein S., Newman J.R., 1967. Empirical Scaling of Common Verbal Phrases Associated with Numerical Probabilities. *Psychonomic Science,* **9**, 563–564.

Light R.J., Pillemer D.B., 1984. *Summing up, the Science of Reviewing Research.* Cambridge, MA: Harvard University Press.

Lijphart A., 1994. *Electoral Systems and Party Systems. A Study of Twenty-Seven Democracies 1945–1990.* Oxford: Oxford University Press.

Lipset S.M., 1986. Historical Traditions and National Characteristics — A Comparative Analysis of Canada and the United States. *Canadian Journal of Sociology,* 11, 113–155.

Lipsey M.W., Wilson D.B., 2001. *Practical Meta-Analysis.* Thousand Oaks, CA: Sage.

Littell R.C., Milliken G.A., Stroup W.W., Wolfinger R.D., 1996. *SAS System for Mixed Models.* Cary, NC: SAS Institute, Inc.

Little T.D., 1997. Mean and Covariance Structures (MACS) Analyses of Cross-Cultural Data: Practical and Theoretical Issues. *Multivariate Behavioral Research,* 32, 53–76.

Lodge M., Tursky B., 1979. Comparisons Between Category and Magnitude Scaling of Political Opinion Employing SRC/CPS Items. *American Political Science Review,* 73, 50–66.

Lodge M., Tursky B., 1981. On the Magnitude Scaling of Political Opinion in Survey Research. *American Political Science Review,* 25, 376–419.

Lodge M., Tursky B., 1982. The Social-Psychological Scaling of Political Opinion. In: *Social Attitudes and Psychophysical Measurement.* Wegener B. (ed.). Hillsdale, NJ: Erlbaum.

Lodge M., Cross D., Tursky B., Tanenhaus J., 1975. The Psychological Scaling and Validation of a Political Support Scale. *American Journal of Political Science,* 19, 611–649.

Lohr S.L., 1999. *Sampling: Design and Analysis.* Duxbury Press.

Long J.S., 1983. *Confirmatory Factor Analysis: A Preface to LISREL.* Beverly Hills, CA: Sage.

Lord F., 1967. A Paradox in the Interpretation of Group Comparisons. *Psychological Bulletin,* 68, 304–305.

Lord F., Novick M.R., 1968. *Statistical Theories of Mental Test Scores.* Reading, MA: Addison-Wesley.

Lutz C., White G.M., 1986. The Anthropology of Emotions. *Annual Review of Anthropology,* 15, 405–436.

Lyberg L.E., 2000. Review of IALS — A Commentary on the Technical Report. In: *Measuring Adult Literacy.* London: Office for National Statistics.

Lyberg L.E., Dean P., 1992. *Methods of Reducing Nonresponse Rates: A Review.* Paper presented at the American Association for Public Opinion Research Conference (AAPOR), May 1992.

Lyberg L.E., Kasprzyk D., 1991. Data Collection Methods and Measurement Error: An Overview. In: *Measurement Errors in Surveys.* Biemer P.P., Groves R.M., Lyberg L.E., Mathiowetz N.A., Sudman S (eds.). New York: Wiley.

Lyberg L.E., Biemer P.P., Collins M., De Leeuw E.D., Dippo C., Schwarz N., Trewin D., 1997. *Survey Measurement and Process Quality.* New York: Wiley.

Lynn P., 1998. The British Crime Survey Sample: A Response to Elliott and Ellingworth. *Sociological Research Online,* 3, 1, http://www.socresonline.org.uk/socresonline/3/1/12.html.

Lynn P., Turner R., Smith P., 1997. The Effect of Complexity of Tone of an Advance Letter on Response to an Interview Surveys. *Survey Methods Centre Newsletter,* 17, 13–17.

MacIntosh R., 1998. A Confirmatory Factor Analysis of the Affect Balance Scale in 38 Nations: A Research Note. *Social Psychology Quarterly,* 61, 83–91.

MacKuen M.B., Turner C.F., 1984. The Popularity of Presidents, 1963–1980. In: *Surveying Subjective Phenomena*. Turner C.F., Martin E. (eds.). New York: Russell Sage Foundation.

Maddens B., Beerten R., Billiet J., 1994. *O Dierbaar Bergië. Het natiebewustzijn van Vlamingen en Walen*. Leuven: ISPO.

Maddens B., Beerten R., Billiet J., 1998. The National Consciousness of the Flemings and the Walloons. An Empirical Investigation. In: *Nationalism in Belgium*. Deprez K., Vos L. (eds.). London: Macmillian.

Maddens B., Billiet J., Beerten R., 2000. National Identity and the Attitude towards Foreigners in Multi-National States: The Case of Belgium. *Journal of Ethnic and Migration Studies*, **26**, 45–60.

Maier M.H., 1991. *The Data Game. Controversies in Social Science Statistics*. Armonk, NY: M.E. Sharpe.

Malmberg L.-E., Wanner B., Sumra S., Little T.D., 2001. Action-Control Beliefs and School Experiences of Tanzanian Primary School Students. *Journal of Cross-Cultural Psychology*, **32**, 577–597.

Marcoulides G.A., Schumacker R.E., 1996. *Advanced Structural Equation Modelling: Issues and Techniques*. Mahwah, NJ: Erlbaum.

Markus H.R., Kitayama S., 1991. Culture and the Self: Implications for Cognition, Emotion, and Motivation. *Psychological Bulletin*, **98**, 224–253.

Marlowe D., Crowne D.P., 1960. *The Approval Motive*. New York: Wiley.

Marín G., Marín B.V., 1989. *Research with Hispanic Populations*. Newbury Park, CA: Sage.

Martin J., Astill St., Boothroyd R., Eliot D., Hickman St., Johnson S., Laux R., Loyd R., Macafee K., Morgan C., Semmence J., Smith P., Teague A., Skinner C., 1996. *Report of the Task Force on Imputation*, GSS Methodology Series No.3. London: Government Statistical Service (GSS)/Office for National Statistics (ONS).

Martin M., Takahashi T., 1999. The Organization of Verbs of Knowing: Evidence for Cultural Communality and Variation in the Theory of Mind. *Memory & Cognition*, **27**, 813–825.

McClosky H., Brill A., 1983. *Dimensions of Tolerance: What Americans Believe about Civil Liberties*. New York: Russell Sage Foundation.

McCrae R.R., 1986. Well-Being Scales Do Not Measure Social Desirability. *Journal of Gerontology*, **41**, 390–392.

McCrae R.R., Costa P.T., 1983. Social Desirability Scales: More Substance than Style. *Journal of Consulting and Clinical Psychology*, **51**, 882–888.

McCrae R.R., Costa P.T., 1997. Personality Trait Structure as a Human Universal. *American Psychologist*, **52**, 509–516.

McCrae R.R., Costa P.T. Jr., del Pilar G.H., Rolland J.P., Parker W.D., 1998. Cross-Cultural Assessment of the Five-Factor Model: The Revised NEO Personality Inventory. *Journal of Cross-Cultural Psychology*, **29**, 171-188.

McCrae R.R., Zonderman A.B., Costa P.T., Bond M.H., Paunonen S.V., 1996. Evaluating Replicability of Factors in the Revised NEO Personality Inventory: Confirmatory Factor Analysis Versus Procrustes Rotation. *Journal of Personality and Social Psychology*, **70**, 522–566.

McGorry S.Y., 2000. Measurement in a Cross-Cultural Environment: Survey Translation Issues. *Qualitative Market Research: An International Journal*, **3**, 74–81.

McDonald C., Vangelder P., 1998. The Changing Context of Research. In: *The ESOMAR Handbook of Market and Opinion Research*. McDonald C., Vangelder P. (eds.). Amsterdam: ESOMAR.

McKay R.B., Breslow M.J., Sangster R.L., Gabbard S.M., Reynolds R.W., Nakamoto J.M., Tarnai J., 1996. Translating Survey Questionnaires: Lessons Learned. *New Directions for Evaluation*, **70**, 93–105.

Meisen G., 2001. Handy-Boom und Internet: Das Ende des Telefoninterviews? In: *Jahrbuch der Markt- und Sozialforschung*. Verband Schweizer Marketing- und Sozialforscher.

Mellenbergh G.J., 1982. Contingency Table Models for Assessing Item Bias. *Journal of Educational Measurement*, **7**, 105–118.

Meloen J.D., Veenman J., 1990. *Het is maar de vraag ..., Onderzoek naar responseffecten bij minderhedensurveys [It is just a Question ..., Research into Response Patterns in minority Surveys]*. Lelyland: Koninklijke Vermande.

Menon G., Raghubir P., Schwarz N., 1995. Behavioral Frequency Judgments: An Accessibility-Diagnosticity Framework. *Journal of Consumer Research*, **22**, 212–228.

Middleton K.L., Jones J.L., 2000. Socially Desirable Response Sets: The Impact of Country Culture. *Psychology & Marketing*, **17**, 149–163.

Miller J., Slomczynski K.M., Schoenberg R., 1981. Assessing Comparability of Measurement in Cross-National Sociocultural Settings. *Social Psychology Quarterly*, **44**, 178–191.

Miller J.G., 2001. The Cultural Grounding of Social Psychological Theory. In: *Blackwell Handbook of Social Psychology, Vol. 1: Intraindividual Processes*. Tesser A., Schwarz N. (eds.). Oxford: Blackwell.

Miller P.V., Cannell C.F., 1982. A Study of Experimental Techniques for Telephone Interviewing. *Public Opinion Quarterly*, **46**, 250–269.

Millsap R., Everson H., 1993. Methodology Review: Statistical Approaches for Assessing Measurement Bias. *Applied Psychological Measurement*, **17**, 297–334.

Mittelstaedt R.A., 1971. Semantic Properties of Selected Evaluative Adjectives: Other Evidence. *Journal of Marketing Research*, **8**, 236–237.

Mochmann E., Oedegaard I.C., Mauer R., 1998. *Inventory of National Election Studies in Europe 1945–1995*. Bergisch Gladbach: Ferger.

Mohler P.Ph., Smith T.W., Harkness J.A., 1998. Respondent's Ratings of Expressions from Response Scales: A Two-Country, Two-Language Investigation on Equivalence and Translation. In: *ZUMA-Nachrichten Spezial No. 3. Cross-Cultural Survey Equivalence*. Harkness J.A. (ed.). Mannheim: ZUMA, 159–184.

Molenaar N.J., 1986. *Formuleringseffecten in survey-interviews*. Amsterdam: VU-uitgenverij.

Moorman R.H., Podsakoff P.M., 1992. A Meta-Analytic Review and Empirical Test of the Potential Confounding Effects of Social Desirability Response Sets in Organizational Behaviour Research. *Journal of Occupational and Organizational Psychology*, **65**, 131–149.

Morton-Williams J., 1993. *Interviewer Approaches*. Brookfield: Dartmouth Publishing.

Moscovici S., 1984. The Phenomenon of Social Representations. In: *Social Representations*. Farr R.M., Moscivici S. (eds.). Cambridge: Cambridge University Press.

Mosier Ch., 1941. A Psychometric Study of Meaning. *Journal of Social Psychology*, **13**, 123–140.

Moum T., 1998. Mode of Administration and Interviewer Effects in Self-Reported Symptoms of Anxiety and Depression. *Social Indicators Research*, **45**, 279–318.

Muniz J., Hambleton R.K., Xing D., 2001. Small Sample Studies to Detect Flaws in Item Translations. *International Journal of Testing*, 1, 115–135.

Muthén B., 1991. Multilevel Factor Analysis of Class and Student Achievement Components. *Journal of Educational Measurement*, 28, 338–354.

Muthén B., 1994. Multilevel Covariance Structure Analysis. *Sociological Methods and Research*, 22, 376–398.

Myers J.H., Warner W.G., 1968. Semantic Properties of Selected Evaluation Adjectives. *Journal of Marketing Research*, 5, 409–412.

Nakanishi M., Johnson K.M., 1993. Implications of Self-Disclosure on Conversational Logics, Perceived Communication Competence, and Social Attraction: A Comparison of Japanese and American Cultures. In: *Intercultural Communication Competence*. Wiseman R.L., Koester J. (eds.). Newbury Park, CA: Sage.

Namenwirth J.Z., Weber R.P., 1987. *Dynamics of Culture*. Boston: Allen & Unwin.

Nápoles-Springer A.M., Stewart A.L., 2001. Use of Health-Related Quality of Life Measures in Older and Ethnically Diverse U.S. Populations. *Journal of Mental Health and Aging*, 7, 173–179.

National Center for Education Statistics (NCES), 1998. *Adult Literacy in OECD Countries: Technical Report on the First International Adult Literacy Survey.*: Washington, DC: NCES.

NESSTAR, 2000. *Final Report*. http://www.nesstar.org/papers/NesstarFinalReport.pdf.

Newby M., Amin S., Diamond I., Naved R.T., 1998. Survey Experience Among Women in Bangladesh. *American Behavioral Scientist*, 42, 252–275.

Newton K., Scarborough E., 1995. *Beliefs in Government*. Oxford: Oxford University Press.

Nisbett R.E., Peng K., Choi I., Norenzayan A., 2001. Culture and Systems of Thought: Holistic Versus Analytic Cognition. *Psychological Review*, 108, 291–310.

Nowak S., 1989. Comparative Studies and Social Theory. In: *Cross-National Research in Sociology*. Kohn M.L. (ed.). Newbury Park, CA: Sage.

O´Muircheartaigh C.A., Gaskell G.D., Wright D.B., 1993. The Impact of Intensifiers. *Public Opinion Quarterly*, 57, 552–565.

O´Muircheartaigh C.A., Krosnick J.A., Helic A., 2000. *Middle Alternatives, Acquiescence, and the Quality of Questionnaire Data*. Working Paper Series 01.3. University of Chicago. http://www.harrisschool.uchicago.edu/pdf/wp_01_3.pdf.

Oakland T., Gulek C., Glutting J., 1996. Children's Test-Taking Behaviors: A Review of Literature, Case Study, and Research of Children. *European Journal of Psychology Assessment*, 12, 240–246.

Office for National Statistics, 1996. *Harmonised Concepts and Questions for Government Social Surveys*. London: Government Statistical Service.

Okazaki S., 2000. Asian American and White American Differences on Affective Distress Symptoms: Do Symptom Reports Differ Across Reporting Methods. *Journal of Cross-Cultural Psychology*, 31, 603–625.

Oksenberg L., Cannell C.F., 1977. Some Factors Underlying the Validity of Response in Self-Report. *International Statistical Bulletin*, 48, 325–346.

Onodera N., 1999. Personal Communication, email, 8/16/1999.

Orren G.R., 1978. Presidential Popularity Ratings: Another View. *Public Opinion*, 1,

Osgood C.E., May W.H., Miron M.S., 1975. *Cross-Cultural Universals in Affective Meaning*. Urbana: University of Illinois Press.

Osgood C.E., Suci G.J., Tannenbaum P.H., 1957. *The Measurement of Meaning*. Urbana: University of Illinois Press.

Ostrom Th.M., Gannon K.M., 1996. Exemplar Generation: Assessing How Respondents Give Meaning to Rating Scales. In: *Answering Questions: Methodology for Determining Cognitive and Communicative Processes in Survey Research*. Schwarz N., Sudman S. (eds.). San Francisco: Jossey-Bass.

Owens L., Johnson T.P., O'Rourke D., 1999. Culture and Item Nonresonse in Health Surveys. In: *Seventh Conference on Health Survey Research Methods*. Cynamon M.L., Kulka R.A. (eds.). Maryland: Hyattsville.

Øyen E., 1990. *Comparative Methodology – Theory and Practice in International Research*. Newbury Park, CA: Sage.

Oyserman D., 2001. Self-Concept and Identity. In: *Blackwell Handbook of Social Psychology, Vol.1: Intraindividual Processes*. Tesser A., Schwarz N. (eds.). Oxford: Blackwell.

Oyserman D., Coon H., Kemmelmeier M., 2002. Rethinking Individualism and Collectivism: Evaluation of Theoretical Assumptions and Meta-Analyses, *Psychological Bulletin*, **128**, 3–72.

Page W.F., Davies J.E., Ladner R.A., Alfassa J., 1977. Urinanalysis Screened vs. Verbally Reported Drug Use: The Identification of Discrepant Groups. *International Journal of the Addictions*, **12**, 439–450.

Parducci A., 1982. Category Ratings: Still More Contextual Effects! In: *Social Attitudes and Psychological Measurement*. Wegener B. (ed.). Hillsdale, NJ: Erlbaum.

Park A., Jowell R., 1997. *Consistencies and Differences in a Cross-National Survey. The International Social Survey Programme (1995)*. Cologne: ZA. (http://www.za.uni-koeln.de/data/en/issp/codebooks/s2880app.pdf)

Park K.B., Upshaw H.S., Koh S.D., 1988. East Asians' Responses to Western Health Items. *Journal of Cross-Cultural Psychology*, **19**, 51–64.

Parker R.N., 1983. Measuring Social Participation. *American Sociological Review*, **48**, 864–873.

Patrick D.L., Hurst B.C., 1998. Migraine-Specific Quality of Life Measure (MSQOL). *Medical Outcomes Trust Bulletin*, **6**, 1, http://www.outcomes-trust.org/bulletin/0198blltn.htm.

Paulhus D.L., Van Selst M., 1990. The Spheres of Control Scale: 10 Years of Research. *Personality and Individual Differences*, **11**, 1029–1036.

Paunonen S.V., Ashton M.C., 1998. The Structured Assessment of Personality Across Cultures. *Journal of Cross-Cultural Psychology*, **29**, 150–170.

Pedhazur E.J., 1982. *Multiple Regression in Behavioral Research: Explanation and Prediction*. 2nd edition. New York: Holt, Rinehart and Winston.

Peterson R.A., 2000. *Constructing Effective Questionnaires*. Thousand Oaks, CA: Sage.

Petty R.E., Cacioppo J.T., 1986. *Communication and Persuasion: Central and Peripheral Routes to Attitude Change*. New York: Springer.

Petty R.E., Wegener D.T., 1999. The Elaboration Likelihood Model: Current Status and Controversies. In: *Dual-Process Theories in Social Psychology*. Chaiken S., Trope Y. (eds.). New York: Guilford Press.

Phillips D.L., Clancy K.J., 1972. Some Effects of "Social Desirability" in Survey Studies. *American Journal of Sociology*, **77**, 921–940.

Pike K.L., 1966. *Language in Relation to a Unified Theory of Human Behavior*. The Hague: Mouton.

Piswanger K., 1975. *Interkulturelle Vergleiche mit dem Matrizentest von Formann [Cross-Cultural Comparisons with Formann`s Matrices Test]*. Vienna: University of Vienna.

Poortinga Y.H., 1989. Equivalence of Cross-Cultural Data: An Overview of Basic Issues. *International Journal of Psychology*, **24**, 737–756.

Poortinga Y.H., Van de Vijver F.J.R., 1987. Explaining Cross-Cultural Differences: Bias Analysis and Beyond. *Journal of Cross-Cultural Psychology*, **18**, 259–282.

Popper K.R., 1974. *Conjectures and Refutations: The Growth of Scientific Knowledge*. 5th edition. London: Routledge & Kegan Paul.

Popper K.R., 1976. The Unended Quest. An Intellecutual Autobiography. Glasgow: Fontana/Collins.

Porst R., Zeifang K., 1987. A Description of the German General Social Survey Test-Retest Study and a Report on the Stabilities of the Sociodemographic Variables. *Sociological Methods and Research*, **15**, 177–218.

Power R., Doran C., Scott D., 1999. *Generating Embedded Discourse Markers from Rhetorical Structure*. Proceedings of the European Workshop on Natural Language Generation, Toulouse, France, 30–38.

Prediger D.J., 1982. Dimensions Underlying Holland's Hexagon: Missing Links Between Interests and Occupations? *Journal of Vocational Behavior*, **15**, 259–287.

Presser S., Blair J., 1994. Survey Pretesting: Do Different Methods Produce Different Results? *Sociological Methodology*, **24**, 73–104.

Presser S., Stinson L., 1998. Data Collection Mode and Social Desirability Bias in Self-Reported Religious Attendance. *American Sociological Review*, **63**, 137–145.

Pretorius T.B., 1993. The Metric Equivalence of the UCLA Loneliness Scale for a Sample of South African Students. *Educational and Psychological Measurement*, **53**, 233–239.

Purdon S., Campanelli P., Sturgis P., 1999. Interviewers' Calling Strategies on Face-to-Face Interview Surveys. *Journal of Official Statistics*, **15**, 199–216.

Przeworski A., Teune H., 1966. Equivalence in Cross-National Research. *Public Opinion Quarterly*, **30**, 551–568.

Przeworski A., Teune H., 1970. *The Logic of Comparative Social Inquiry*. New York: Wiley.

RAC (Response Analysis Newsletter), 1984. Quantifies the Vast Majority. *The Sampler from Response Analysis*, **31**,

Radloff L.S., 1977. A CES-D Scale: A Self-Report Scale for Research in the General Population. *Applied Psychological Measurement*, **1**, 385–401.

Ramsey J.O., 1986. *Multiscale II Manual*. Montreal: McGill University, Department of Psychology.

Rao J.N.K., Wu C.F.J., 1988. Resampling Inference with Complex Survey Data. *Journal of American Statistical Association*, **83**, 231–241.

Rasbash J., 2000. *A User's Guide to MLwiN*. London: University of London, Institute of Education.

Rattenbury J., 1974. *Data Processing in the Social Sciences with Osiris*. Ann Arbor, MI: ISR.

Raudenbush S.W., 1994. Random Effects Models. In: *The Handbook of Research Sysnthesis*. Cooper H., Hedges L.V. (eds.). New York: Russell Sage Foundation.

Raudenbush S.W., Bryk A.S., Cheong Y.F., Congdon R., 2000. *HLM 5.0. Hierarchical Linear and Nonlinear Modeling*. Chicago: Scientific Software International.

Raven J.C., 1938. *Progressive Matrices: A Perceptual Test of Intelligence*. London: Lewis.

Raven J.C., 1956. *Coloured Progressive Matrices, Sets A, Ab and B*. London: Lewis.

Reese S.D., Danielson W.A., Shoemaker P., Chang T., Hsu H.-L., 1986. Ethnicity-of-Interviewer Effects among Mexican-Americans and Anglos. *Public Opinion Quarterly*, 50, 563–572.

Rensvold R.B., Cheung G.W., 1998. Testing Measurement Models for Factorial Invariance: A Systematic Approach. *Educational and Psychological Measurement*, 58, 1017–1034.

Richards J.M., Gottfredson D.C., Gottfredson G.D., 1991. Units of Analysis and Item Statistics for Environmental Assessment Scales. *Current Psychology: Research and Reviews*, 9, 407–413.

Richman W.L., Kiesler S., Weisband S., Drasgow F., 1999. A Meta-Analytic Study of Social Desirability Distortion in Computer-Administered Questionnaires, Traditional Questionnaires and Interviews. *Journal of Applied Psychology*, 84, 754–775.

Robinson W.S., 1950. Ecological Correlations and the Behavior of Individuals. *American Sociological Review*, 15, 351–357.

Rockwood T., Sangster R.L., Dillman D.A., 1997. The Effect of Response Categories on Survey Questionnaires: Context and Mode Effects. *Sociological Methods and Research*, 26, 118–140.

Rodgers W.L., Andrews F.M., Herzog A.R., 1992. Quality of Survey Measures: A Structural Modelling Approach. *Journal of Official Statistics*, 8, 251–275.

Rogers H.J., Swaminathan H., 1993. A Comparison of Logistic Regression and Mantel-Haenszel Procedures for Detecting Differential Item Functioning. *Applied Psychological Measurement*, 17, 105–116.

Rokkan S., 1955. Comparative Cross-National Research. *International Sociological Science Bulletin*, 7, 622–641.

Rokkan S., 1969. Cross-National Survey Research: Historical, Analytical and Substantive Contexts. In: Rokkan S., Verba S., Viet J., Almasy E. (eds.). Paris: Mouton.

Rokkan S., 1993. Cross-Cultural, Cross-Societal and Cross-National Research. *Historical Social Research*, 18, 6–54.

Rokkan S., Verba S., Viet J., Almasy E. (eds.). 1969. *Comparative Survey Analysis*. Paris: Mouton.

Rosenberg S., Nelson C., Vivekenanthan P.S., 1968. A Multi-Dimensional Approach to the Structure of Personality Impressions. *Journal of Personality and Social Psychology*, 9, 283–294.

Rosenthal R., 1984. *Meta-Analytic Procedures for Social Research*. Newbury Park, CA: Sage.

Rosenthal R., Rosnow R.L., Rubin D.R., 2000. *Contrast and Effect Sizes in Behavioral Research. A Correlational Approach*. Cambridge: Cambridge University Press.

Ross C.E., Mirowsky J., 1984. Socially-Desirable Response and Acquiescence in a Cross-Cultural Survey of Mental Health. *Journal of Health and Social Behavior*, 25, 189–197.

Roth G., 1971. Max Weber's Comparative Approach and Historical Typology. In: *Comparative Methodology on Sociology — Essays on Trends and Applications*. Vallier I. (ed.). Berkeley: University of California Press.

Rounds J., Tracey T.J., 1996. Cross-Cultural Structural Equivalence of RIASEC Models and Measures. *Journal of Counseling Psychology*, 43, 310–329.

Rubin D.B., 1987. *Multiple Imputation for Nonresponse in Surveys*. New York: Wiley.

Rumelhart D.E., 1980. Schemata: The Building Blocks of Cognition. In: *Theoretical Issues in Reading and Comprehension*. Spiro R.J., Bruce B., Brewer W.F. (eds.). Hillsdale, NJ: Erlbaum.

Russell J.A., 1980. A Circumplex Model of Affect. *Journal of Personality and Social Psychology*, **39**, 1161–1178.

Russell J.A., 1983. Pancultural Aspects of the Human Conceptual Organization of Emotions. *Journal of Personality and Social Psychology*, **45**, 1281–1288.

Russell J.A., 1991. Culture and the Categorization of Emotions. *Psychological Bulletin*, **110**, 426–450.

Russell J.A., Lewicka M., Niit T., 1989. A Cross-Cultural Study of a Circumplex Model of Affect. *Journal of Personality and Social Psychology*, **57**, 848–856.

Ryder N.B., 1965. The Cohort as a Concept in the Study of Social Change. *American Sociological Review*, **30**, 843–861.

Särndal C.-E., Swensson B., Wretman J., 1992. *Model Assisted Survey Sampling*. New York: Springer.

Sampson R.J., Lauritsen J.L., 1997. Racial and Ethnic Disparities in Crime and Criminal Justice in the United States. In: *Ethnicity, Crime, and Immigration: Comparative and Cross-National Perspective*. Tonry M. (ed.). Chicago: University of Chicago Press.

Sanders D., 1994. Methodological Considerations in Comparative Cross-National Research. *International Social Science Journal*, **46**, 513–521.

Saris W.E., 1986. *Variation in Response Functions: A Source of Measurement Error in Attitude Research*. Amsterdam: SRF.

Saris W.E., 1990. The Choice of a Model for Evaluation of Measurement Instruments. In: *Evaluation of Measurement Instruments by Meta-Analysis of Multitrait-Multimethod Studies*. Saris W.E., Van Meurs A. (eds.). Amsterdam: North-Holland.

Saris W.E., 1998. The Effects of Measurement Error in Cross-Cultural Research. In: *ZUMA-Nachrichten Spezial No.3. Cross-Cultural Survey Equivalence*. Harkness J.A. (ed.). Mannheim: ZUMA.

Saris W.E., Andrews F.M., 1991. Evaluation of Measurement Instruments Using a Structural Modelling Approach. In: *Measurement Errors in Surveys*. Biemer P.P., Groves R.M., Lyberg L.E., Mathiowetz N.A., Sudman S. (eds.). New York: Wiley.

Saris W.E., Hagenaars J.A., 1997. Mode Effects in the Standard Eurobarometer Questions. In: *ZUMA-Nachrichten Spezial No. 2. Eurobarometer. Measurement Instruments for Opinions in Europe*. Saris W.E., Kaase M. (eds.). Mannheim: ZUMA.

Saris W.E., Kaase M. (eds.), 1997. *ZUMA-Nachrichten Spezial No.2. Eurobarometer. Measurement Instruments for Opinions in Europe*. Mannheim: ZUMA.

Saris W.E., Meurs A., 1990. *Evaluation of Measurement Instruments by Meta-Analysis of Multitrait Multimethod Studies*. Amsterdam: North-Holland.

Saris W.E., Münnich A. (eds.), 1995. *The Multitrait-Multimethod Approach to Evaluate Measurement Instruments*. Budapest: Eötvös University Press.

Saris W.E., Veenhoven R., Scherpenzeel A.C., Bunting B., 1996. *A Comparative Study of Satisfaction with Life in Europe*. Budapest: Eötvös University Press.

Sasaki M., 1995. Research Design of Cross-National Attitude Surveys. *Behaviormetrica*, **22**, 99–114.

Schaeffer N.C., 1991. Hardly Ever or Constantly? Group Comparisons Using Vague Quantifiers. *Public Opinion Quarterly*, **55**, 395–423.

Scheepers P., Felling A., Peters J., 1990. Social Conditions, Authoritarianism and Ethnocentrism: A Theoretical Model of the Early Frankfurt School Updated and Tested. *European Sociological Review*, **6**, 15–29.

Scheepers P., Felling A., Peters J., 1992. Anomie, Authoritarianism, and Ethnocentrism: Update of a Classic Thema and an Empirical Test. *Politics and the Individual*, **2**, 29–42.

Scherpenzeel A.C., 1995. Meta Analysis of a European Comparative Study. In: *The Multitrait-Multimethod Approach to Evaluate Measurement Instruments*. Saris W.E., Münnich A. (eds.). Budapest: Eötvös University Press, 225–243.

Scherpenzeel A.C., 2000. *Mode Effects in Panel Surveys: A Split-Ballot Comparison of CAPI and CATI*. http://www.unine.ch/psm/file/methods/capi_cati.pdf.

Scherpenzeel A.C., Saris W.E., 1997. The Validity and Reliability of Survey Questions: A Meta-Analysis of MTMM Studies. *Sociological Methods and Research*, **25**, 341–383.

Scheuch E.K., 1968. The Cross-Cultural Use of Sample Surveys: Problems of Comparability. In: *Comparative Research Across Cultures and Nations*. Rokkan S. (ed.). Paris: Mouton.

Scheuch E.K., 1973. Entwicklungsrichtlinien bei der Analyse sozialwissenschaftlicher Daten. In: *Handbuch der empirischen Sozialforschung, Vol. 1, Geschichte und Grundprobleme*. König R. (ed.). 3rd edition. Stuttgart: F. Enke, 161–226.

Scheuch E.K., 1989. Theoretical Implications of Comparative Survey Research: Why the Wheel of Cross-Cultural Methodology Keeps on Being Reinvented. *International Sociology*, **4**, 147–167.

Scheuch E.K., 1990. From a Data Archive to an Infrastructure for the Social Sciences. *International Social Science Journal*, **42**, 1, 93–112.

Scheuch E.K., Rokkan S., 1963. Data Archives in the Social Sciences. *Social Science Information*, **2**, 109–114.

Schiffman S.S., Reynolds M.L., Young F.W., 1981. *Introduction to Multidimensional Scaling: Theory, Methods, and Applications*. Orlando, FL: Academic Press.

Schimmack U., 1996. Cultural Influences on the Recognition of Emotion by Facial Expressions: Individualistic or Causasian Cultures? *Journal of Cross-Cultural Psychology*, **27**, 37–50.

Schnell R., 1991. Wer ist das Volk? Undercoverage, Schwererreichbare und Nichtbefragbare bei "allgemeinen Bevölkerungsumfragen". *Kölner Zeitschrift für Soziologie und Sozialpsychologie*, **43**, 106–137.

Schober M.F., 1999. Making Sense of Questions: An Interactional Approach. In: *Cognition and Survey Research*. Sirken M.G., Herrmann D.J., Schechter S., Schwarz N., Tanur J., Tourangeau R. (eds.). New York: Wiley.

Schooler C., Diakite C., Vogel J., Mounkoro P., Caplan L., 1998. Conducting a Complex Sociological Survey in Rural Mali: Three Points of View. *American Behavioral Scientist*, **42**, 252–275.

Schoua A., 1985. *An English / Spanish Test of Decentering for the Translation of Questionnaires*. Unpublished dissertation. Chicago: Northwestern University.

Schoua-Glusberg A., 1988. *A Focus-Group Approach to Translating Questionnaire Items*. Paper presented at the Annual Meeting of the American Association for Public Opinion Research, Toronto.

Schoua-Glusberg A., 1992. *Report on the Translation of the Questionnaire for the National Treatment Improvement Evaluation Study*. Chicago: National Opinion Research Center.

Schriesheim Ch., Schriesheim J., 1974. Development and Empirical Verification of New Response Categories to Increase the Validity of Multiple Response Alternatives Questionnaires. *Educational and Psychological Measurement*, **34**, 877–884.

Schuman H., Presser S., 1981. *Questions and Answers in Attitude Surveys: Experiments on Question Form, Wording, and Context*. New York: Academic Press.

Schuman H., Steeh Ch., Bobo L., Krysan M., 1997. *Racial Attitudes in America: Trends and Interpretations*. Revised edition. Cambridge, MA: Harvard University Press.

Schwanenflugel P.J., Martin M., Takahashi T., 1999. The Organization of Verbs of Knowing: Evidence for Cultural Communality and Variation in the Theory of Mind. *Memory & Cognition*, 27, 813–825.

Schwanenflugel P.J., Fabricius W.V., Noyes C.R., Bigler K.D., Alexander J.M., 1994. The Organization of Mental Verbs and Folk Theories of Knowing. *Journal of Memory & Language*, 33, 376–395.

Schwartz S.H., 1992. Universals in the Content and Structure of Values: Theoretical Advances and Empirical Tests in 20 Countries. In: *Advances in Experimental Social Psychology*, Vol. 26. Zanna M. (ed.). Orlando: Academic Press.

Schwartz S.H., Sagiv L., 1995. Identifying Culture-Specifics in the Content and Structure of Values. *Journal of Cross-Cultural Psychology*, 26, 92–116.

Schwarz N., 1994. Judgment in a Social Context: Biases, Shortcomings, and the Logic of Conversation. In: *Advances in Experimental Social Psychology, Vol. 26.* Zanna M. (ed.). San Diego: Academic Press.

Schwarz N., 1995. What Respondents Learn from Questionnaires: The Survey Interview and the Logic of Conversation. (The 1993 Morris Hansen Lecture). *International Statistical Review*, 63, 153–177.

Schwarz N., 1996. *Cognition and Communication: Judgmental Biases, Research Methods and the Logic of Conversation.* Hillsdale, NJ: Erlbaum.

Schwarz N., 1999a. Cognitive Research into Survey Measurement: Its Influence on Survey Methodology and Cognitive Theory. In: *Cognition and Survey Research.* Sirken M.G., Herrmann D.J., Schechter S., Schwarz N., Tanur J., Tourangeau R. (eds.). New York: Wiley.

Schwarz N., 1999b. Self-Reports: How the Questions Shape the Answer. *American Psychologists*, 54, 93–105.

Schwarz N., 1999c. Self-Reports of Behaviors and Opinions: Cognitive and Communicative Processes. In: *Cognition, Aging, and Self-Reports.* Schwarz N., Park D., Knäuper B., Sudman S. (eds.). Philadelphia: Psychology Press.

Schwarz N., Hippler H.-J., 1995. The Numeric Values of Rating Scales: A Comparison of Their Impact in Mail Surveys and Telephone Interviews. *International Journal of Public Opinion Research*, 7, 72–74.

Schwarz N., Sudman S., 1994. *Autobiographical Memory and the Validity of Retrospective Reports.* Heidelberg: Springer.

Schwarz N., Sudman S., 1996. *Answering Questions — Methodology for Determining Cognitive and Communicative Processes in Survey Research.* San Francisco: Jossey-Bass.

Schwarz N., Garyson C.E., Knäuper B., 1998. Formal Features of Rating Scales and the Interpretation of Question Meaning. *International Journal of Public Opinion Research*, 10, 177–183.

Schwarz N., Knäuper B., Parl D., 1999. *Aging, Cognition, and Context Effects: How Differential Context Effects Invite Misleading Conclusions about Cohort Differences.* St. Petersburg Beach, FL: American Association for Public Opinion Research.

Schwarz N., Strack F., Mai H.P., 1991. Assimilation and Contrast Effects in Part-Whole Question Sequences: A Conversational Logic Analysis. *Public Opinion Quarterly*, 55, 3–23.

Schwarz N., Hippler H.-J., Deutsch B., Strack F., 1985. Response Scales: Effects of Category Range on Reported Behavior and Comparative Judgements. *Public Opinion Quarterly,* **49**, 388–395.

Schwarz N., Knäuper B., Hippler H.-J., Noelle-Neumann E., Clark L., 1991a. Rating Scales: Numeric Values May Change the Meaning of Scale Labels. *Public Opinion Quarterly,* **55**, 570–582.

Schwarz N., Strack F., Hippler H.-J., Bishop G.F., 1991b. The Impact of Administration Mode on Response Effects in Survey Measurement. *Applied Cognitive Psychology,* **5**, 193–212.

Schwarzer R., 1989. *Meta-Analysis Programs. Program Manual.* Berlin: Institut für Psychologie, Freie Universität Berlin.

Segall M.H., Dasen P.R., Berry J.W., Poortinga Y.H., 1990. *Human Behavior in Global Perspective. An Introduction to Cross-Cultural Psychology.* New York: Pergamon Press.

Shaver P., Schwartz J., Kirson D., O'Connor C., 1987. Emotion Knowledge: Further Exploration of a Prototype Approach. *Journal of Personality and Social Psychology,* **52**, 1061–1086.

Shweder R.A., 1972. *Semantic Structure and Personality Assessment.* Cambridge, MA: Harvard University Press.

Sigelman L., 1990. Answering the 1,000,000-Person Question: The Measurement and Meaning of Presidential Popularity. *Research in Micropolitics,* **3**, 209–226.

Signorella M.L., Jamieson W., 1986. Masculinity, Feminity, Androgyny, and Cognitive Performance: A Meta-Analysis. *Psychological Bulletin,* **100**, 207–228.

Silberstein A.S., Scott S., 1991. Expenditure Diary Surveys and Their Associated Errors. In: *Measurement Errors in Surveys.* Biemer P.P., Groves R.M., Lyberg L.E., Mathiowetz N.A., Sudman S. (eds.). New York: Wiley.

Simpson R.H., 1944. The Specific Meanings Certain Terms Indicating Differing Degrees of Frequency. *Quarterly Journal of Speech,* **30**, 328–330.

Sinaiko H.W., Brislin R.W., 1973. Evaluating Language Translations: Experiments on Three Assessment Methods. *Journal of Applied Psychology,* **57**, 328–334.

Singer E., Presser S., 1989. The Interviewer. In: *Survey Research Methods.* Singer E., Presser S. (eds.). Chicago: University of Chicago Press.

Singer E., 2001. Incentives in Survey Research. In: *Survey Nonresponse.* Groves R.M., Eltinge J.L., Dillman D.A., Little R.J.A. (eds.). New York: Wiley.

Singer E., Van Hoewyk J., Gebler N., Raghunathan T.E., McGonagle K.A., 1999. The Effect of Incentives on Response Rates in Interviewer-Mediated Surveys. *Journal of Official Statistics,* **15**, 217–230.

Sirken M.G., Herrmann D.J., Schechter S., Schwarz N., Tanur J.M., Tourangeau R., 1999. *Cognition and Survey Research.* New York: Wiley.

Skinner C., Holt D., Smith T.M.F. (eds.), 1989. *Analysis of Complex Surveys.* New York: Wiley.

Skinner J.K., 2001. Acculturation: Measures of Ethnic Accommodation to the Dominant American Culture. *Journal of Health and Aging,* **7**, 41–52.

Smeeding T.M., Ward M., Castles I., Lee H., 2000. *Cross-Country Comparisons.* http://lisweb.ceps.lu/links/canberra/luxembourg/chapter8.pdf.

Smelser N.J., 1972. The Methodology of Comparative Studies. In: *Comparative Research Methods.* Warwik D.P., Osherson S. (eds.). Englewood Cliffs, NJ: Prentice-Hall.

Smith E.R., 1998. Mental Representation and memory. In: *The Handbook of Social Psychology. Volume 1.* Gilbert D.T., Fiske S.T., Lindzey G. (eds.). 4th edition. New York: Random House.

Smith M.L., Glass G.V., Miller T.I., 1980. *The Benefits of Psychotherapy.* Baltimore: Johns Hopkins University Press.

Smith P.B., Bond M.H., 1998. *Social Psychology Across Cultures.* 2nd edition. Boston: Allyn & Bacon.

Smith P.B., 2001. The End of the Beginning? *International Journal of Cross-Cultural Management,* 1, 21–24.

Smith T.W., 1979. Happiness: Time Trends, Seasonal Variations, Intersurvey Differences, and Other Mysteries. *Social Psychology Quarterly,* 42, 18–30.

Smith T.W., 1984. Nonattitudes: A Review and Evaluation. In: *Surveying Subjective Phenomenon.* Turner C.F., Martin E. (eds.). New York: Russell Sage Foundation.

Smith T.W., 1988. *The Ups and Downs of Cross-National Survey Research.* GSS Cross-National Report No. 8. Chicago: NORC.

Smith T.W., 1990. Social Inequality in Cross-National Perspective. In: *Attitudes to Inequality and the Role of Government.* Becker J.W., Davis J.A., Ester P., Mohler P.Ph. (eds.). Rijswijk: Sociaal en Cultureel Planbureau.

Smith T.W., 1991a. *An Analysis of Missing Income Information on the General Social Survey.* GSS Methodological Report No. 71. Chicago: NORC.

Smith T.W., 1991b. Context Effects in the General Social Survey. In: *Measurement Errors in Surveys.* Biemer P.P., Groves R.M., Lyberg L.E., Mathiowetz N.A., Sudman S. (eds.). New York: Wiley.

Smith T.W., 1992. Thoughts on the Nature of Context Effects. In: *Context Effects in Social and Psychological Research.* Schwarz N., Sudman S. (eds.). New York: Springer.

Smith T.W., 1994. *An Analysis of Response Patterns to the Ten-Point Scalometer.* American Statistical Association. Proceedings of the Section on Survey Research Methods, Alexandria, VA.

Smith T.W., 1996. *Environmental and Scientific Knowledge Around the World.* GSS Cross-National Report No. 16. Chicago: NORC.

Smith T.W., 1997. *Improving Cross-National Survey Response by Measuring the Intensity of Response Categories.* GSS Cross-National Report No. 78. Chicago: NORC.

Smith T.W., Heaney K., 1995. *Who, What, When, Where, and Why: An Analysis of Usage of the General Social Survey.* GSS Project Report No. 19. Chicago: NORC.

Snell-Hornby M., 1988. *Translation Studies. An Integrated Approach.* Amsterdam: John Benjamins.

Snijkers G., Luppes M., 2000. The Best of Two Worlds: Total Design Method and New Kontiv Design. An Operational Model to Improve Respondant Co-Operation. *Netherlands Official Statistics,* 15, 4–10.

Snijkers G., Hox J.J., De Leeuw E.D., 1999. Interviewers 'Tactics for Fighting Survey Nonresponse. *Journal of Official Statistics,* 15, 185–198.

Sommer R., Unholzer G., Wiegand E., 1999. *Standards for Quality Assurance in Market and Social Research.* Frankfurt: ADM Arbeitskreis Deutscher Markt- und Sozialforschungsinstitute e. V.

Spector P.E., 1976. Choosing Response Categories for Summated Rating Scales. *Journal of Applied Psychology,* 61, 374–375.

Sperber D., Wilson D., 1986. *Relevance: Communication and Cognition.* Oxford: Blackwell.

Spielberger C.D., Gorsuch R.L., Lushene R.E., 1970. *Manual for The State-Trait Anxiety Inventory*. Palo Alto, CA: Consulting Psychologists Press.

Srole L., 1996. Social Integration and Certain Corollaries: An Exploratory Study. *American Sociological Review*, **21**, 709–716.

Stacey M., 1969. *Comparability in Social Research*. London: Heineman.

Statistisches Bundesamt, 1999. *Demographische Standards*. 3rd edition. Wiesbaden: Statistisches Bundesamt.

Stevens J., 1996. *Applied Multivariate Statistics for the Social Sciences*. 3rd edition. Mahwah, NJ: Erlbaum.

Stevenson H.C., 1994. The Psychology of Sexual Racism and AIDS: An Ongoing Saga of Distrust and the "Sexual Other". *Journal of Black Studies*, **25**, 62–80.

Stocké V., 2001. *Socially Desirable Response Behavior as Rational Choice: The Case of Attitudes Towards Foreigners*. Sonderforschungsbereich 504, No. 01–22. Mannheim: Universität Mannheim.

Stocké V., 2002. *Form oder Inhalt? Die unterschiedlichen Ursachen für Framing-Effekte*. München: Oldenbourg.

Stocks J.Th., 1999. *Quota Sampling*. Web-based Course at the School of Social Work, Michigan State University. http://www.msu.edu/user/sswwebed/.

Strack F., Martin L.L., 1987. Thinking, Judging, and Communicating: A Process Account of Context Effects in Attitude Surveys. In: *Social Information Processing and Survey Methodology*. Hippler H.-J., Schwarz N., Sudman S. (eds.). New York: Springer.

Strack F., Martin L.L., Schwarz N., 1988. Priming and Communication: The Social Determinants of Information Use in Judgments of Life-Satisfaction. *European Journal of Social Psychology*, **18**, 429–442.

Strack F., Schwarz N., Wänke M., 1991. Semantic and Pragmatic Aspects of Context Effects in Social and Psychological Research. *Social Cognition*, **9**, 111–125.

Strahan R., Gerbasi K.C., 1973. Semantic Style Variance in Personality Questionnaires. *Journal of Psychology*, **85**, 109–118.

Strosahl K.D., Chiles J.A., Linehan M.M., 1984. Will the Real Social Desirability Please Stand Up? Hopelessness, Depression, Social Desirability, and the Prediction of Suicidal Behavior. *Journal of Consulting and Clinical Psychology*, **52**, 449–457.

Strube M.J., 1981. Meta-Analysis and Cross-Cultural Comparison: Sex Differences in Child Competitiveness. *Journal of Cross-Cultural Psychology*, **12**, 3–20.

Sudman S., Bradburn N.M., 1974. *Response Effects in Surveys: A Review and Synthesis*. Chicago: Aldine.

Sudman S., Bradburn N.M., 1982. *Asking Questions: A Practical Guide to Questionaire Design*. San Francisco: Jossey-Bass.

Sudman S., Bradburn N.M., Schwarz N., 1996. *Thinking About Answers: The Application of Cognitive Processes to Survey Methodology* . San Francisco: Jossey-Bass.

Sullivan J.L., Piereson J.E., Marcus G.E., 1992. *Political Tolerance and American Democracy*. Chicago: University of Chicago Press.

Summer Wolfson K., Pearce W.B., 1983. A Cross-Cultural Comparison of the Implications of Self-Disclosures in Conversational Logics. *Communication Quarterly*, **31**, 249–256.

Sumner W.G., 1906. *Folkways. A Study of the Sociological Importance of Usages, Manners, Customs, Mores and Morals*. New York: Mentor Books.

Sykes W., Collins M., 1988. Effects of Mode of Interview: Experiments in the UK. In: *Telephone Survey Methodology*. Groves R.M., Biemer P.P., Lyberg L.E., Massey J.T., Nicholls II W.L., Wakesberg J. (eds.). New York: Wiley.

Sykes W., Hoinville G., 1985. *Telephone Interviewing on a Survey of Social Attitudes: A Comparison with Face-to-Face Procedures*. London: Social and Community Planning Research.

Szalai A., 1972. *The Use of Time*. The Hague: Mouton.

Szalai A., 1993. The Organization and Execution of Cross-National Survey Research Projects. *Historical Social Research*, **18**, 139–171.

Türk Smith S., Smith K.D., Seymour K., 1993. Social Desirability of Personality Items as a Predictor of Endorsement: A Cross-Cultural Analysis. *Journal of Social Psychology*, **133**, 43–52.

Tabachnick B.G., Fidell L.S., 2001. *Using Multivariate Statistics*. 4th edition. New York: Allyn & Bacon.

Tajfel H., 1981. *Human Groups and Social Categories. Studies in Social Psychology*. Cambridge: Cambridge University Press.

Tan B.C.Y., Wie K., Watson R.T., Clapper D.L., McLean E., 1998. Computer-Mediated Communication and Majority Influence: Assessing the Impact of an Individualistic and a Collectivistic Culture. *Management Science*, **44**, 1263–1278.

Tanaka-Matsumi J., Marsella A.J., 1976. Cross-Cultural Variations in the Phenomenological Experience of Depression: I. Word Association Studies. *Journal of Cross-Cultural Psychology*, **7**, 379–396.

Tanur J., 1992. *Questions about Questions: Inquiries into the Cognitive Bases of Surveys*. New York: Russell Sage Foundation.

Tanzer N.K., 1995. Cross-Cultural Validity of Likert-Type Scales: Perfect Matching Factor Structures and Still Biased? *European Journal of Psychology Assessment*, **11**, 194–201.

Tanzer N.K., forthcoming. Developing Tests for Use in Multiple Languages and Cultures: A Plea for Simultaneous Development. In: *Adapting Educational and Psychological Test for Cross-Cultural Assessment*. Hambleton R.K., Merenda P.F., Spielberger C.D. (eds.). Hillsdale, NJ: Erlbaum.

Tanzer N.K., Sim C.Q.E., 1991. *Test Anxiety in Primary School Students: An Empirical Study in Singapore*. Research Report 1991/6. Graz: Department of Psychology, University of Graz.

Tanzer N.K., Gittler G., Ellis B.B., 1995. Cross-Cultural Validation of Item Complexity in a LLTM-Calibrated Spatial Ability Test. *European Journal of Psychology Assessment*, **11**, 170–183.

Tanzer N.K., Ellis B.B., Zhang H., Sim C.Q.E., Broer M., Gittler G., 1997. *Cross-Cultural Decentering of Test Instructions in a Letter-Cancellation Test: A Field Test of the ITC Guidelines for Test Adaptions*. Paper presented at the 5th European Congress of Psychology, Dublin.

Tateneni K., Browne M.W., Cudeck R., Mels G., 1998. *CEFA: Comprehensive Exploratory Factor Analysis*. http://quantrm2.psy.ohio-state.edu/browne.

Taylor T.R., Boeyens J.C., 1991. The Comparability of the Scores of Blacks and Whites of the South African Personality Questionnaire: An Exploratory Study. *South-African Journal of Psychology*, **21**, 1–11.

Ten Berge J.M.F., 1986. Rotatie naar perfecte congruentie en de multipele groep methode. *Nederlands Tijdschrift voor de Psychologie*, **41**, 219–225.

Teresi J.A., 2001. Statistical Methods for Examination of Differential Item Functioning (DIF) with Applications to Cross-Cultural Measurement of Functional, Physical and Mental Health. *Journal of Mental Health and Aging,* 7, 31–40.

Teresi J.A., Holmes D., Ramírez M., Gurland B.J., Lantigua R., 2001. Performance of Cognitive Tests Among Different Racial/Ethnic and Education Groups: Findings of Differential Item Functioning and Possible Item Bias. *Journal of Mental Health and Aging,* 7, 79–90.

Ting-Toomey S., 1999. *Communicating across Nations.* New York: Guilford Press.

Tourangeau R., 1999. Context Effects to Answers to Attitude Questions. In: *Cognition and Survey Research.* Sirken M.G., Herrmann D.J., Schechter S., Schwarz N., Tanur J.M., Tourangeau R. (eds.). New York: Wiley.

Tourangeau R., Rasinski K.A., 1988. Cognitive Processes Underlying Context Effects in Attitude Measurement. *Psychological Bulletin,* 103, 299–314.

Tourangeau R., Rips L.J., Rasinski K.A., 2000. *The Psychology of Survey Response.* Cambridge: Cambridge University Press.

Tourangeau R., Smith T.W., 1996. Asking Sensitive Questions. The Impact of Data Collection Mode, Question Format, and Question Context. *Public Opinion Quarterly,* 60, 275–304.

Tränkle U., 1987. Auswirkung der Gestaltung der Antwortskala auf Quantitative Urteile. *Zeitschrift für Sozialpsychologie,* 18, 88–99.

Treiman D.J., 1977. *Occupational Prestige in Comparative Perspective.* New York: Academic Press.

Trentini G., Muzio G.B., 1995. Values in a Cross-Cultural Perspective: A Further Analysis. In: *Life Roles, Values, and Careers: International Findings of the Work Importance Study.* Super D.E., Sverko B. (eds.). San Francisco: Jossey-Bass.

Triandis H.C., 1995. *Individualism and Collectivism.* Boulder, CO: Westview Press.

Triandis H.C., Marín G., 1983. Etic Plus Emic versus Pseudoetic: A Test of a Basic Assumption of Contemporary Cross-Cultural Psychology. *Journal of Cross-Cultural Psychology,* 14, 489–500.

Triandis H.C., Martin G., Lisansky J., Betancourt H., 1984. Simpatia as a Cultural Script for Hispanics. *Journal of Personality and Social Psychology,* 47, 1363–1375.

Tsushima W.T., 1969. Responses of Irish and Italians of Two Social Classes on the Marlowe-Crowne Social Desirability Scale. *Journal of Social Psychology,* 77, 215–219.

Tucker L.R., 1951. *A Method for Synthesis of Factor Analysis Studies.* Personnel Research Section Report No. 984. Washington, DC: Department of the Army.

Tukey J.W., 1977. *Exploratory Data Analysis.* Reading, MA: Addison-Wesley.

Turner J.C., 1982. Towards a Cognitive Redefinition of the Social Group. In: *Social Identity and Intergroup Relations.* Tajfel H. (ed.). Cambridge: Cambridge University Press.

Uher R., 2000. The International Social Survey Programme (ISSP). In: *Schmollers Jahrbuch: Journal of Applied Social Science Studies, Volume 4.* Wagner G.G., Burkhauser R.V., Hauser R., Petzina D., Riphahn R.T., Smeeding T.M., Wagner J. (eds.). Berlin: Duncker und Humboldt.

UNESCO, 1997. *International Standard Classification of Education. ISCED 1997.* Paris: UNESCO.

Usunier J.C., 1999. *Marketing Across Cultures.* 3rd edition. New York: Prentice-Hall.

Vallier I. (ed.), 1971. *Comparative Methodology on Sociology — Essays on Trends and Applications.* Berkeley: UCA-Press.

Van Dam D., 1996. *Blijven we buren in België? Vlamingen en Walen over Vlamingen en Walen.* Leuven: Van Halewijck.

Van de Vijver F.J.R., 1997. Meta-Analysis of Cross-Cultural Comparisons of Cognitive Test Performance. *Journal of Cross-Cultural Psychology,* **28**, 678–709.

Van de Vijver F.J.R., 1998. Towards a Theory of Bias and Equivalence. In: *ZUMA-Nachrichten Spezial No.3. Cross-Cultural Survey Equivalence.* Harkness J.A. (ed.). Mannheim: ZUMA.

Van de Vijver F. J. R,. 2001. Récents développements dans l'evaluation interculturelle. *Psychologie Française,* **46**, 251–257.

Van de Vijver F.J.R., 2002. Inductive Reasoning in Zambia, Turkey, and The Netherlands: Establishing Cross-Cultural Equivalence. *Intelligence,* **30**, 313–351.

Van de Vijver F.J.R., Hambleton R.K., 1996. Translating Tests: Some Practical Guidelines. *European Psychologist,* **1**, 89–99.

Van de Vijver F.J.R., Leung K., 1997a. Methods and Data Analysis of Comparative Research. In: *Handbook of Cross-Cultural Psychology.* Berry J.W., Poortinga Y.H., Pandey J. (eds.). 2nd edition. Boston: Allyn & Bacon.

Van de Vijver F.J.R., Leung K., 1997b. *Methods and Data Analysis for Cross-Cultural Research.* Newbury Park, CA: Sage.

Van de Vijver F.J.R., Poortinga Y.H., 1994. Methodological Issues in Cross-Cultural Studies on Parental Rearing Behavior and Psychopathology. In: *Parental Rearing and Psychopathology.* Perris C., Arrindell W.A., Eisemann M. (eds.). Chicester, England: Wiley.

Van de Vijver F.J.R., Poortinga Y.H., 1997. Towards an Integrated Analysis of Bias in Cross-Cultural Assessment. *European Journal of Psychology Assessment,* **13**, 29–37.

Van de Vijver F.J.R., Poortinga Y.H., 2002. Structural Equivalence in Multilevel Research. *Journal of Cross-Cultural Psychology,* **33**, 141–156.

Van de Vijver F.J.R., Tanzer N.K., 1997. Bias and Equivalence in Cross-Cultural Assessment: An Overview. *European Review of Applied Psychology,* **41**, 263–279.

Van de Vijver F.J.R., Watkins D., in review. *Structural Equivalence at Individual and Country Levels: A Multilevel Factor Analysis of a Measure of the Independent-Interdependent Self.*

Van de Vijver F.J.R., Daal M., Van Zonneveld R., 1986. The Trainability of Abstract Reasoning: A Cross-Cultural Comparison. *International Journal of Psychology,* **21**, 589–615.

Van der Linden W.J., Hambleton R.K. (eds.), 1997. *Handbook of Multilevel Research.* New York: Springer.

Van der Zouwen J., 1976. A Conceptual Model for the Auxiliary Hypotheses Behind the Interview. In: *Annals of Systems Research. Volume 4.* Van Rootselaar B. (ed.). Dordrecht: Kluwer Academic Publisher.

Van der Zouwen J., 2000. An Assessment of the Difficulty of Questions Used in the ISSP-Questionnaires, the Clarity of Their Wording, and the Comparability of the Responses. *ZA-Information,* **46**, 96–114.

Van Deth J.W., 1998a. Equivalence in Comparative Political Research. In: *Comparative Politics: The Problem of Equivalence.* Van Deth J.W. (ed.). London: Routledge.

Van Deth J.W. (ed.), 1998b. *Comparative Politics. The Problem of Equivalence.* London: Routledge.

Van Deth J.W., Scarbrough E. (eds.), 1995. *The Impact of Values*. Oxford: Oxford University Press.

Van Dijk T.A., 1993. *Elite Discource and Racism*. Newbury Park, CA: Sage.

Van Haaften E.H., Van de Vijver F.J.R., 1996. Psychological Consequences of Environmental Degradation. *Journal of Health Psychology*, **1**, 411–429.

Van Hemert D.D.A., Van de Vijver F.J.R., Poortinga Y.H., Georgas J., in press. *Structure and Score Levels of the Eysenck Personality Questionnaire Across Individuals and Countries. Personality and Individual Differences*. Tilburg: Tilburg University Press.

Van Herk H., 2000. *Equivalence in a Cross-National Context: Methodological and Empirical Issues in Marketing Research*. Tilburg, The Netherlands: Tilburg University Press.

Vehovar V., 1995. Field Substitutions in Slovene Public Opinion Survey. In: *Contributions to Methodology and Statistics, Metodoloski zvezki, 10*. Ferligoj A., Kramberger A. (eds.). Ljubljana: FDV, 39–66.

Vehovar V., 1999. Field Substitution and Unit Nonresponse. *Journal of Official Statistics*, **15**, 335–350.

Verba S., 1971. Cross-National Survey Research: The Problem of Credibility. In: *Comparative Methodology on Sociology — Essays on Trends and Applications*. Vallier I. (ed.). Berkeley: University of California Press.

Verba S., Lehman Scholzmann K., Brady H.E., 1995. *Voice and Equality. Civic Voluntarism in American Politics*. Cambridge, MA: Harvard University Press.

Verma V., 1992. *Household Surveys in Europe: Some Issues in Comparative Methodology*. Paper presented at the seminar on International Comparison of Survey Methodologies, Athens, March/April 1992.

Verma V., Gabilondo L.G., 1993. *Family Budget Surveys in the EC: Methodology and Recommendations for Harmonization*. Statistical Document 3E. Luxembourg: Eurostat.

Verma V., Le T., 1995. *Sampling Errors for the DHS Surveys*. Bulletin of the International Statistical Institute. Proceeding of the 50th Session, Tome LVI, Beijing, China, 1995, 839–859.

Vermunt J.K., 1995. *Loglinear Event History Analysis. A General Approach with Missing Data, Latent Variables and Unobserved Heterogeneity*. Tilburg, The Netherlands: WORC.

Vermunt J.K., 1997. *LEM: A General Program for the Analysis of Categorial Data*. Tilburg, The Netherlands: Tilburg University Press.

Vidali J.J., 1975. Context Effects on Scales Evaluatory Adjective Meaning. *Journal of the Market Research Society*, **17**, 21–25.

Von Alemann U., Tönnesmann W., 1995. Grundriss: Methoden in der Politikwissenschaft. In: *Politikwissenschaftliche Methoden. Grundriss für Studium und Forschung*. Von Alemann U. (ed.). Opladen: Westdeutscher Verlag.

Voogt R.J.J., forthcoming. *Mixed Mode Design — A Solution for the Problem of Unlisted Telephone Numbers in Survey Research*. Amsterdam: University of Amsterdam.

Voss K.E., Stem D.E. Jr., Johnson L.W., Arce C., 1996. An Exploration of the Comparability of Semantic Adjectives in Three Languages: A Magnitude Estimation Approach. *International Marketing Review*, **13**, 44–58.

Wallsten Th.S., Budescu D.V., Rapoport A., Zwick R., Forsyth B., 1986. Measuring the Vague Meanings of Probability Terms. *Journal of Experimental Psychology*, **115**, 348–365.

Ward L.M., 1977. Multidimensional Scaling of the Molar Physical Environment. *Multivariate Behavioral Research*, **12**, 23–42.

Ware J.E., Gandek B., 1998. Overview of the SF-36 Health Survey and the International Quality of Life Assessment (IQOLA) Project. *Journal of Clinical Epidemiology*, **51**, 903–912.

Warnecke R., Ferrans C.E., Johnson T.P., Chapa-Resendez G., O'Rourke D., Chavez N., Dudas S., Smith E.D., Schallmoser L.M., Hand R.P., Lad T., 1996. Measuring Quality of Life in Culturally Diverse Populations. *Journal of the National Cancer Institute Monographs*, **20**, 29–38.

Warnecke R., Johnson T.P., Chavez N., Sudman S., O'Rourke D., Lacey L., Horm J., 1997. Improving Question Wording in Surveys of Culturally Diverse Populations. *Annals of Epidemiology*, **7**, 334–342.

Watten R.G., 1996. Coping Styles in Abstainers from Alcohol. *Psychopathology*, **29**, 340–346.

Wechsler D., 1981. *Wechsler Adult Intelligence Scale — Revised*. San Antonio, TX: The Psychological Corporation.

Wechsler D., 1989. *Wechsler Preschool and Primary Scale of Intelligence — Revised*. New York: The Psychological Corporation.

Wechsler D., 1991. *Wechsler Intelligence Scale for Children*. 3rd edition. San Antonio, TX: The Psychological Corporation.

Weech-Waldonado R.W., Weidmer B., Morlaes L.S., Hays R.D., 2001. *Cross Cultural Adaption of Survey Instruments*. Proceedings of the Seventh Conference in Health Survey Research Methods, DHHS, Hyattsville, MD, 2001.

Weidmer B., Brown J., Garcia L., 1999. Translating the CAHPS 1.0 Survey Instruments into Spanish. *Medical Care*, **37**, 89–96.

Welkenhuysen-Gybels J., Van de Vijver F.J.R., in review. *Methods for the Evaluation of Construct Equivalence in Studies Involving Many Groups*.

Welte J.W., Russell M., 1993. Influence of Socially Desirable Responding in a Study of Stress and Substance Abuse. *Alcoholism: Clinical and Experimental Research*, **17**, 758–761.

Werner O., Campbell D.T., 1970. Translating, Working Through Interpreters, and the Problem of Decentering. In: *A Handbook of Cultural Anthropology*. Naroll R., Cohen R. (eds.). New York: American Museum of Natural History.

White G.M., 1978. Ambiguity and Ambivalence in A`ara Personality Descriptors. *American Ethnologist*, **5**, 334–360.

White G.M., 1980. Conceptual Universals in Interpersonal Language. *American Anthropologist*, **82**, 759–781.

Wierzbicka A., 1992. *Semantics, Culture and Cognition: Universal Human Concepts in Culture-Specific Configurations*. Oxford: Oxford University Press.

Wiggins J.S., 1979. A Psychological Taxonomy of Trait-Descriptive Terms: The Interpersonal Domain. *Journal of Personality and Social Psychology*, **20**, 395–412.

Wilcox C., Sigelman L., Cook E., 1989. Some Like It Hot: Individual Differences in Responses to Group Feeling Thermometers. *Public Opinion Quarterly*, **53**, 246–257.

Wildt A.R., Mazis M.B., 1978. Determinants of Scale Response: Label Versus Position. *Journal of Marketing Research*, **15**, 261–267.

Williams J.E., Satterwhite R.C., Saiz J.L., 1998. *The Importance of Psychological Traits*. New York: Plenum Press.

Wilson D., Cutts J., Lees I., Mapungwana S., Maunganidze L., 1992. Psychometric Properties of the Revised UCLA Loneliness Scale and Two Short-Form Measures of Loneliness in Zimbabwe. *Journal of Personality Assessment,* **59**, 72–81.

Wilson E.C., 1958. Problems of Survey Research in Modernizing Areas. *Public Opinion Quarterly,* **22**, 230–234.

Wilss W., 1996. *Knowledge and Skills in Translator Behavior.* Amsterdam: John Benjamins.

Windle M., Iwawaki S., Lerner R.M., 1988. Cross-Cultural Comparability of Temperament among Japanese and American Preschool Children. *International Journal of Psychology,* **23**, 547–567.

Wiseman R.L., Koester J. (eds.), 1993. *Intercultural Communication Competence.* Newbury Park, CA: Sage.

Witherspoon S., Mohler P.Ph., Harkness J.A., 1995. *Report on Research into Environmental Attitudes and Perceptions (REAP) to the European Commission.* Mannheim: ZUMA. http://intraweb.zuma-mannheim.de/en/cooperation/research/compass/reap.htm.

Wolter K.M., 1985. *Introduction to Variance Estimation.* New York: Springer.

Worcester R.M., Burns T.R., 1975. A Statistical Examination of the Relative Precision of Verbal Scales. *Journal of the Market Research Society,* **17**, 181–197.

Wortman P.M., 1994. Judging Research Quality. In: *The Handbook of Research Synthesis.* Cooper H., Hedges L.V. (eds.). New York: Russell Sage Foundation.

Wortman P.M., Bryant F.B., 1985. School Desegregation and Black Achievement: An Integrative Review. *Sociological Methods and Research,* **13**, 289–324.

Wright D.L., Aquilino W.S., Rasinski K.A., 1998. A Comparison of Computer-Assisted and Paper-and-Pencil Self-Administrative Questionnaires in a Survey on Smoking, Alcohol, and Drug Use. *Public Opinion Quarterly,* **62**, 331–353.

Wrigley C.F., Neuhaus J.O., 1955. The Matching of Two Sets of Factors. *American Psychologist,* **10**, 418–419.

Yang K.S., Bond M.H., 1990. Exploring Implicit Personality Theories with Indigenous or Imported Constructs: The Chinese Case. *Journal of Personality and Social Psychology,* **58**, 1087–1095.

Yates F., 1949. *Sampling Methods for Censuses and Surveys.* London: Griffin.

Young C.A., 1999. *What We Know About 'I Don't Know': An Analysis of the Relationship between 'Don't Know' and Education.* Paper presented to the American Association of Public Opinion Research, St. Petersburg Beach, FL, May.

Zarkovich S.S., 1983. Some Problems of Sample Work in Underdeveloped Countries. In: *Social Research in Developing Countries.* Bulmer M., Warwick D.P. (eds.). New York: Wiley.

Zentralarchiv für Empirische Sozialforschung, 1991. *ISSP 1988 — Family and Changing Gender Roles I. Codebook. ZA Study 2620.* Cologne: ZA.

Zentralarchiv für Empirische Sozialforschung, 1993. *ISSP CD-ROM 1985–1992.* Cologne: ZA.

Zentralarchiv für Empirische Sozialforschung, 1996. *ISSP 1994 — Family and Changing Gender Roles II. Codebook. ZA Study 2620.* Cologne: ZA.

AUTHOR INDEX

SUBJECT INDEX

(Terms with an asterisk are explained in the Glossary)

WILEY SERIES IN SURVEY METHODOLOGY
Established in Part by WALTER A. SHEWHART AND SAMUEL S. WILKS

Editors: *Robert M. Groves, Graham Kalton, J. N. K. Rao, Norbert Schwarz, Christopher Skinner*

Wiley Series in Survey Methodology covers topics of current research and practical interests in survey methodology and sampling. While the emphasis is on application, theoretical discussion is encouraged when it supports a broader understanding of the subject matter.

The authors are leading academics and researchers in survey methodology and sampling. The readership includes professionals in, and students of, the fields of applied statistics, biostatistics, public policy, and government and corporate enterprises.

*Now available in a lower priced paperback edition in the Wiley Classics Library.